Fundamentals of Deep Excavations

Fundamentals of Deep Excavations

Chang-Yu Ou

CRC Press
Taylor & Francis Group
Boca Raton London New York

CRC Press is an imprint of the
Taylor & Francis Group, an **informa** business

Cover image: A strut-free deep excavation case history with a depth of
13.2 m and a size of 123 x 105 m. The excavation system was designed
by the author and his team with a combination of cross walls and buttress
walls as introduced in Chapter 11.

CRC Press/Balkema is an imprint of the Taylor & Francis Group, an informa business

© 2022 Taylor & Francis Group, London, UK

Typeset by codeMantra

Library of Congress Cataloging-in-Publication Data
Names: Ou, Chang-Yu, 1954- author.
Title: Fundamentals of deep excavations / Chang-Yu Ou.
Description: Boca Raton : CRC Press, 2021. | Includes bibliographical
references and index.
Subjects: LCSH: Excavation. | Soil stabilization.
Classification: LCC TA730 .O925 2021 (print) | LCC TA730 (ebook) |
DDC 624.1/52—dc23
LC record available at https://lccn.loc.gov/2021009937
LC ebook record available at https://lccn.loc.gov/2021009938

Published by: CRC Press/Balkema
            Schipholweg 107C, 2316 XC Leiden, The Netherlands
            e-mail: Pub.NL@taylorandfrancis.com
            www.crcpress.com – www.taylorandfrancis.com

ISBN: 978-0-367-42601-9 (Hbk)
ISBN: 978-0-367-42608-8 (Pbk)
ISBN: 978-0-367-85385-3 (eBook)

DOI: 10.1201/9780367853853

Contents

Preface

The author published a monograph entitled *Deep Excavation: Theory and Practice* in 2006. Since then, the book has been extensively used as a textbook at many famous universities in the world, and it has also become an important reference work for engineers to deal with deep excavation problems. However, recently soil constitutive models and geotechnical software developments have made great progress, and many new research results have been obtained. During this period, the author and his team also published many articles in international journals and published conference papers at some important international conferences. The author feels it is time to update the book.

The most important updates include the construction of diaphragm wall, fundamentals of effective stress undrained analysis and total stress undrained analysis, advanced constitutive soil models, allowable settlement of buildings, and control of ground settlement with cross walls and buttress walls. The failure mechanism of excavations is also elucidated in detail. The stability analysis methods are therefore presented in relation to the failure mechanism as discussed in the book, which are very different from that in the previous version and that in other journal papers.

Most of the content of this book comes from the research results of my supervised domestic and international graduate students. I am gratified that these students are successful in the industry and academia. In addition, some of the concepts and contents are also derived from international and domestic papers and case histories in Taiwan. Many of the figures are drawn by my graduated students, and I would like to thank them here. Finally, I hope that if you find any errors in this book, or need further discussion or clarification, please feel free to give me advice and/or contact me by sending an email **(ou@mail.ntust.edu.tw)** to me or visit my website **(deepexcavation.net)**. The solution to the problems at the end of the chapters can be found on the website as well.

Chang-Yu Ou
January 2021

About the author

Chang-Yu Ou, Ph.D., is a Chair Professor of the Department of Civil and Construction Engineering at the National Taiwan University of Science and Technology, Taipei, Taiwan. He received his doctoral degrees from Stanford University in 1987. He is the chair of the ATC6 of the International Society for Soil Mechanics and Geotechnical Engineering (ISSMGE). He was the Dean of the College of Engineering at the National Taiwan University of Science and Technology and the President of the Taiwan Geotechnical Society.

His areas of interest are deep excavations, soil behavior, soft ground tunneling, and ground improvement. He has published more than 200 journal and conference papers. He has received three outstanding research awards and received a research fellow award from the National Science Council of Taiwan. He was also awarded many journal paper awards. In addition to publishing the book "Deep Excavation: Theory and Practice" with Taylor & Francis, he has also published three deep excavation books in the Chinese language. Because of his outstanding research results, he has participated in many large-scale projects related to deep excavations as a reviewer, designer, or analyst.

Chapter 1

Introduction to the analysis and design of excavations

When Terzaghi (1943) first considered the stability of excavations, he defined those whose excavation depths were smaller than their widths as shallow excavations while those with depths larger than their widths were deep excavations. Years later, Terzaghi and Peck (1967) and others, including Peck et al. (1977), revised that excavations whose depths were less than 6m could be defined as shallow excavations and those deeper than that as deep excavations, considering that the use of sheet piles or soldier piles grows uneconomical once the excavation depth goes beyond 6m. Generally speaking, the analysis methods for shallow excavations are comparatively simple. In fact, more and more excavation projects are located in urban areas. To avoid damage to adjacent properties caused by excavation, diaphragm walls are commonly used as retaining walls. What's more, computer programming has done most of the analysis and design, which applies to all depths, following the same theories. Therefore, it isn't meaningful to distinguish between deep and shallow excavations any more.

Analysis of deep excavations is usually required before going into design. Deep excavation analysis is a typical soil–structure interaction problem. Soil is a nonlinear, inelastic, and anisotropic material. Its behavior is normally affected by water contents. Some types of soils have the characteristics of consolidation and creep. Theoretically, analysis of deep excavation involves simulations of elastoplastic behavior of soil, interface behavior between soil and retaining walls, and the excavation process. A reasonable excavation analysis in practice should make use of conventional soil mechanics and simple structural mechanics, along with appropriate modifications according to field observation. For a detailed discussion of excavation analyses, please see Chapters 5–8.

A complete deep excavation design includes a retaining system, a strutting system, a dewatering system, excavation procedure, a monitoring system, and building protection. Figure 1.1 illustrates the general course of deep excavation design, which will be paraphrased as follows:

1.1 GEOLOGICAL INVESTIGATION AND SOIL TESTS

Deep excavation projects include the construction of building basements and subway stations, whose depths may range from several meters to 30 or 40m. The process of an excavation may encounter different kinds of soils underneath the same excavation site—from soft clay to hard rocks. The closer the construction site to a hillside, the more complicated the geological condition. The geological condition determines the

DOI: 10.1201/9780367853853-1

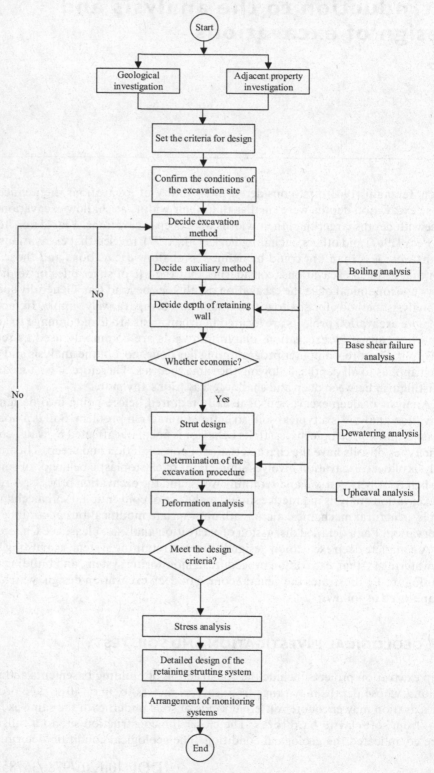

Figure 1.1 Flowchart of analysis and design of an excavation.

type and construction of retaining walls and greatly influences the excavation behavior as well. In addition to the geological condition, the distribution of groundwater also contributes to the excavation behavior. For example, it may fall below the hydrostatic water pressure in an urban area because of the long-term overuse of groundwater. On hillsides, there might exist an artesian aquifer, which has a rather high pressure. In seaside areas, seawater may permeate into the soils and tides will make it fluctuate daily. To sum up, the geological investigation of an excavation project aims at the soil conditions underneath the construction site and the distribution of groundwater.

There are many soil tests for deep excavations. These include tests of basic soil behavior, such as unit weight, specific weight, water content, and Atterberg limit, and tests of mechanical behavior, such as consolidation and strength. According to the information from the soil tests, engineers can judge whether the soil is a drained or an undrained material. Since the strength differs significantly between drained and undrained materials, the choice of analysis methods and retaining walls varies accordingly. The more precise the results of soil tests, the more reasonable the analysis results and the more economical are the retaining and excavation systems.

1.2   CONDITIONS OF THE ADJACENT PROPERTIES

From the perspective of mechanics, deep excavation necessarily gives rise to movement of the soils near the excavation site. However, if the movement or settlement is too large, it will damage neighboring buildings or public facilities. Some buildings or facilities are especially sensitive to settlement, a little of which may bring about cracks in beams or columns, while others can stand more settlement. The allowable settlement of a building or a facility is highly correlated with its foundation type, construction material, structural type, and age. Therefore, investigation of the condition of adjacent properties and public facilities before designing an excavation project is required to determine the allowable settlement, which in turn determines the type of retaining and strutting systems and the selection of auxiliary methods.

1.3   CONFIRMATION OF THE CONDITIONS OF AN EXCAVATION SITE

According to the shape, area, and elevation of the excavation site, along with geological conditions, the distribution of groundwater, and conditions of neighboring properties, we can decide on a provisional retaining method and an excavation method. Therefore, it is necessary to have a thorough understanding of the conditions of the excavation site.

1.4   DESIGN CRITERIA

Whether an excavation is successful is significant to the lives and properties of many people. Thus, an appropriate design criterion must be selected before design.

A deep excavation design criterion should include at least the method of stability analysis, the methods of simplified and advanced deformation and stress analyses, a dewatering scheme, the design of structural components, and property protection.

This book explores theories, as well as their application to deep excavation, from both the theoretical and practical perspectives. Getting familiar with the contents should not only help the reader to understand excavation behavior and design and analyses, but also help develop a suitable excavation design criterion.

1.5 COLLECTING CASE HISTORIES OF THE NEARBY EXCAVATIONS

The first job of excavation design is to decide the type of retaining wall and the excavation method. Though we can choose the most reasonable methods based on geotechnical theories, geological conditions, and neighboring property conditions, an excavation analysis doesn't always predict the excavation behavior exactly because geological investigation may not cover all kinds of soils to be encountered during excavation and because the simulation of excavation process may not be complete. A case history of nearby excavation, equivalent to a full-scale excavation experiment, helps design the excavation project no matter if it was successful or not in the end.

1.6 AUXILIARY METHODS

A deep excavation may have difficulty meeting design criteria, due to poor geological conditions or deteriorated adjacent buildings. Even if it reaches the criterion, it may be very expensive. Auxiliary methods can help solve the dilemma. These include soil improvement, buttress walls, cross walls, micropiles, and underpinning. Please see Chapter 11 for the design of auxiliary methods.

1.7 EXCAVATION ANALYSES

Excavation analyses consist of stability analyses, deformation analyses, and stress analyses. Stability analyses, including base shear failure analyses, sand boiling analyses, and upheaval analyses, aim at avoiding failure or collapse. Base shear failure analyses and sand boiling analyses can determine how deeply the retaining wall should penetrate into the soil. Upheaval analyses can decide on dewatering schemes at different stages. For stability analyses of excavation, please see Chapter 5. For dewatering analyses, see Chapter 9.

Deformation analyses are to find the lateral deformation of retaining walls, the heave of the excavation bottom, and the settlement of the soil outside the excavation zone. The lateral deformation and the settlement of the soils affect not only the safety of the retaining wall but also the adjacent properties. As to the heave of the excavation bottom, it is correlated with the capacity of the strutting system.

Stress analyses involve those of strut load and of bending moment and shear of retaining walls. The data on the strut load are necessary for the detailed design of struts or anchors, while those of the bending moment and the shear are relevant to the choice of the appropriate type and dimension of retaining walls, and sometimes to the design of reinforcements. For methods of stress and deformation analyses of excavations, please refer to Chapters 6–8.

1.8   LAYOUT OF THE STRUTTING SYSTEM

A strutting system comprises either horizontal struts or anchors, which contribute to the resistance to the lateral earth pressure generated by excavation. Stability analyses determine the penetration depth of a retaining wall. After finishing the stability analyses, the tentative locations and vertical distances of the struts can be determined. The procedure of installation and of the later removal of the struts for the construction of floor slabs basically determines the locations and vertical distances of the struts. After analyzing deformation and stress, the type and size of the struts are accordingly decided. For the detailed design of strutting systems, please see Chapter 10.

1.9   MONITORING SYSTEM

In spite of thorough geological investigations, soil tests, rigorous analyses, and design before excavation, excavation theories, based on many hypotheses, can hardly cope with many uncertainties of geological conditions. Therefore, excavation has to be carried out along with monitoring instruments, which tell, immediately, changes in stress and displacement generated by excavation. The engineers in charge can thereby check the safety of excavation at any time. For large-scale excavations, the geological uncertainty increases and a monitoring system is urgently required. For the items and design of monitoring systems for deep excavation, please see Chapter 12.

1.10   PROTECTION OF NEIGHBORING PROPERTIES

Due to the unbalanced earth pressures on the two sides of a retaining wall, excavation produces displacement of retaining walls and settlement of the ground. The buildings and public facilities within the range of the settlement may have differential settlement. When settlement goes beyond the allowable amount, the nearby buildings or public facilities may be damaged. The damages may turn out to be structural or non-structural. To avoid such damages, prediction of the settlement is necessary to decide whether and how to take protecting measures. There are many ways to protect neighboring properties. Some of them can be deduced from theoretical analyses. Others rely on engineering experience and empirical data. Chapter 11 introduces various measures to protect adjacent properties and related analyses.

Chapter 2

Engineering properties of soils and geotechnical analysis

2.1 INTRODUCTION

The chapter will introduce the engineering properties of soils. Though basic soil properties are many, we will concentrate on those that are directly relevant to deep excavations. Some representative values of the engineering properties of urban soils in the world are collected as references for beginners to have an idea about the characteristics of urban soils and for engineers to help design excavation projects. To save space, this chapter won't go into the defining of soil properties, which can be found in detail in books on soil mechanics and foundation engineering.

Moreover, deformation behavior of soils is introduced, which includes yield behavior, Young's modulus, and Poisson's ratio. Those parameters are highly related to finite element analysis of deep excavation problems. When studying Chapter 8, readers are advised to refer to their definition and characteristics.

2.2 ENGINEERING PROPERTIES

This section will investigate the soil properties related to analyses and designs of deep excavations, such as the specific gravity, the unit weight, the water content, the Atterberg limits, the permeability, the compression and swelling indices, and the strength characteristics. To help comprehend the engineering properties of soils and to provide information for excavation analysis and design, this section introduces the range of soil properties in some urban areas in the world, such as San Francisco bay mud (Bonaparte and Mitchell, 1979), Boston blue clay (Lambe and Whitman, 1969; Ladd and Foote, 1974; Pestana, 2001), Chicago silty clay (Finno et al, 2002a; 2002b), London clay (Dawson et al., 1996; Simpson, 1992; Higgin et al., 1996; Ng, 1992), Mexico City clay (Lo, 1962; Mesri et al., 1975; Diaz-Rodriguez et al., 1992), Bangkok clay (Bergado et al., 1991; Moh et al., 1969), Singapore marine clay (Chu et al., 2002; Tan et al., 2002; Lee et al., 1998), Shanghai clay (Ou, 2001), and Taipei silty clay and sand (Wu, 1987; Moh et al., 1989). Comparing the basic properties of soils with these various urban areas, one can develop a further understanding of the engineering characteristics of soils.

DOI: 10.1201/9780367853853-2

2.2.1 Specific gravity

Except for some types such as peat and highly organic soil, the specific gravities (G_s) of soils don't vary significantly. Table 2.1 lists the magnitudes of the specific gravities of some general types of soils. Table 2.2 gives those of some well-known urban soils. Since specific gravities don't range widely, using the data from Table 2.1 or 2.2 is acceptable when the test results are not available before designing.

2.2.2 Unit weight and water content

The unit weight and the water content of soils correlate with their degrees of denseness, degrees of saturation, mineral compositions, and overburden depth. The degrees of denseness of sandy soils have something to do with their grain sizes and gradations; those of clayey soils, however, are affected by their overconsolidation ratios (OCRs).

Table 2.3 shows the saturated unit weights and the water contents for some urban soils. Theoretically, the deeper the overburden layer, the larger the unit weight, and the smaller the water content. Near-ground clays are often overconsolidated because of desiccation and the fluctuation of groundwater. Under the same conditions, the unit weights of overconsolidated soils are larger than those of normally consolidated soils, while their water contents are smaller. Though Table 2.3 doesn't handle factors such as

Table 2.1 Specific Gravities of Some Soils

Soil Type	G_s
Gravel	2.65–2.68
Sand	2.65–2.68
Clay	2.62–2.68
Clay (inorganic)	2.68–2.75
Clay (organic)	2.58–2.75
Peat	1.30–1.90

Table 2.2 Typical Values of Specific Gravity for Some Urban Soils

Soil Type	G_s
San Francisco bay mud (CH) mud(CH)	2.69–2.73
Boston blue clay (CL)	2.76–2.79
Chicago silty clay (CL)	2.63–2.76
London clay (CH)	2.72–2.76
Mexico City clay (MH)	2.35–2.65
Bangkok clay (CH)	2.59–2.74
Singapore marine clay (CH)	2.62–2.78
Shanghai clay (CL)	2.72–2.74
Taipei silty clay (CL)	2.69–2.71
Taipei silty sand	2.64–2.73

Table 2.3 Typical Values of Saturated Unit Weight and
Saturated Water Content for Some Urban Soils

Soil Type	w (%)	γ_{sat} (kN/m^3)
San Francisco bay mud (CH)	88–96	14.4–14.8
Boston blue clay (CL)	34–38	18.2–18.7
Chicago silty clay (CL)	21–33	18.1–19.0
London clay (CH)	23–27	19.9–20.3
Mexico City clay (MH)	439–574	11.0–14.4
Bangkok clay (CH)	66–92	14.4–15.9
Singapore marine clay (CH)	66–92	15.5–16.5
Shanghai clay (CL)	28–42	17.2–18.6
Taipei silty clay (CL)	18–42	17.7–19.3
Taipei silty sand	14–36	18.2–20.0

overburden depth, it is still useful to understand the possible range of the unit weights and water contents, as these are also important references for preliminary design.

2.2.3 Atterberg limit

The commonly used Atterberg limits include the liquid limit (LL), the plastic limit (PL), and the plastic index (PI). Table 2.4 lists the values of LL, PL, and PI for some urban soils. For the convenience of comparison, the LL and PL values of typical clay minerals such as Kaolinite, Illite, and Montmorillonite are also included.

The LL, the PL, and the water content can be used to estimate the degree of consolidation of soils. This is because the density of an overconsolidated soil is usually higher in the natural state than shown in the Atterberg limit test (remolded state). Thus, considering that the porosity or natural water content of the saturated overconsolidated soil is lower than that in the normally consolidated state, we can conclude reasonably the following:

Table 2.4 Atterberg Limits for Some Soils

Soil Type	LL	PL	PI
Kaolinite	35–100	25–35	–
Illite	50–100	30–60	–
Montmorillonite	100–800	50–100	–
San Francisco bay mud (CH)	85–88	35–44	43–54
Boston blue clay (CL)	40–49	20–25	18–23
Chicago silty clay (CL)	30–42	17–22	14–20
London clay (CH)	70–80	24–29	47–70
Mexico City clay (MH)	425–550	57–150	300–493
Bangkok clay (CH)	76–102	29–47	23–27
Singapore marine clay (CH)	75–100	22–30	52–62
Shanghai clay (CL)	28–42	17–24	10–19
Taipei silty clay (CL)	25–40	17–24	8–16

If w_n is about LL%, the soil is normally consolidated.
If w_n is smaller than PL%, the soil is highly overconsolidated.
If w_n is between LL% and PL%, the soil is overconsolidated.
If w_n is larger than LL%, the soil is unstable.

When the natural water content (w_n) is larger than LL%, it follows that the undrained shear strength is low and the soil is susceptible to collapse or becoming mobile once it is disturbed. The reason why the soil can remain solid and stable is the vertical overburden stress or the intergranular cementation. The above explanation can be summed up by the following expression of the liquid index (LI):

$$LI = \frac{w_n - PL}{LL - PL} \qquad\qquad (2.1)$$

Combining the equation with the above discussion, we can conclude that when LI $\approx$ 1.0, the soils are normally consolidated; when LI < 1.0, the soils are overconsolidated; and when LI > 1.0, the soils are soft or sensitive.

2.2.4 Permeability

Generally speaking, the coefficient of permeability is not necessarily required in deep excavation analysis and design. An advanced analysis of deep excavation sometimes has to consider the dissipation of pore water pressure or consolidation and therefore has to take advantage of coefficient of permeability.

Table 2.5 gives the possible range of the coefficients of permeability and could be useful for reference when designing an excavation. Table 2.6 offers some typical coefficients of permeability for some urban soils.

2.2.5 Compression and swelling

The compression index (C_c) of clayey soils is the slope of the straight-line segment on the $e - \log p$ curve, which is produced in the primary consolidation stage of the consolidation test. Table 2.7 lists the typical values of C_c for some urban soils. Many investigators have established the relationships between basic soil properties and C_c through regression analyses, as shown in Table 2.8. To select the most suitable empirical formula for local types of soils, a further laboratory test is necessary.

Table 2.5 Coefficients of Permeability
of Some Soils

Soil Type	k (cm/s)
Clean gravel	$>10^{-1}$
Coarse-to-fine soil	10^{-1} to 10^{-3}
Fine sand, silty sand	10^{-3} to 10^{-5}
Silt, silty clay	10^{-4} to 10^{-6}
Clay	$<10^{-7}$

Table 2.6 Typical Values of Permeability for Some Urban Soils

Soil Type	k (cm/s)
San Francisco bay mud (CH)	$(2.0-100) \times 10^{-9}$
Boston blue clay (CL)	$(1.0-100) \times 10^{-9}$
Chicago silty clay (CL)	$< 10^{-7}$
London clay (CH)	$(0.5-100) \times 10^{-8}$
Mexico City clay (MH)	$(1.0-100.0) \times 10^{-9}$
Bangkok clay (CH)	$(4.0-60) \times 10^{-7}$
Singapore marine clay (CH)	$(1.0-7.0) \times 10^{-9}$
Shanghai clay (CL)	–
Taipei silty clay (CL)	$(0.5-2.0) \times 10^{-7}$
Taipei silty sand	$(0.5-6.0) \times 10^{-4}$

Table 2.7 Compression Indices of Some Soils

Soil Type	C_c
Normally consolidated, medium-sensitive clay	0.2–0.5
Peat	10.0–15.0
Organic clay	>4.0
San Francisco bay mud (CH)	0.8–1.8
Boston blue clay (CL)	0.3–0.4
London clay (CH)	0.23
Chicago silty clay (CL)	0.18–0.25
Mexico City clay (MH)	7–10
Bangkok clay (CH)	0.5–1.52
Singapore marine clay (CH)	0.77–0.97
Shanghai clay (CL)	0.25–0.35
Taipei silty clay (CL)	0.33–0.50

Table 2.8 Some Empirical Equations for C_c (after Rendon-Herrero, 1980)

Equation	Regions of Applicability
$C_c = 0.009(LL - 10)$[a]	Normally consolidated clays
$C_c = 0.007(LL - 7)$[b]	Remolded clays
$C_c = 0.156e_0 + 0.0107$	All clays
$C_c = 1.15(e_0 - 0.35)$	All clays
$C_c = 0.30(e_0 - 0.27)$	Inorganic cohesive soil; silt, silty clay
$C_c = 0.75(e_0 - 0.50)$	Soils with low plasticity
$C_c = 0.01w_n$	Chicago clays
$C_c = 0.208e_0 + 0.0083$	Chicago clays
$C_c = 0.0115w_n$	Organic soils; peats; organic silt and clay
$C_c = 0.0046(LL - 9)$	Brazilian clays
$C_c = 1.21 + 1.055(e_0 - 1.87)$	Motley clays from Sao Paulo City
$C_c = 0.015(w_n - 8)$ or $C_c = 0.54(e_0 - 0.23)$[c]	Taipei silty clay

[a] Terzaghi and Peck (1967).
[b] Skempton (1944).
[c] Moh, et al. (1989).

Table 2.9 Typical Values of Compression Index and Swelling Index for Some Urban Soils

Soil Type	C_s/C_c
San Francisco bay mud (CH)	0.08–0.15
Boston blue clay (CL)	0.24–0.33
London clay (CH)	0.046
Chicago silty clay (CL)	0.12–0.17
Mexico City clay (MH)	0.04–0.05
Bangkok clay (CH)	0.13–0.22
Singapore marine clay (CH)	0.16–0.21
Shanghai clay (CL)	0.08–0.21
Taipei silty clay (CL)	0.15–0.32

The swelling index $\left(C_s\right)$ is the slope of the unloading curve. The value of the swelling index can also be obtained through laboratory tests. However, if test data are not available, the C_s/C_c ratio can be assumed to be between 0.05 and 0.1. Table 2.9 provides some typical C_s/C_c values for some urban soils for analysis and design references.

2.3 PRINCIPLE OF EFFECTIVE STRESS

The stresses acting at a point in soil can be differentiated as the total stress (σ), pore water pressure (u), and the effective stress (σ'), which have the following relation:

$$\sigma = \sigma' + u \tag{2.2}$$

The total stress can also be called the body stress, which is produced by the gravitational force acting on the mass. Thus, the total stress acting at a point in soil is the sum of the unit weights of all the materials (including solids and water) above the point. When the pore water pressure in soil changes, the unit weight of soil remains unchanged, and therefore, the total stress remains unchanged. As shown in Figure 2.1a, the total stress at a depth of $H_1 + H_2$ is:

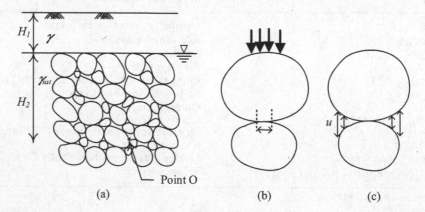

Figure 2.1 Principle of effective stress. (a) Soil profile, (b) contact area between particles, and (c) effect of pore water pressure on the effective stress.

$$\sigma = \gamma H_1 + H_2 \gamma_{sat} \tag{2.3}$$

where γ = unit weight of soil above the groundwater table and γ_{sat} = saturated unit weight of soil.

The contact stress between soil particles is the intergranular contact force divided by contact area. Because the contact area between soil particles is very small, the contact stress is enormous, as shown in Figure 2.1b. The effective stress is defined as the average contact stress on a unit cross-sectional area, which consists of solid areas and pore areas. However, the effective stress is not the true contact stress between soil particles but can represent the intergranular contact force between soil particles. The larger the effective stress, the larger the contact stress.

Since the effective stress represents the intergranular contact forces over a unit gross cross-sectional area, judging from the mechanical point of view, the intergranular friction (i.e., shear strength) increases with the increase in the effective stress. Moreover, the larger the effective stress, the larger the soil particles subject to the contact force, and the greater the soil compressibility. Obviously, the shear strength (drained or undrained) and compressibility of soil relate only to the effective stress rather than the total stress.

The change in the stress environment may cause the pore water pressure in soil to change. However, the change in the pore water pressure does not necessarily cause the total stress to change. As long as the unit weight of soil and water above the point does not change, the total stress remains unchanged. From Eq. 2.2, it can be seen that under the condition of constant total stress, the increase in pore water pressure will lead to the decrease in effective stress. It can also be seen from Figure 2.1c that when the pore water pressure increases, the particles in contact with each other tend to be pushed away; therefore, the contact stress between the particles decreases, and the effective stress decreases. The shear strength of the soil (called the particle friction between) reduces as a result.

Assuming the pore water pressure comes from only the weight of water (e.g., there is no seepage or excess pore water pressure), then it is called the hydrostatic water pressure. The pore water pressure at a point O in Figure 2.1a is:

$$u = H_2 \gamma_w \tag{2.4}$$

where γ_w = unit weight of water.

Then, the effective stress at O is:

$$\sigma' = \sigma - u = H_1 \gamma + H_2 (\gamma_{sat} - \gamma_w) = H_1 \gamma + H_2 \gamma' \tag{2.5}$$

where γ' = the submerged unit weight of the soil or the effective unit weight.

If the pore water pressure at point O is not in the hydrostatic condition, the pore water pressure is:

$$u = H_2 \gamma_w + u_e \tag{2.6}$$

where u_e is the excess pore water pressure, implying the pore water pressure beyond the hydrostatic water pressure.

The excess pore water pressure may be positive or negative. According to Eq. 2.2, we can understand that when the positive excess pore water pressure is generated in the soil, the effective stress decreases and the shear strength decreases as a result, and vice versa.

2.4   FAILURE OF SOILS

2.4.1   Mohr–Coulomb failure theory

Mohr (1900) predicts that when the major (σ_1) and the minor (σ_3) principal stresses acting on a point in a material have a specific functional relation, the material will fail. The relation can be expressed as:

$$(\sigma_1 - \sigma_3) = f(\sigma_1 + \sigma_3) \tag{2.7}$$

where f indicates the specific function, which varies with the type of the material.

Figure 2.2a shows a Mohr's circle depicted according to Eq. 2.7. Different combinations of σ_1 and σ_3 that engender failure will form different Mohr's circles, from which a Mohr failure envelope can be obtained, as shown in Figure 2.2b. The function that represents the Mohr failure envelope is f in Eq. 2.7, indicating that a specific function at failure exists between σ_1 and σ_3.

Coulomb (1776) suggests that the shear strength of soils can be represented as follows:

$$\tau_f = c + \sigma \tan\phi \tag{2.8}$$

where τ_f = the shear strength of soil; σ = the normal stress on the failure surface; c = the cohesion; and ϕ = the angle of the internal friction.

The cohesion and internal friction angle are also called the strength parameters of soil.

Though stress can be distinguished into total stress and effective stress, only the latter is relevant to the engineering behavior (such as strength and deformation), as explained by the principle of effective stress. Thus, Eq. 2.8 should be rewritten as follows:

$$\tau_f = c' + \sigma' \tan\phi' \tag{2.9}$$

where c' and ϕ' denote the effective cohesion and the effective angle of internal friction. σ' represents the effective normal stress on the failure surface.

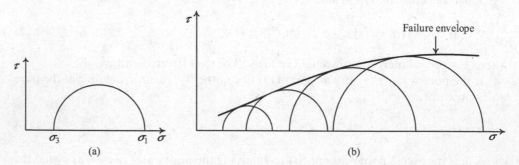

Figure 2.2 Mohr's theory. (a) Mohr's circle and (b) envelope of Mohr's circles.

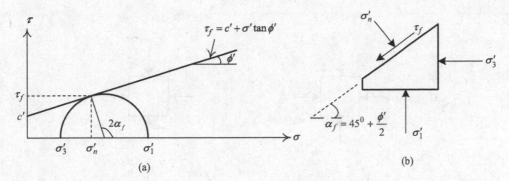

Figure 2.3 Failure angle determined by Mohr's circle. (a) Mohr's circle and its envelope and (b) stresses on the failure surface.

In soil mechanics, it is assumed that the Mohr failure envelope, as illustrated in Figure 2.2b, can be represented by Coulomb's equation, i.e., Eq. 2.9. As Figure 2.3 shows, with the above assumption, according to the theory of Mohr's circles, the angle $\left(\alpha_f\right)$ between the failure surface and the direction of the major principal stress is:

$$\alpha_f = 45^0 + \frac{\phi'}{2} \tag{2.10}$$

It should be noted that the above ϕ' is the effective friction angle because soil failure is related to the effective stress and has nothing to do with the total stress according to the principle of effective stress.

In addition to the Mohr–Coulomb failure criterion, some other commonly used criteria are von Mises, Tresca, and Drucker–Prager criteria. However, many studies (e.g., Lade and Duncan, 1973) have shown that the Mohr–Coulomb failure criterion is the most accurate criterion available to predict soil failures.

2.4.2 Commonly used laboratory shear strength tests

The triaxial test, direct shear test, and the direct simple shear (DSS) test are the commonly used laboratory shear strength tests to obtain required strength parameters of soil. The details of the tests can refer to relevant literatures.

There are at least three types of triaxial tests, depending on the drained conditions: the consolidated drained (CD) test, the consolidated undrained (CU) test, and the unconsolidated undrained (UU) test. The loading and drained conditions at the first and second stages in the tests are shown in Figure 2.4. The above CD and CU tests usually go through the consolidation stage under isotropic compression; i.e., the consolidation stresses in each direction at the first stage are made equal (i.e., $\sigma_1 = \sigma_2 = \sigma_3$). Thus, the CD and CU tests are also designated as CID and CIU tests where "I" indicates "isotropic." Besides, to simulate the K_0 consolidation condition of the in situ soil before shearing, we sometimes have specimens consolidated under the K_0 condition

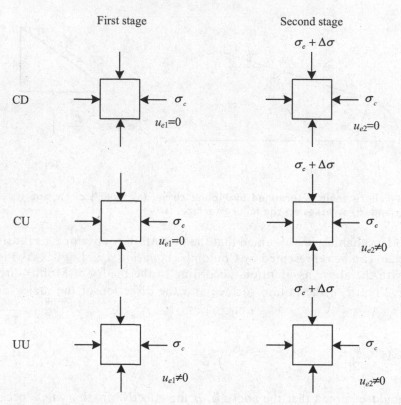

Figure 2.4 Loading patterns and drainage conditions of the CD, CU, and UU tests (u_{e1} and u_{e2} are excess pore water pressures).

and then carry out the shear test. Thus, the CD and CU tests are also called CK_0D and CK_0U tests.

Theoretically, the triaxial test is applicable to all kinds of soil. However, a tube specimen has to be trimmed before putting it to the triaxial test. With no cohesion, however, trimming is impossible for sand. Therefore, triaxial tests are not suitable for the strength testing of sand unless a special testing method is adopted. In addition, the traditional triaxial cell is too small for gravelly soil and a larger-size triaxial cell is usually adopted.

The soil specimen of the direct shear test can be directly pressed into the shear box from a thin tube. Therefore, if the soil specimen cannot stand on its own (such as sand), the direct shear test can still be performed. Sandy soil has high permeability, and the excess pore water pressure is always equal to 0 during testing. Therefore, the drained test (so-called CD test) is usually conducted to obtain the strength parameters of the soil (c', ϕ'). Therefore, although the direct shear test cannot control the drained condition, it can still be carried out to obtain the strength parameters of sandy soil. On the other hand, the permeability of sand is high, and the direct shear test cannot control its drained condition, so the CU and UU tests of sand cannot be performed with the direct shear test.

Theoretically, we can perform CD, CU, and UU tests using the direct shear test for clayey soils by controlling the testing rate. According to the principle of the CD test, it is workable to carry out the CD test on clayey soils using the direct shear test. However, the procedures are not simpler than those of the triaxial test, and therefore, the latter is more often adopted. Since the direct shear test can't control the drained condition and can't measure the excess pore water pressure during CU testing, the testing results are meaningless, even though CU tests can be carried out to obtain the total strength parameters by manipulating the test rate. What's more, as will be explained in Section 2.6.1, the undrained shear strength of a clayey soil is the shear strength on the failure surface by definition. The failure surface of the direct shear test is forced to be the horizontal plane. Thus, the results of the UU test on clayey soils from the direct shear test are questionable. Besides, the inability to control drainage of the direct shear test makes it a less favored choice for most people conducting the UU test on clayey soils.

Considering many weaknesses of the direct shear test, especially the inability to be applied to clayey soils, the direct simple shear (DSS) test apparatus was developed by modifying the direct shear apparatus. The specimen of the direct simple test can be either cylindrical or cubic, which is confined by a wire-reinforced rubber membrane to prevent it from lateral deformation. Shear acts on the top of the specimen, and a pure shear condition is thus generated.

Without lateral deformation produced, soil specimens in direct shear simple tests are subject to pure shear stress and are all in the state of plane strain. The DSS test overcomes most of the weaknesses of the direct shear test: the failure surface forced to be a horizontal plane and the inability to control drainage. The problem of the uneven distribution of stresses is also basically solved. Thus, the DSS test is widely adopted in Europe and America.

2.4.3 Stress path and stress path tests

Traditional triaxial CD and CU tests are to apply isotropic consolidation pressure and then apply axial pressure to generate shear stress in the soil specimen. Since the strength of the soil is related to the way the pressure acts to the test specimen, sometimes in order to obtain the shear strength of the in site soil reasonably, the pressure applied in the strength test must simulate the process (path) of the in situ soil subject to loading (or unloading). The applied pressure is not necessarily in the axial (or vertical) direction. Those types of tests are called the stress path tests. During the stress path test, the stress state of the soil specimen at any stage can be expressed by the following equation (Figure 2.5):

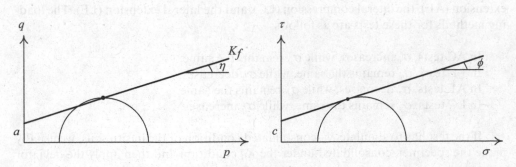

Figure 2.5 K_f line and Mohr and Coulomb's failure envelope.

$$q = \frac{\sigma_v - \sigma_h}{2} \tag{2.11}$$

$$p = \frac{\sigma_v + \sigma_h}{2} \tag{2.12}$$

where σ_v = total vertical stresses and σ_h = total horizontal stresses.

Depict the p and q during shearing on the p-q diagram, which denotes a serial of shearing. The path of shearing is designated as the total stress path (TSP). Eqs. 2.11 and 2.12 can also be rewritten in terms of the effective stress:

$$q = \frac{\sigma_v' - \sigma_h'}{2} \tag{2.13}$$

$$p' = \frac{\sigma_v' + \sigma_h'}{2} \tag{2.14}$$

Plot the values of p' and q at each stage, and we have the path that is designated as the effective stress path (ESP). The Mohr failure envelope can be transformed onto the p-q diagram, which is called K_f line. Comparing the geometric relationship between the K_f line and the Mohr failure envelope, there exists the following relationship (refer to Figure 2.5):

$$\sin\phi = \tan\eta \tag{2.15}$$

$$c = \frac{a}{\cos\phi} \tag{2.16}$$

where a and η are respectively the intercept on the q-axis and the angle of the K_f line with respect to the horizontal.

Depict the TSP and the ESP for an undrained test onto the same p-q diagram, and the variation of the pore water pressure during the test can be easily observed, which will help one to understand the strength behavior of soils.

The stress path test can be carried out using special testing apparatus (the DSS testing apparatus, for example) or by increasing or decreasing the horizontal (confining pressure) and the vertical pressures in the triaxial test. Generally speaking, the commonly used triaxial stress path tests include the axial compression (AC), the axial extension (AE), the lateral compression (LC), and the lateral extension (LE). The loading methods for these tests are as follows:

In AC tests, σ_v increases, while σ_h remains the same.
In LE tests, σ_v remains the same, while σ_h decreases.
In AE tests, σ_v decreases, while σ_h remains the same.
In LC tests, σ_v remains the same, while σ_h increases.

If the test has to simulate K_0-consolidated condition of the in situ soils, we usually make the specimen consolidated under the K_0 condition and then apply the deviator

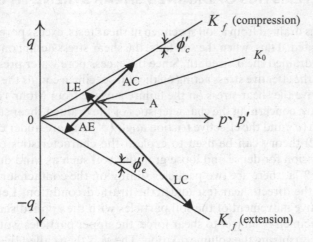

Figure 2.6 Drained stress paths for normally consolidated soils (or total stress paths).

Table 2.10 Effective Friction Angle of Undisturbed Clay from CK_0U Compression Test and Extension Test (Ladd et al., 1977)

Soil Type	LL PI	Compression Test		Extension Test		Reference
		ε_f (%)	ϕ_c' (deg)	ε_f (%)	ϕ_e' (deg)	
Undisturbed sensitive Haney clay	44 18	0.4	25.2	10.5	34.3	Vaid and Campanella (1974)
Reconstituted Boston blue clay	41 21	0.4	29	4.3	≤40	Ladd et al. (1971)
Undisturbed AGS CH clay	71 40	1.0	36.5	8.2	36.5	Unpublished data from the University of British Columbia (1975)
San Francisco bay mud	88 45	3.6	38	10.2	35	Duncan and Seed (1966)

ε_f: failure strain.

stress onto it. Figure 2.6 illustrates the TSPs of the K_0-consolidated specimens for AC, AE, LC, and LE tests.

Ou et al. (1995) carried out a series of triaxial stress path tests on the Taipei silty clay. The stress paths included the above-mentioned four paths: AC, AE, LE, and LC. The test types included CD and CU tests. Results show that the effective friction angles are the same despite the differences in stress paths and drained conditions as long as the deformation directions of the specimens are the same. For example, the effective friction angles of the AC and LE tests are the same, while those of the AE and LC tests are equal also. Table 2.10 provides the results of the CK_0U tests under the plane strain condition on four different soils by Ladd et al. (1977). Their study shows that the effective friction angle obtained from AC tests (ϕ_c') is larger than that from the AE test (ϕ_e'). However, the differences lessen with the increase in the plasticity of soils.

2.5 CHARACTERISTICS OF DRAINED SHEAR STRENGTH OF SOILS

If the pore water is drained from a soil specimen in shear tests, excess pore water pressure will not be generated. Thus, when the soil fails, the shear stress acting on the failure surface is called the drained shear strength. Since the excess pore water pressure in the soil specimen is zero, the effective stress acting on the soil specimen equals the total stress and we can easily derive the shear stress on the failure surface from Mohr's circle as shown in Figure 2.3. Thus, concerning the characteristics of the drained shear strength, it is the effective cohesion (c') and the effective friction angle (ϕ') that are under consideration.

Rowe's (1962) theory can be used to explain the characteristics of volume dilation and compression for dense and loose granular soil such as sand during shear. As shown in Figure 2.7a, there are two particles sliding on the contact surface in a dense granular soil in the direct shear test under the drained condition. Let ψ denote the direction of relative movement of the soil particles with the applied shear force; then, when the soil specimen is subject to shear force, the upper particles would move along the sliding surface, causing the volume dilation. The ψ is then called the dilation angle. The applied shear force must overcome the work done by the volume dilation before it can continue to produce shear strain.

When the upper particles "climb" up to the top of the lower particles, the shear force acting on the soil achieves a peak, and the corresponding friction angle is called the peak friction angle (ϕ'). After that, the shear force acting on the soil and dilation gradually decrease as the soil particles continue to be sheared until the volume no longer changes, and the acted shear force no longer reduces and reaches constant. At this time, the shear strength of the soil almost entirely comes from the frictional resistance between the surfaces of the soil particles. The corresponding friction angle is called the critical state friction angle, and the corresponding void ratio is called the critical state void ratio. At this time, the soil can be said to be at a critical state, and the dilation angle ψ is equal to 0.

Draw the forces acting on the contact point and form a force polygon diagram as shown in Figure 2.7b, and then, we can obtain $T = N \tan(\phi'_\mu + \psi)$ or $\tau_f = \sigma \tan(\phi'_\mu + \psi)$.

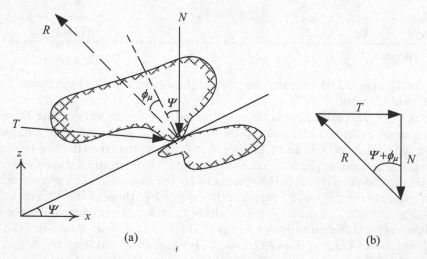

(a)   (b)

Figure 2.7 Rowe's theory. (a) Forces at the contact point and (b) force polygon diagram.

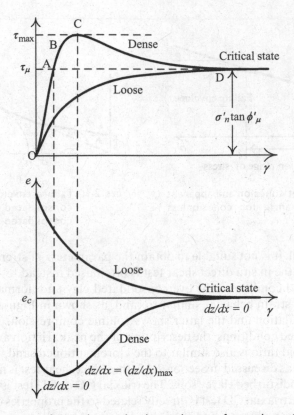

Figure 2.8 Stress–strain relationship and volume change of granular soils.

Then, it is understandable that $\phi' = \phi'_u + \psi$. The complete stress–strain behavior of dense soil is shown in Figure 2.8.

Figure 2.8 also shows that loose sand is compressed all the way to the critical state. The friction angle of loose sand at the critical state is equal to that of dense sand because at this state the shear strength only comes from the friction resistance between the surfaces of soil particles and has nothing to do with the density of soil.

As discussed in Section 2.4.2, for undisturbed sands, c' and ϕ' are usually obtained from the direct shear test. Though completely undisturbed specimens of sands are not easily obtained, the c' and ϕ', which are basically related to the friction properties of material, are less affected by the disturbance of specimens. On the other hand, the Mohr failure envelope for cohesionless soils is usually a curve over a wide range of stress, and one has to derive the curve from regression using the Coulomb equation within the range of stress interested to us. As shown in Figure 2.9, the thus-acquired c' and ϕ' are not the true c' and ϕ' of the soil. They simply represent the parameters proper for the segment within a range of stress which interests to us and are designated separately as the apparent cohesion (c') and the apparent friction angle (ϕ'). As a matter of fact, there is no cohesion between particles for cohesionless soils. When calculating the lateral earth pressure using the apparent cohesion and the apparent friction angle, we may have a negative earth pressure, which means that the soil is subjected to tension forces and thus is unreasonable.

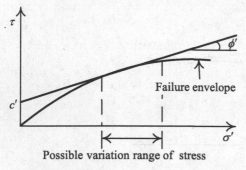

Figure 2.9 Apparent cohesion and apparent friction angle for cohesionless soils.

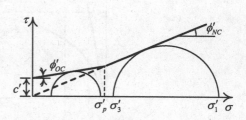

Figure 2.10 Failure envelopes for the over-consolidated clay and normally consolidated clay.

For gravel soil, it is not suitable to obtain the parameters of strength from a laboratory test. Thus, the in situ direct shear test is often used instead.

Under drained conditions, the overconsolidated clay and normally consolidated clay have a stress–strain behavior similar to sand, as shown in Figure 2.8. The former exhibits volume dilation and the latter shows volume compression. Therefore, in the case of fully drained conditions, the description of the peak friction angle, critical friction angle, and void ratio is also similar to the aforementioned sand.

Furthermore, as discussed in Section 2.4.2, the direct shear test is not suitable to obtain c' and ϕ' of undisturbed clayey soils. The triaxial CD or CU test is usually adopted. Basically, since the triaxial CD test is directly related to the properties of friction among particles of clay, it can obtain c' and ϕ' of soils directly. As to the triaxial CU test, a type of undrained shear test, its results are also affected by the effective stress acting on the soils. If the pore water pressure is measured during the test, the effective stress on soil can be known. The testing results can thus be expressed in terms of the effective stress, and c' and ϕ' are thus obtained. Since the effective stress represents the intergranular contact stress among particles, the results of the triaxial CU test conducted following the above principles are also related to the properties of intergranular friction, and thus, the c' and ϕ' should not be different from those obtained from the CD test.

The c' of normally consolidated soils equals 0. The c' of overconsolidated soils, having been compressed, doesn't equal 0. However, when the consolidation pressure or confining pressure in the CD/CU test is greater than the preconsolidation pressure of the soil, the consolidation pressure or confining pressure becomes the maximum pressure which has acted on the soil. Under such conditions, the soil reverts to the normal consolidation and $c' = 0$ as shown in Figure 2.10. The typical c' and ϕ' of clays for some urban areas are also listed in Table 2.11.

2.6 CHARACTERISTICS OF UNDRAINED SHEAR STRENGTH OF SATURATED COHESIVE SOILS

2.6.1 Principle of undrained shear strength

Let the pore water not be drained from a soil specimen during the shear strength test. Excess pore water pressure is thus generated. Therefore, when the soil comes to failure,

Table 2.11 Typical Values of Effective Cohesion and Effective
Angle of Friction for Some Urban Soils

Soil Type	c' (kN/m^2)	ϕ' (degree)
San Francisco bay mud (CH)	0	32.5–35[a]
Boston blue clay (CL)	0	32–34[a]
Chicago silty clay (CL)	0	24–29
London clay (CH)	10	25
Mexico City clay (MH)	0	34–47
Bangkok clay (CH)	0	24–26
Singapore marine clay (CH)	0	22–25
Shanghai clay (CL)	0	30–35
Taipei silty clay (CL)	0	29–32[a]
Taipei silty sand	0	32.5–34

[a] Parameters at normally consolidated state.

the shear stress acting on the failure surface is called the undrained shear strength, as illustrated by τ_f in Figure 2.11a. The circle B in Figure 2.11c is the corresponding total stress Mohr's circle.

According to the principle of effective stress, in the undrained state, soils fail only when its effective stress Mohr's circle is tangent to the effective failure envelope, and it has nothing to do with the total stress. As illustrated in Figure 2.11c, circle A is the effective Mohr's circle tangent to the effective failure envelope. Due to the inability of the pore water in soils to bear the shear stress, the shear stress on the failure plane in

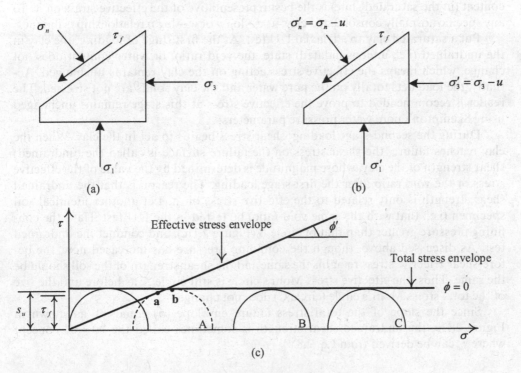

Figure 2.11 Basic principle of undrained shear strength.

terms of the effective stress equals that in terms of the total stress, as shown by τ_f in Figure 2.11b. Since τ_f is identical in both the total stress and effective stress Mohr's circles, the size of the effective Mohr's circle of should be the same as that of the total stress Mohr's circle. Known from Figure 2.11c, the shear stress on the failure plane (point **a** in Figure 2.11c) can be expressed in the following form:

$$\tau_f = \frac{\sigma_1' - \sigma_3'}{2}\cos\phi' = \frac{\sigma_1 - \sigma_3}{2}\cos\phi' \tag{2.17}$$

where τ_f denotes the shear stress on the failure plane, which can be defined as the undrained shear strength of the soil.

Observing the nearness of the top point (**b** in Figure 2.11c) to the tangent point (**a** in Figure 2.11c) of the Mohr's circle and, on the other hand, considering the inconvenience to have the c' and ϕ' of a clay through tests for the calculation of its undrained shear strength using Eq. 2.17, the value of the shear strength at the top point of the Mohr's circle is adopted to calculate the undrained shear strength of soils as follows:

$$s_u = \frac{\sigma_1 - \sigma_3}{2} \tag{2.18}$$

Though Eq. 2.18 gives some error in the computation of the undrained shear strength of clay, it has the merit of leaving out c' and ϕ' and is therefore frequently used in engineering practice. From the above discussions, we know that the undrained shear strength of clay relates only to the effective stress acting on it and has nothing to do with the total stress. As far as normally consolidated clay is concerned, the void ratio or water content (in the saturated state) is the best representative of the effective stress on it. To any specific normally consolidated clay, its $e - \log p$ or $w - \log p$ relationship is unique.

Put a saturated clay to a triaxial UU test. At the first stage of loading, the clay in the undrained (i.e., unconsolidated) state, the void ratio, or water content does not change, which means the effective stress acting on the clay remains unchanged. Actually, the load acts totally on the pore water and the clay solids are not stressed. The reader is recommended to prove the effective stress at this stage remains unchanged using Skempton's pore water pressure parameters.

During the second-stage loading, shear stress begins to act in the clay. When the clay reaches failure, the shear stress on the failure surface is called the (undrained) shear strength of the clay, whose magnitude is determined by the value of the effective stress or the void ratio after the first-stage loading. The reason is that the undrained shear strength is only related to the effective stress on it. Let another identical soil specimen (i.e., that with the same void ratio) be tested in the UU test. Have the confining pressure greater than that of circle B (Figure 2.11c), and conduct the undrained test. As discussed above, though the confining pressure has increased now, the before-shear effective stress remains the same and the shear strength of the soil should be the same. Thus, the effective stress Mohr's circle is still circle A as before and the size of the total stress Mohr's circle (circle C) does not change.

Since the slope of the total stress failure envelope (ϕ) equals 0, as shown in Figure 2.11c, the undrained shear strength is sometimes called the "$\phi = 0$" concept where s_u can be derived from Eq. 2.8.

However, there is a pitfall that needs to be explicated. Though the total stress failure envelope can be represented by Eq. 2.8, it seems workable to substitute $\phi = 0$ into the equation to obtain the undrained shear strength, which will be the same as that computed by Eq. 2.18. The procedure is wrong according to the principle of effective stress. Eq. 2.8 should not be used to compute the undrained shear strength.

2.6.2   Characteristics of undrained shear strength

The undrained shear strength of soil has the following characteristics:

1.   Strength–depth relation

As discussed in the above section, the undrained shear strength of the soil relates only to the effective stress acting on it and has nothing to do with the total stress. For a normally consolidated clayey deposit, the effective stress increases with depth. As a result, the undrained shear strength increases with depth. However, for clayey soils near the ground surface, frequently influenced by the change of groundwater level, the desiccation, and the capillarity, the soil normally exhibits overconsolidated behavior. The void ratio of overconsolidated soil, having been compressed, is smaller than that of normally consolidated soil under the same effective stress. Thus, the undrained shear strength of overconsolidated soil is greater than that of normally consolidated soil under the same effective stress. Figure 2.12 displays the typical relationship between the undrained shear strength and the depth.

The average value of the undrained shear strengths within the range of the stress influence is often used for analysis. As shown in Figure 2.12, to analyze the bearing capacity of an individual footing with a width of B, the average undrained shear strength $\left(s_{u,\mathrm{avg}}\right)$ within the range equal to the foundation width from the foundation bottom is used to compute the bearing capacity, using bearing capacity equations.

The variation of excess pore water pressure in underconsolidated soil is usually uncertain. As a result, the effective stress may not increase with depth. That is, the undrained shear strength of soil does not necessarily increase with the depth, either. Since the effective stress in underconsolidated clay is low, the undrained shear strength of underconsolidated clay might also be low and often exhibit constant value, not increasing with depth.

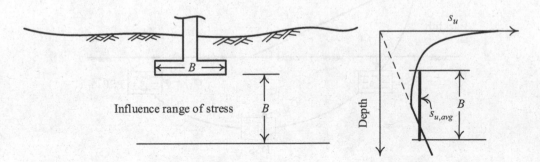

Figure 2.12 Profile of undrained shear strength and bearing capacity of footings.

2. Influence of sampling disturbance

According to many studies (Ladd and Lambe, 1963; Seed, et al., 1964), the undrained shear strength of clay is particularly affected by sampling disturbance. The main source of disturbance is the relief of the total stress during the sampling process, which brings about a negative value of the pore water pressure within soil (due to the undrained condition, as illustrated in Example 2.4). The absolute value of the negative pore water pressure might decrease because of vibration or draining and decreases, in turn, the effective stress. The undrained shear strength obtained by the test might be smaller than the true in situ value.

3. Anisotropic behavior

Take an example of an embankment on a clay deposit, as shown in Figure 2.13, and each point below the embankment on the failure surface is subject to a different stress path. The directions of the major principal stress at points **a**, **b**, and **c**, as shown in Figure 2.13, are different. Thus, when conducting laboratory tests, letting the directions of loading be consistent with those of the major principal stress at points **a**, **b**, and **c**, respectively, the resulting undrained shear strengths will be different accordingly. Since the conventional boring and loading directions are both vertical, the results from conventional triaxial test procedure (vertical sampling and vertical loading) can represent the undrained shear strength at point **a** in Figure 2.13, where the obtained undrained shear strength is greater than that obtained by other stress path tests (**b** and **c** in Figure 2.13).

4. Strain rate dependent

According to many studies (e.g., Bjerrum, 1971), when shearing in the undrained state, the slower the strain rate, the smaller the undrained shear strength. The reason is that, with a low strain rate, creep is produced within soil and the excess pore water pressure increases, which in turn decreases the undrained shear strength. Because the laboratory triaxial tests (UU and CU tests) have high strain rates, the specimens usually fail within 1 hour. However, the in situ undrained shearing strain rate at failure (as shown in Figure 2.13) is often much smaller than that applied in laboratory tests. Thus, the differences between the strain rate in the field and that in laboratory tests should be considered.

5. Normalized soil behavior

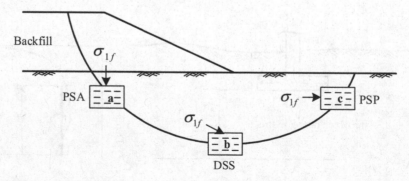

Figure 2.13 Failure of soft soil under an embankment.

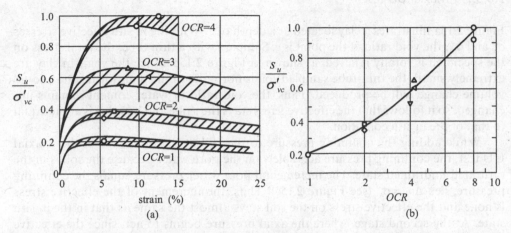

Figure 2.14 Normalized behavior of normally consolidated and overconsolidated Boston blue clay (Ladd and Foote, 1974) (τ_h = shear stress; σ'_{vc} = vertical consolidation stress; and s_u = undrained shear strength).

According to Ladd and Foote (1974), saturated clay with the same OCR may have similar stress–strain behavior though the effective stresses acting on the soil, which refers to the consolidated stress at the first loading stage in a CU test or the effective overburden stress on the specimen before sampling in a UU test, are different. If we normalize the above stress–strain relation with respect to the effective stress, the normalized stress–strain relation will fall on a narrow band, like a unique stress–strain relation, which is often called the normalized soil behavior. Ladd and Foote also found that clays usually have the normalized behavior, except for those with a high degree of structure, such as quick clays or naturally cemented clays.

Figure 2.14a illustrates the normalized stress–strain behavior of Boston blue clay by the direct simple shear CK_0U test. The strength and stiffness of clay with the same OCR can be represented by the normalized parameters, such as the normalized undrained shear strength (s_u/σ') and the normalized stiffness (E_u/σ'), where σ' is the average principal stress (i.e., $\sigma' = (\sigma'_1 + \sigma'_2 + \sigma'_3)/3$) or the vertical effective overburden pressure (σ'_v).

6. Plane strain

Basically, the failure pattern of the soil under an embankment, as shown in Figure 2.13, is in the plane strain condition. Thus, the stress states at point **a** are σ'_1, $\sigma'_2(= K_0\sigma'_1)$, and σ'_3, while those in the conventional triaxial test are σ'_1, $\sigma'_2(= \sigma'_3)$, and σ'_3. According to Ladd and Foote (1974), the undrained shear strength obtained from a triaxial test is smaller than that from a plane strain test under the same conditions. The definition of plane strain can be referred to Appendix C.

2.6.3 Methods to obtain the undrained shear strength

The undrained shear strength of a saturated soil can be determined by the following methods:

2.6.3.1   Triaxial UU test

Figure 2.15a illustrates a clay located at a depth of z. With the in situ effective stresses σ'_v and σ'_h, the void ratio at the point is e. Sampled with a thin tube, the total stress on the specimen is totally relieved, as shown in Figure 2.15b. Since the voids in clay are extremely small, the thin-tube sampled soil is normally in the undrained state and no volume change will be produced. Thus, the void ratio or water content remains unchanged, so it follows that the effective stress stays the same and its value is about equal to that of the in situ condition.

While adding the confining pressure to the soil at the first stage in the triaxial UU test, the confining pressure acts solely on the pore water because the soil is in the saturated undrained state. The incremental pore water pressure equals the confining pressure, i.e., $\Delta u_1 = \sigma_c$ (see Figure 2.15c). Thus, the increment of the effective stress is none and the effective stress on the soil stays almost the same as that in the in situ state. At the second stage, where the axial pressure begins to act, since the effective stress on the soil specimen before axial loading is close to that of the in situ soil, the obtained strength can therefore represent the undrained shear strength of the in situ soil. The change in the pore water pressure and the effective stress due to sampling is as illustrated in Example 2.4.

Theoretically, to obtain the in situ undrained shear strength by way of lab tests, a completely undisturbed soil specimen is necessary. Besides, the stress condition, the stress path, and the strain rate of the in situ soil must be simulated in laboratory tests. Since soil specimens are often disturbed during sampling, the results of triaxial UU tests are largely affected by sampling disturbance. What's more, soil specimens are subjected to vertical loading and triaxial UU tests have a larger strain rate than the

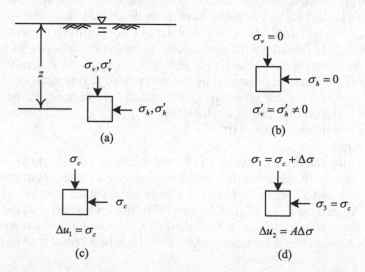

Figure 2.15 Undrained shear strength of saturated clay obtained from triaxial UU tests. (a) State of stresses at a depth of z; (b) state of stresses of the soil specimen after sampling; (c) loading at the first stage of the UU test; and (d) loading at the second stage of the UU test. (Note: A is a Skempton pore water pressure parameter.)

in situ soil at failure. With so much difference, how can the results of triaxial UU tests represent the undrained shear strength of the in situ soil? The reason is that the discrepancy can be made up for by self-compensation. The higher strain rate and the vertical loading of triaxial UU tests bring about a larger undrained shear strength. On the other hand, the triaxial stress state and sampling disturbance lead to a lower value. The two groups of factors are compensated for each other in such a way that test results can suitably represent those of the in situ soil.

The unconfined compression (UC) test is a special triaxial UU test with no confining pressure. As shown in Figure 2.11, the effective stress Mohr's circle obtained from the UC test on a saturated soil specimen is identical with that obtained from the triaxial UU test as long as the specimens have the same effective stress (or void ratio). Without being held by confining pressure, the UC test specimen is capable of standing by itself because a negative pore water pressure is generated in the specimen, which renders the effective stress on the specimen equivalent to that of the in situ soil. Nevertheless, if the specimen has been disturbed, or its pores are relatively large, or a thin layer of permeable material exists in the specimen or there are fine cracks in the specimen, the negative pore water would dissipate, which in turn reduces the effective stress. Moreover, because the specimen is not restrained by the confining pressure, the undrained shear strength obtained from the UC test tends to be smaller.

The self-compensation effect of the triaxial UU test renders the obtained undrained shear strength capable of representing the in situ undrained shear strength as discussed above. Nevertheless, the effect is not always reliable: the extra amount does not necessarily equal the lowered amount. It counts on either the testing or sampling quality, or both.

2.6.3.2 CU test

The triaxial CU test produces a series of total stress Mohr's circles, which is able to form a failure envelope whose intercept on the y-axis c_T and slope ϕ_T can be determined, similar to the UU test results (Figure 2.16). As discussed earlier, the Mohr–Coulomb equation represents only the effective failure envelope. If we substitute c_T and ϕ_T into Eq. 2.8, an erroneous result will follow.

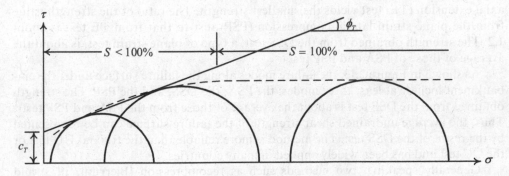

Figure 2.16 UU triaxial tests on saturated and unsaturated clays.

Table 2.12 Undrained Shear Strength of Boston Blue Clay from Various Strength Tests (Ladd and Foote, 1974)

Soil Type	s_u / σ'_{vc}	γ_f	$s_u / s_{u,TC}$	Mark
CK_0U(PSA)	0.34	0.8%	1.03	Plane strain compression test
CK_0U(TC)	0.33	0.5%	1.00	Conventional triaxial compression test
CK_0U(DSS)	0.20	6%	0.61	Direct simple test
CK_0U(PSP)	0.19	8.5%	0.57	Plane strain lateral compression test
CK_0U(TE)	0.155	15%	0.47	Conventional triaxial extension test

γ_f: shear strain at failure; σ'_{vc}: effective vertical pressure; definition of plane strain refers to Appendix E.

The shear stress on the failure surface derived from the triaxial CU test or the shear stress at the apex of the effective Mohr's circle is called the undrained shear strength, as shown by point **a** or **b** in Figure 2.11c. Since the triaxial CU test undergoes consolidation process, the undrained shear strength relates more to the consolidation pressure of the test than to the effective overburden pressure of in situ soil. If the consolidation pressure is greater than the effective overburden pressure, it follows that the undrained shear strength relates only to the consolidation pressure and no longer correlates with the effective overburden pressure of in situ soil. Besides, whether the CU test can represent the undrained shear strength of the in situ soil depends on whether the testing can fully simulate the in situ stressed condition. Similar to the UU triaxial test discussed earlier, the CU test has some problems to be resolved: the rate of strain, the plane strain, the anisotropy of strength, and the sampling disturbance.

The strain rate of the CU test should be capable of simulating the in situ condition. The rate of strain can be determined by referring to case histories before conducting the CU test. If the soil failure exhibits the characteristics of plane strain and strength anisotropy (e.g., the arc failure surface under the embankment in Figure 2.13), the CU test should simulate the condition of plane strain and the variation of strength along the failure surface. Table 2.12 provides the values of the undrained shear strength obtained from the CU stress path test on normally consolidated Boston blue clay according to Ladd and Foote's study (1974). The plane strain axial compression (PSA) test results in the largest strength. The triaxial compression (TC) test results, though a little smaller, are close to those of the PSA test. The triaxial extension (TE) test yields the smallest strength. The ratio of the strength value from the plane strain lateral compression (PSP) test to that from TE test is about 1.2. The strength obtained from the DSS test, a type of plane strain test, is about the average of those of PSA and PSP tests.

As shown in Figure 2.13, the failure modes along the failure surface under the embankment include at least three modes: the PSA, the DSS, and the PSP. The strength obtained from the DSS test is about the average of those from the PSA and PSP tests. Thus, the average undrained shear strength of the failure surface can be represented by the result of the DSS test. The method is more reliable than the triaxial UU test or the UC test and has been widely applied in many countries.

Generally speaking, two methods such as recompression (Bjerrum, 1973) and Stress History and Normalized Soil Engineering Properties (SHANSEP) (Ladd and Foote, 1974) can be used to eliminate the disturbance of the specimen during sampling.

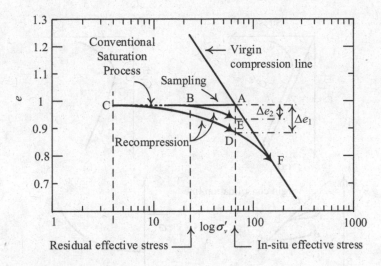

Figure 2.17 Recompression method.

The recompression method is first to recompress a specimen to its stress state close to the in situ condition and then to conduct the CU test. As shown in Figure 2.17, point A represents the in situ soil at the K_0-consolidated state. After sampling, first estimate the amount of retained effective stress, that is, residual negative pore water pressure, and then apply back pressure to the interior of the specimen and simultaneously confining pressure to saturate the specimen. The difference between the applied confining pressure and the backwater pressure is approximately equal to the effective stress retained in the specimen. At this time, the state of the specimen is at point B in Figure 2.17. Then, the soil specimen is compressed or consolidated to the in situ stress, as shown in point E in Figure 2.17. Finally, the undrained shear is conducted on the soil specimen at point E. Note that the traditional saturation procedure is to impose approximately equal confining pressure and back pressure. At this time, the net effective stress acting on the specimen is approximately 0 (point C), and the soil is in an overconsolidated state. When it is recompressed to the in situ state (point D in Figure 2.17), a relatively larger compression will be induced, which is unable to reproduce the stress state at the in situ condition. Estimation of the retained effective stress or residual pore water pressure can refer to relevant literatures (e.g., Cho and Finno, 2010; Teng et al., 2014).

The SHANSEP approach is another method of eliminating sampling disturbance. The SHANSEP approach (see Figure 2.18) can be described as follows (Ladd and Foote, 1974):

a. Divide the soil into several layers according to the results of subsurface exploration.
b. Establish the profile of the effective overburden pressure (σ'_{vo}) and the preconsolidation pressure (σ'_{vp}), as shown in Figure 2.18a, on the basis of the unit weight, the in situ pore water pressure, and the consolidation test.
c. According to the potential failure modes of the in situ soil, choose a suitable testing method which can best represent the failure modes of the in situ soil to conduct the CU test (PSA, PSP, or DSS, for example).

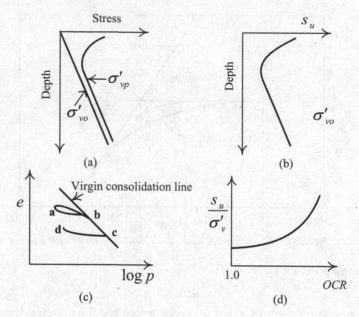

Figure 2.18 SHANSEP approach.

d. Examine whether the soil has normalized properties by seeing whether it exhibits a similar behavior type, as shown in Figure 2.14a.
e. At the first loading stage of the CU test (i.e., the consolidation stage), consolidate the specimen to the virgin consolidation curve and then unload it to the predetermined value of OCR before proceeding to the second stage (i.e., the shear test). As shown in Figure 2.18c, point **a** represents the initial state after sampling. The soil is consolidated to the virgin consolidation curve (point **c**) and then unloaded to point **d** where the CU test is going to be conducted. In this way, various values of normalized strength $\left(s_u/\sigma_v' \right)$ with the corresponding OCR value can be obtained, as shown in Figure 2.18d.
f. With the OCR values and the effective overburden pressures (σ_{vo}') at various depths (Figure 2.18a) and referring to Figure 2.18d, we can then compute the strengths at various depths, as shown in Figure 2.18b.

2.6.3.3   Field vane shear test

The device for the field vane shear test (FV test) consists of a connecting shaft and four vanes as shown in Figure 2.19a. The most commonly used vane is rectangular to obtain the average undrained shear strength in different directions. ASTM also suggests other different shapes of vanes to obtain anisotropic undrained shear strength for specific directions.

As shown in Figure 2.19b, to conduct the FV test, first bore to a depth 50 cm above the test depth and stay till the dirt is washed away, and then press the shaft and the

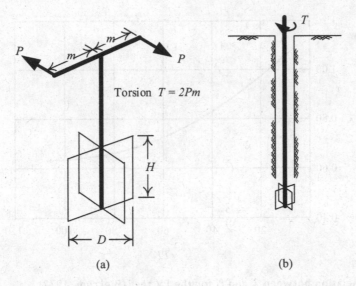

Torsion $T = 2Pm$

Figure 2.19 Field vane shear test (FV test). (a) FV test apparatus and (b) FV test apparatus in a boring hole.

vanes to the test depth. Install the torsion apparatus onto the connecting shaft and then have the torsion act on vanes with a constant rate to make them swirl till the soil within the vanes and outside them separates. Thus, the torsion (T) at failure is obtained.

The soil within the vanes is column-shaped at failure. As shown in Figure 2.19a, the torsion should equal the sum of the resistant moments of the soil strength on the top of the column (M_T), on the bottom of the column (M_B), and on the side of the column and can be expressed as follows:

$$T = M_T + M_B + M_S \tag{2.19}$$

Assume that the soil on the top, the bottom, and the side of the column all fail with equal undrained shear strength, which implies that the undrained shear strengths in all directions are the same. Thus,

$$M_S = \pi D H \frac{D}{2} s_u \tag{2.20}$$

$$M_T = M_B = \frac{\pi D^2}{4} \frac{2}{3} \frac{D}{2} s_u \tag{2.21}$$

Therefore,

$$T = s_u \left[\left(\pi D H \frac{D}{2} \right) + 2 \left(\frac{\pi D^2}{4} \frac{2}{3} \frac{D}{2} \right) \right] \tag{2.22}$$

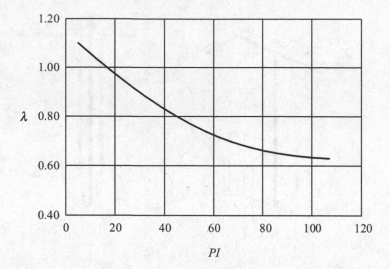

Figure 2.20 Relation between λ and *PI* for the FV test (Bjerrum, 1972).

Then,

$$s_u = \frac{T}{\pi\left(D^2 H/2 + D^3/6\right)} \tag{2.23}$$

where s_u = undrained shear strength of soil; D = diameter of the vanes; and H = height of the vanes.

As discussed above, the FV test causes little disturbance to the in situ soil. The testing results also represent the average strength of soils in various directions. However, its result often overestimates the strength of in situ soil, according to Bjerrum (1972). Therefore, Bjerrum suggested that the result of the FV test be modified to be applied to the design as follows:

$$s_{u,d} = \lambda s_{u,\text{FV}} \tag{2.24}$$

where $s_{u,d}$ = undrained shear strength for design; $s_{u,\text{FV}}$ = undrained shear strength obtained from the FV test; and λ = modification factor, related to the *PI* values of soil, which can be determined by referring to Figure 2.20.

2.6.3.4 Cone penetration test

The cone penetration test (CPT) is to push into soil a drilling rod with a mechanical or electrical cone and take the cone end resistance and side friction measurements. Some CPTs can also measure the pore water pressure, in addition to end resistance and side friction. Such an apparatus is called a CPTU. From these data, the type of the soil and its parameters of strength can be estimated.

Figure 2.21 illustrates the basic configuration of a CPTU apparatus. According to Robertson and Campanella (1989), the undrained shear strength of clayey soils can be obtained by the following equation:

$$s_u = \frac{q_t - \sigma_{vo}}{N_k} \qquad (2.25)$$

$$q_t = q_c + u(1 - a) \qquad (2.26)$$

where s_u = undrained shear strength of soil; q_c = cone end resistance; u = pore water pressure; σ_{vo} = total overburden pressure; a = area ratio, which equals D^2 / d^2 (see Figure 2.21); D = diameter at the top of the CPTU cone; d = diameter at the place where the pore water pressure is measured; and N_k = an empirical value that correlates with the value of OCR, the sensitivity, and the stiffness of the soil collectively. Usually, $N_k = 15 \pm 5$.

Since the N_k-value relates to the type of soil, to obtain the in situ undrained shear strength, a correlation between the result of the FV test (or the UU triaxial test) and that of the CPT is required to derive the N_k-value. Then, the value is used to obtain s_u for other CPTs.

Because of the continuity of CPT results, the results will not miss any soil layer no matter how thin it is. Thus, CPTs are used not only to obtain the parameters of strength, but also to classify layers of soil. The procedures are also simple. As a result, CPTs have been widely applied in recent years.

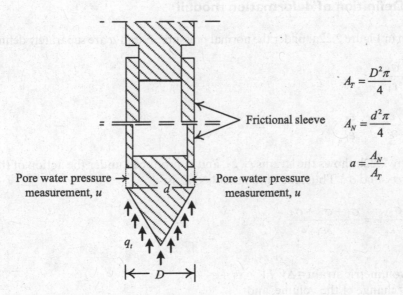

Figure 2.21 Configuration of a CPTU.

2.6.3.5   Other methods and empirical formulas

The undrained shear strength of saturated soils can also be obtained by a pressurometer or laboratory vane shear test. To save space, this book will pass them over. Interested readers may refer to relevant literatures.

2.7   UNDRAINED SHEAR STRENGTH OF UNSATURATED COHESIVE SOILS

Unsaturated soils refer to those whose voids are not fully filled by water or where the degree of saturation is less than 100%. Unsaturated soils possess suction, which pulls particles together tightly and increases their effective stress and strength accordingly. Therefore, with the suction in the soil known, the undrained shear strength of unsaturated soils can be computed according to the effective stress. The literature relating to failure theories of unsaturated soils based on effective stress can be found, for example, Fredlund (1997). Nevertheless, since the measurement of suction requires a special instrument, the application of the above failure theories is comparatively inconvenient. UU triaxial tests are more commonly adopted to obtain the undrained shear strength.

Figure 2.16 illustrates a typical test result from UU triaxial tests where the intercept (c) and the slope (ϕ) of the failure envelope are the parameters of strength. Due to the unsaturated state, ϕ is apparently not 0. The strength of the soil on the failure surface can be calculated using Eq. 2.18. Though the equation is expressed in total stress and doesn't conform to the principle of effective stress, some small discrepancy is allowed considering that not too much unsaturated soil is encountered in most of the deep excavation problems.

2.8   DEFORMATION CHARACTERISTICS OF SOILS

2.8.1   Definition of deformation moduli

As shown in Figure 2.22a, under the normal pressure, E and μ are separately defined as:

$$E = \frac{\sigma_1}{\varepsilon_1} \tag{2.27}$$

$$\mu = -\frac{\varepsilon_2}{\varepsilon_1} = -\frac{\varepsilon_3}{\varepsilon_1} \tag{2.28}$$

Figure 2.22b shows the strains ε_1, ε_2, and ε_3 produced under the action of the pressures σ_1, σ_2, and σ_3. Thus, the bulk modulus, K, is:

$$K = \frac{\sigma_{avg}}{\varepsilon_v} = \frac{\sigma_1 + \sigma_2 + \sigma_3}{3\varepsilon_v} \tag{2.29}$$

where
    ε_v = volumetric strain = $\Delta V / V \approx \varepsilon_1 + \varepsilon_2 + \varepsilon_3$;
    ΔV = change of the volume; and
    V = volume.

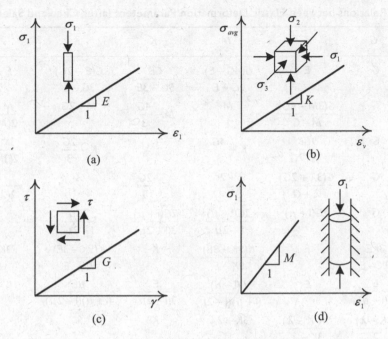

Figure 2.22 Definition of various deformation moduli. (a) Young's modulus, (b) bulk modulus, (c) shear modulus, and (d) constrained modulus.

Figure 2.22c shows the shear strain produced under the action of the shear stress, τ. Thus, the shear modulus is:

$$G = \frac{\tau}{\gamma} \tag{2.30}$$

where γ = shear strain.

Figure 2.22d displays the axial pressure, σ_1, on the material while the lateral strain is restrained. That is, the lateral strain is 0. On the other hand, we can say that the material is subject to one-dimensional compression. The constrained modulus, M or E_{oed}, is:

$$M = \frac{\sigma_1}{\varepsilon_1} \tag{2.31}$$

According to the theory of elasticity, the relationship between the deformation moduli E, μ, G, K, and M (or E_{oed}) can be derived, as shown in Table 2.13.

2.8.2 Various forms of Young's modulus and Poisson's ratio

Although deformation modulus includes Young's modulus, shear modulus, bulk modulus, and Poisson's ratio, for simplicity, this section only discusses Young's modulus and Poisson's ratio.

Table 2.13 Relations between Elastic Deformation Parameters (after Chen and Saleeb, 1982)

	G	E	M	K	λ	μ
G, E	G	E	$\dfrac{G(4G-E)}{3G-E}$	$\dfrac{GE}{9G-3E}$	$\dfrac{G(E-2G)}{3G-E}$	$\dfrac{E-2G}{2G}$
G, M	G	$\dfrac{G(3M-4G)}{M-G}$	M	$M-\dfrac{4G}{3}$	$M-2G$	$\dfrac{M-2G}{2(M-G)}$
G, K	G	$\dfrac{9GK}{3K+G}$	$K+\dfrac{4G}{3}$	K	$K-\dfrac{2G}{3}$	$\dfrac{3K-2G}{2(3K+G)}$
G, λ	G	$\dfrac{G(3\lambda+2G)}{\lambda+G}$	$\lambda+2G$	$\lambda+\dfrac{2G}{3}$	λ	$\dfrac{\lambda}{2(\lambda+G)}$
G, μ	G	$2G(1+\mu)$	$\dfrac{2G(1-\mu)}{1-2\mu}$	$\dfrac{2G(1+\mu)}{3(1-2\mu)}$	$\dfrac{2G\mu}{1-2\mu}$	μ
E, K	$\dfrac{3KE}{9K-E}$	E	$\dfrac{K(9K+3E)}{9K-E}$	K	$\dfrac{K(9K-3E)}{9K-E}$	$\dfrac{3K-E}{6K}$
E, μ	$\dfrac{E}{2(1+\mu)}$	E	$\dfrac{E(1-\mu)}{(1+\mu)(1-2\mu)}$	$\dfrac{E}{3(1-2\mu)}$	$\dfrac{\mu E}{(1+\mu)(1-2\mu)}$	μ
K, λ	$\dfrac{3(K-\lambda)}{2}$	$\dfrac{9K(K-\lambda)}{3K-\lambda}$	$3K-2\lambda$	K	λ	$\dfrac{\lambda}{3K-\lambda}$
K, M	$\dfrac{3(M-K)}{4}$	$\dfrac{9K(M-K)}{3K+M}$	M	K	$\dfrac{3K-M}{2}$	$\dfrac{3K(2M-1)+M}{3K(2M+1)-M}$
K, μ	$\dfrac{3K(1-2\mu)}{2(1+\mu)}$	$3K(1-2\mu)$	$\dfrac{3K(1-2\mu)}{1+\mu}$	K	$\dfrac{3K\mu}{1+\mu}$	μ

λ is the Lame parameter. For its definition, please see books on theory of elasticity. It is, though, rarely used in geotechnical analyses.

Consider the stress–strain behavior of a soil subject to a triaxial test as shown in Figure 2.23. Basically, Young's modulus is the ratio of stress increment to strain increment over a certain range of stress. Since soil is a highly nonlinear and plastic material, as the soil is subject to loading to a state and then unloading, from point B to point C, for example, a permanent or plastic strain will occur. However, according to the explanation in Section 2.8.3, as long as the stress state is on the unloading/reloading line, the behavior is elastic. Therefore, the slope (E_{ur}) of the unloading/reloading line is called the elastic Young's modulus.

The tangent Young's modulus is the slope of a line tangent to the stress–strain curve at a certain stress state. Figure 2.23 shows the initial Young's modulus (E_i) and tangent Young's modulus (E_t). It should be noted that it is unsuitable to derive E_i from the tangent line at the origin because the E_i highly depends on the resolution of the measurement device. For example, the smallest strain, 10^{-3} or 10^{-4}, that can be measured in conventional triaxial tests will result in different E_i values if it is derived directly by drawing a tangent line at the origin. The author recommends to adopt Duncan and Chang's (1970) approach to derive the E_i. The secant Young's modulus (E_s) is the slope of a line connecting the origin and a certain stress, point E, for example. If point E is at the 50% failure strength, $(\sigma_1 - \sigma_3)_f$, the E_s is usually expressed by E_{50}.

Furthermore, the Young's modulus obtained from the stress–strain curve with the drained shear test is called drained (or effective) Young's modulus. Similarly, the Young's modulus obtained from the stress–strain curve with the undrained shear test and expressed in terms of total stress is called undrained (or total) Young's modulus. Therefore, the E_s, E_{ur}, and E_i have the distinction between the effective and undrained E_s, E_{ur}, and E_i. See Section 8.5 for more discussion.

Table 2.14 gives a possible range of the E_s for various soils. The ranges shown in the table are wide because the values of Young's modulus of a soil are related to the effective stress on the soil, drainage condition of the soil, stress paths, and the depth of the soil. More discussions and estimation of Young's modulus for excavation problems can also refer to Section 8.5.

Poisson's ratio also has a distinction of μ_i and μ_t, which have exactly the same meaning as the initial Young's modulus and tangent Young's modulus. Similar to Young's modulus, the magnitude of Poisson's ratio is also related to the effective stress on the soil, drainage condition of the soil, stress paths, and the depth of the soil, but its range of variation is relatively narrow. Table 2.15 gives a possible range of the μ for

Table 2.14 Ranges of Young's Modulus for Various Soils (Bowles, 1988)

Soil Type	E_s (MPa)
Very soft clay	2–15
Soft clay	5–25
Medium-stiff clay	15–50
Stiff clay	50–100
Sandy clay	25–250
Silty sand	5–20
Loose sand	10–25
Dense sand	50–81
Loose gravel	50–150
Dense gravel	100–200
Shale	150–5000
Silt	2–20

[a] The table only lists the possible range of E_s for soils. E_s of in situ soils is related to water content, density, stress history, etc.
[b] $MPa = 1 \times 10^6 N/m^2 = 100 t/m^2$.

Table 2.15 Ranges of Poisson's Ratio for Various Soils

Soil Type	μ_s
Saturated clay (undrained)	0.5
Unsaturated clay (undrained)	0.35–0.4
Silty sand	0.3–0.4
Sand, gravel	0.15–0.35
Silt	0.3–0.35
Rock	0.1–0.4 (depending on the type of rock)
Ice	0.36
Concrete	0.15

various soils. Poisson's ratios also have a distinction of the effective and undrained Poisson's ratios, which will be further discussed in Section 8.5. Note that saturated clay has no volume change under undrained conditions, so the undrained $\mu = 0.5$.

2.8.3   Yield and yield stress

Figure 2.24 shows the stress and strain behavior of a metal material subject to a one-dimensional loading. With the loading, the stress in metal increases along line AB starting from point A. If the loading stops and reduces to zero, the stress–strain relation would return to point A along line BA. No permanent or plastic strain is generated, and the metal is in an elastic state with the slope (E) of the stress–strain relation. When the stress exceeds point B, the load continues to increase to point C and then reduces to 0, and the stress–strain relation will follow line CD to point D. Compared with the initial point A, the material will produce permanent or plastic strain AD. Therefore, for the stress state at any point on line AB, point B is the yield point and $\sigma_{a,YB}$ is the corresponding yield stress; if the stress exceeds $\sigma_{a,YB}$, the material will produce permanent strain or plastic strain. If the loading starts at any point on the unloading line (e.g., point F), the stress–strain behavior is elastic as long as the increased stress does not exceed point C. As the stress is increased beyond the point C, a permanent or plastic strain will occur again. Therefore, for the stress state at any point on the unloading line CD, the stress at point C is the new yield stress and the corresponding yield stress $\sigma_{a,YC}$. Under such a condition, $\sigma_{a,YB}$ is no longer the yield stress, so it is called the initial yield stress.

In general, a yield point refers to a state that as long as the stress is less than the stress at the yield point, its behavior is elastic. As the stress exceeds the yield point, the stress, point C, for example, is the new yield stress. If the stress continues to be applied, plastic strain will occur.

It can be understood from the above explanation that unloading/reloading (line CD) is an elastic behavior, exactly the same as the initial loading (line AB). Therefore, lines CD and AB should be parallel to each other.

If the metal material is subjected to two-dimensional loading, in both the **a** and **b**-directions, at the same time (Figure 2.25), the yield stress in the a direction will be

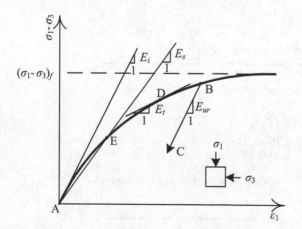

Figure 2.23 Various definitions of Young's modulus.

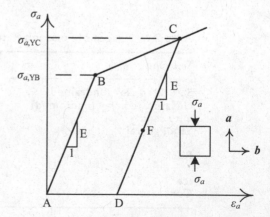

Figure 2.24 Yield and yield stress in one-dimensional loading.

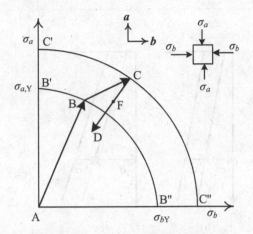

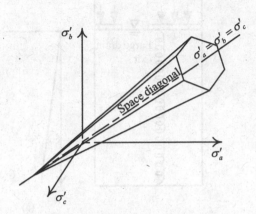

Figure 2.25 Yield and yield surface in two-dimensional loading.

Figure 2.26 Yield surface in three-dimensional loading.

affected by the stress in the **b**-direction; the same yield stress in the **b**-direction also depends on the magnitude of the stress in the **a**-direction. Therefore, in a two-dimensional stress system, the yield stress is not a single stress, but a combination of σ_a and σ_b. Connect all the stress states on the two-dimensional system that reach the yield to form a yield line or surface; the stress path inside the yield line or surface is elastic; the stress state at any point on the yield line or surface, if the stress continues to be applied, will produce plastic strain. For the stress state at any point on the stress path AB in Figure 2.25, B′B″ is the yield line or surface, similar to the yield point (point B) in Figure 2.24. When the stress is increased from point B to point C, plastic strain will occur. If it is unloading to point D (similar to the CF line in Figure 2.24), the stress–strain will be elastic on the stress path CD; if the stress reloads from any point on the CD line (e.g., point F), as long as the stress does not exceed the C′C″ line (similar to point C in Figure 2.24), it is an elastic behavior. Therefore, for the stress state at any point on the CD line, C′C″ is the new yield line or yield surface (similar to point C in Figure 2.24), and B′B″ is called the initial yield line or yield surface (similar to point B in Figure 2.24).

Similarly, if the metal material is subjected to stresses in three directions, the yield stress can be represented by a yield surface, as shown in Figure 2.26. Similarly, the behavior of the stress path inside the yield surface is elastic; on the yield surface, the behavior of the stress path outward is elastoplastic. In geotechnical engineering, the soil is normally subject to stresses in three directions. Therefore, even when dealing with two-dimensional problems (plane strain), we often use the term "yield surface" and rarely use the term "yield lines."

2.9 GEOTECHNICAL ANALYSIS METHOD

2.9.1 Drained behavior and undrained behavior

As a soil is stressed, the ease of the pore water flow out will affect the soil deformation behavior significantly. Figure 2.27 shows a schematic diagram of saturated coarse-grained soil subject to a normal pressure, $\Delta\sigma$. The spring

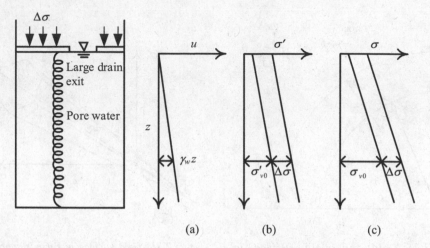

Figure 2.27 Drained condition for saturated coarse granular soil subject to loading.

represents the coarse-grained soil, which will deform as it is stressed. There-fore, the stress on the spring represents effective stress in soil. The orig-inal pore water pressure is in the hydrostatic condition, that is, $u = \gamma_w z$. The original effective stress and total stress are denoted by σ'_0 and σ_{v0}, respectively. As a pressure $\Delta\sigma$ acts on a piston, causing the piston moving down, the water would flow out rapidly and the volume thus decreases. The spring is there-fore stressed. Under such a condition, the excess pore water pressure is nil and the water pressure remains the same as the hydrostatic pressure (Figure 2.27a). The applied pressure $\Delta\sigma$ directly transmits to the soil, causing the effective stress in the soil being increased to $\sigma'_{v0} + \Delta\sigma$ (Figure 2.27b). The total pressure is increased to $\sigma_{v0} + \Delta\sigma$ (Figure 2.27c). Such a kind of deformation behavior of the soil is called drained behavior.

Figure 2.28 shows a schematic diagram of saturated fine-grained soil, i.e., clay, subject to a normal pressure, $\Delta\sigma$. Similarly, $\gamma_w z$, σ'_{v0}, and σ_{v0} represent the original pore water pressure, effective stress, and the original total stress, respectively. In this condition, when a pressure $\Delta\sigma$ acts on a piston, the water flows out very slowly. There-fore, at the initial stage of loading, the pressure $\Delta\sigma$ is borne by water entirely and the spring is unstressed. The stress in the spring or the effective stress in the soil remains unchanged (Figure 2.28b). This condition is called undrained behavior or short-term behavior. The pore water pressure is thus the sum of the hydrostatic water pressure and excess pore water pressure, while the excess pore water pressure is equal to $\Delta\sigma$ (Figure 2.28a). If there exists a shear stress-induced pore water pressure, then the ex-cess pore water pressure will be generated by $\Delta\sigma$ as well as the shear stress.

However, as shown in Figure 2.28, the water can still flow out eventually. The pis-ton moves down eventually and the volume change still occurs. The excess pore water would decrease or dissipate to 0 and the effective stress increases accordingly, but the total stress remains unchanged. When the excess pore water pressure is nil, the stress

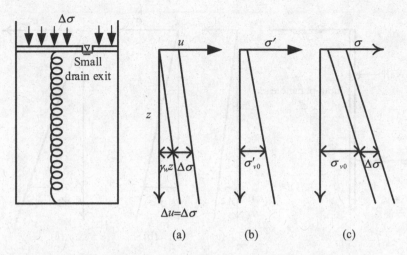

Figure 2.28 Undrained condition for saturated fine soil subject to loading.

in the spring is equal to $\sigma'_{v0} + \Delta\sigma$. This condition is also called drained behavior or long-term behavior.

In the short-term condition, the materials with drained behavior are called drained materials, while the materials with undrained behavior are called undrained materials. When conducting geotechnical analysis, one should firstly determine whether the target soil belongs to drained materials or undrained materials, based on the soil classification test results of the target soil and the description of the site investigation, and then perform drained analysis or undrained analysis, respectively.

2.9.2 Drained analysis

The excess pore water pressure is equal to 0 for the soil under drained shear condition. Therefore, during analysis, it is necessary to assume a zero excess pore water pressure, and the pore water pressure below the groundwater level is thus equal to the hydrostatic water pressure. This type of analysis is called drained analysis. The drained analysis is a type of effective stress analysis because all the stresses in computation are expressed in terms of effective stress. Since the pore water pressure is known, the effective stress can be calculated in an easy way, which is equal to the total stress minus the pore water pressure. The required strength parameters of soil for drained analysis are effective cohesion (c') and effective friction angle (ϕ'). Figure 2.29 shows the lateral earth pressure on a retaining wall for drained conditions. The effective active earth pressure should be computed using effective strength parameters (c', ϕ'), the submerged (or effective) unit weight, and effective deformation modulus (E', μ'). The pore water pressure is equal to $\gamma_w z$ if seepage does not occur. The total active earth pressure is the sum of $K_a \gamma' z$ and $\gamma_w z$ where $K_a = \tan^2\left(45^o + \phi'/2\right)$.

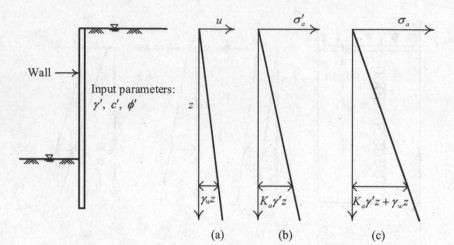

Figure 2.29 Lateral earth pressure of coarse granular soil obtained from drained analysis.

2.9.3   Undrained analysis

Undrained materials such as clay will generate excess pore water pressure during undrained shear, which in turn affects the magnitude of effective stress. The shear strength under undrained condition will be affected accordingly. It is therefore necessary to conduct undrained analysis. The undrained analysis can be categorized into the effective stress undrained analysis and total stress undrained analysis, which will be introduced below.

2.9.3.1   Effective stress undrained analysis

The force and deformation of soil are only related to the effective stress, based on the principle of effective stress. If the soil and pore water in soil are treated as a two-phase material in finite element computation and the material property parameters of the soil and water such as Young's modulus are given separately, which are implemented in the finite element formulation, we can find the effective stress of the soil and pore water pressure simultaneously. Since this type of undrained analysis is based on the effective stress and the required parameters are effective strength parameters (c', ϕ'), the submerged (or effective) unit weight, and effective deformation modulus (E', μ'), it is called the effective stress undrained analysis. Figure 2.30 shows the active earth pressure (σ_a') and pore water pressure (u) on a retaining wall using the effective stress undrained analysis where the pore water pressure is the sum of the hydrostatic water pressure and excess pore water pressure. Note that $\sigma_a' \neq K_a \times \gamma'z$ because $u \neq \gamma_w z$.

The effective stress undrained analysis only can be performed with the finite element method. The analysis accuracy is highly related to soil constitutive laws. Since advanced soil constitutive models have been developed recently and implemented in some professional geotechnical software, effective stress undrained analysis has been widely applied to geotechnical analysis.

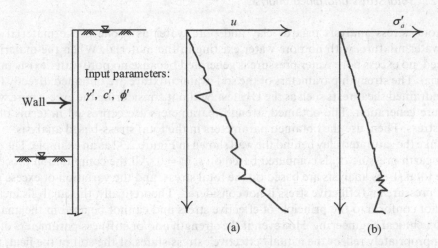

Figure 2.30 Lateral earth pressure of cohesive soil obtained from effective stress undrained analysis.

In the effective stress undrained analysis, the pore water pressure should be set according to the actual pore water pressure in clay deposits. This is because the pore water pressure will influence the magnitude of effective stress, which in turn affects the deformation modulus such as Young's modulus and Poisson's ratio. Take the second soil layer (note: clay) shown in Figure 2.31 as an example. If the effective stress undrained analysis is adopted when long-term dewatering is encountered, the pore water pressure distribution is different from the hydrostatic condition. The initial deformation moduli would be different from that of hydrostatic condition in analysis.

More discussion on the effective stress undrained analysis and its finite element formulation can refer to Section 8.3.

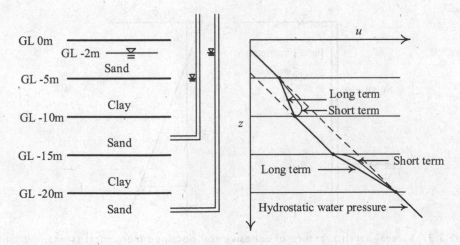

Figure 2.31 Non-hydrostatic water pressure distribution in the sand and clay deposit.

2.9.3.2 Total stress undrained analysis

The total stress analysis regards clay and pore water as a one-phase material or a soil–water mixture, with no pore water existing in the material. When the material is stressed, no excess pore water pressure is generated because no pore water exists in this material. The strength parameters of the soil–water mixture are obtained directly from the undrained shear tests such as the UU test, without considering the excess pore water pressure generation. The obtained strength parameters are expressed in terms of the total stress. Then, use the obtained parameters in the total stress-based analysis.

Take the saturated clay behind the wall shown in Figure 2.32 as an example. The total active earth pressure σ_a is computed based on s_u, $\phi = 0$. All the computation processes in the total stress analysis are based on the total stress, and the variation of excess pore water pressure and effective stress is not considered. Theoretically, this analysis method does not conform to the principle of effective stress and cannot be used in the analysis of geotechnical engineering. However, if the strength and/or stiffness parameters of soil can appropriately reflect the actual "effective" stress states of the soil in the field, then total stress analysis can result in good analysis results. Therefore, the keys to the rationality of the total stress undrained analysis lie in the selection of those parameters.

In the total stress undrained analysis, the effect of groundwater has been reflected in the saturated unit weight of the soil and the undrained shear strength parameters. Therefore, when performing total stress undrained analysis, the pore water pressure and/or groundwater level of the clay layers must be assumed to be 0; in the finite element analysis, the groundwater level (or piezometric water level) of the layers can be set below the soil layers to prevent the water pressure in the soil layers from affecting the results of the analysis. Take the second soil layer (note: clay) shown in Figure 2.31 as an example again. If the total stress undrained analysis is adopted to examine the behavior of the clay after dewatering, the pore water pressure in the clay should be assumed to be equal to 0 or the groundwater level of the clay should be set below GL-10 m. The parameters needed for analysis are obtained by performing the triaxial UU test on the soil specimen sampling at the site or conducting the FV test. The soil

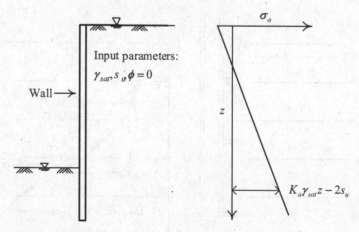

Figure 2.32 Lateral earth pressure of cohesive soil obtained from total stress undrained analysis.

specimen used in the triaxial UU test or FV test results has been influenced by the effective stress after long-term dewatering. Therefore, the obtained undrained shear strength can reflect the actual state of stress in the field.

More discussion on the total stress undrained analysis and its finite element formulation can refer to Section 8.3.

2.10   STRESS PATHS IN EXCAVATIONS

It is generally believed that in deep excavations, due to the excavation of the soil in front of the wall, the wall moves forward, so the soil in front of the wall is in a state of vertical unloading and lateral compression. Furthermore, due to the forward movement of the wall, the soil behind the retaining wall is in a state of constant vertical pressure and lateral unloading. Nevertheless, the soil in front of and behind the wall is affected by many factors and presents a rather complex stress state. These factors include the friction of the wall, the settling of the wall, the uplift of the excavation bottom, and drainage states.

Figure 2.33 shows an idealized stress path in deep excavations. For granular soil, Figure 2.33 represents the change of total stress and effective stress where the influence of the retaining wall friction is not considered (Becker, 2008).

As stated in Chapter 4, the soil in front of and behind the wall will be affected not only by the release of overburden pressure, but also by shear stress, especially near the wall. The closer to the wall, the larger the impact. Therefore, in the undrained state, the excess pore water pressure induced by excavation will be affected by both. The author and his team employed the effective stress undrained analysis method to explore the ESP of the soil in front of and behind the wall in the Taipei National Enterprise Center (TNEC) excavation. The geological and excavation profiles of the TNEC excavation are shown in Appendix B.

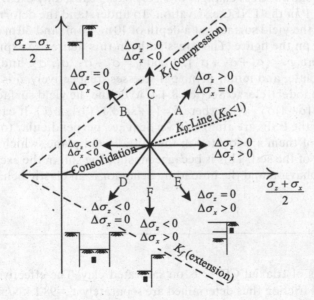

Figure 2.33 Total stress paths in excavations in clay.

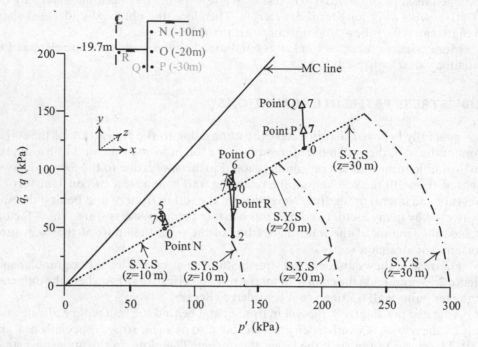

Figure 2.34 Effective stress paths of some representative soils in the TNEC excavation case (note: S.Y.S denotes yield surface).

Figure 2.34 shows stress changes or stress paths caused by excavation at points R, Q, N, O, and P in the TNEC excavation. To understand the deformation characteristics of soil, the yield surfaces at a depth of 10 m, 20 m, and 30 m at those points are also plotted on the figure. The stress paths in this figure are expressed in terms of p' and q where $p' = (\sigma_1' + \sigma_2' + \sigma_3')/3$ and $q = \sigma_1' - \sigma_3'$. σ_1', σ_2', and σ_3' denote the major, intermediate, and minor principal stresses, respectively. $\tilde{q}$ is the parameter used by the HS model (refer to Section 8.4.4) to define the yield surface, which is defined as $\tilde{q} = \sigma_1' + (\delta - 1)\sigma_2' - \delta\sigma_3'$, where $\delta = (3 + \sin\phi')/(3 - \sin\phi')$. It can be seen from Figure 2.34 that the ESPs are almost all vertical i.e., perpendicular to the horizontal axis, and most of them are located in the initial yield plane, which represents the elastic behavior of the soil. This is because most of the soil in the excavation area is an unloading behavior, and the unloading behavior is an elastic behavior, as stated in Section 2.8.3.

Example 2.1

Conduct a series of triaxial CD tests on saturated clay. The effective cohesion and angle of internal friction thus determined are separately $c' = 98.1$ kN/m² and $\phi' = 30°$. Let another sample of the same clay undergo a triaxial CU test. We then obtain the following total stress-related results: $c_T =58.8$ kN/m^2 and $\phi_T =15°$. What is the angle of the failure surface of the soil at failure?

Answer

According to the principle of effective stress, the strength or deformation of a soil relates only to the effective stress, whereas it has nothing to do with the total stress. c' and ϕ' are effective strength parameters, while c_T and ϕ_T are the parameters of total stress. Thus, using Eq. 2.10, the angle between the failure surface and the horizontal can be obtained as follows:

$$\alpha_f = 45° + \frac{\phi'}{2} = 45° + \frac{30°}{2} = 60°$$

Example 2.2

The friction angle of a saturated normally consolidated clay $\phi' = 30$ degree. The major and minor principal stresses at failure of the UU test on the clay $\sigma_3 = 98.1$ kN/m^2 and $\sigma_1 = 490.5$ kN/m^2. If the apex of the Mohr's circle is adopted as the undrained shear strength (i.e., $s_u = (\sigma_1 - \sigma_3) / 2$), what will be the error for the method?

Answer

Referring to Figure 2.11, in the triaxial UU test, the effective Mohr's circle at failure should be tangent to the effective failure envelope. What they are tangent to represents the undrained shear strength. Use Eq. 2.17, and the shear stress on the failure surface $\left(\tau_f\right)$ can be computed as follows:

$$\tau_f = \frac{\sigma_1' - \sigma_3'}{2}\cos\phi' = \frac{\sigma_1 - \sigma_3}{2}\cos\phi' = \frac{490.5 - 98.1}{2}\cos 30° = 169.9 \text{kN/m}^2$$

The shear strength at the apex of the Mohr's circle $\left(s_u\right)$ is:

$$s_u = \frac{490.5 - 98.1}{2} = 196.2 \text{kN/m}^2$$

$$\text{Error} = \frac{196.2 - 169.9}{169.9} = 15.5\%$$

Example 2.3

As shown in Figure 2.35, the ground is mainly composed of saturated clay. Suppose an embankment is to be constructed on the ground in two stages. In the first stage of construction, the embankment consolidates the soil completely. In the second stage,

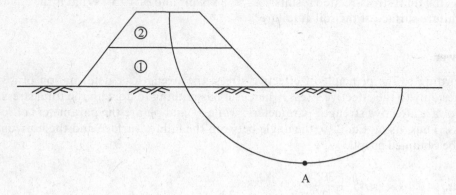

Figure 2.35 Embankment constructed in two stages.

the embankment is constructed at such a speed that the soil through the whole process stays in the undrained state. Before construction (the first stage), what parameters are necessary for the stability analysis of the embankment in the second stage?

Answer

Before the first stage of construction, use the SHANSEP approach discussed in Section 2.6.3.2 and conduct a series of undrained CK_0UDSS tests on the soil specimen sampled at point **A** in Figure 2.35. Establish the s_u / σ_v'–OCR relation, as illustrated in Figure 2.18d. Then, calculate the value of vertical stress (σ_v') at **A** after the first stage of construction. According to the OCR at **A**, the value of s_u / σ_v' at **A** can be obtained, referring to the corresponding figure, such as one like Figure 2.18d. Substitute σ_v' into s_u / σ_v' to obtain s_u, which can be used as the average s_u along the failure surface.

Example 2.4

Figure 2.36 shows the $w_f - p_f' - q_f$ and $w_0 - p_0'$ relations of Weald Clay where the subscript "f" refers to the failure condition and "0" the initial condition. According to the principle of effective stress, the relation of water content (or void ratio) and stress state (e.g., p' and q) for a particular type of normally consolidated soil is unique. Assume that a ground is composed of normally consolidated Weald Clay whose groundwater table is on the ground surface. $\gamma_{sat} = 17.66$ kN/m³, $K_0 = 0.5$, $c' = 0$, and $\phi' = 30°$. Sample the clay at the depth of 10 m. Due to sampling disturbance, a 15% of negative pore water pressure loss is estimated. Put the trimmed specimen to the UU triaxial test. When the confining pressure (σ_3) is equal to 78.5 kN/m², compute the undrained shear strength of the soil and the pore water pressure at failure.

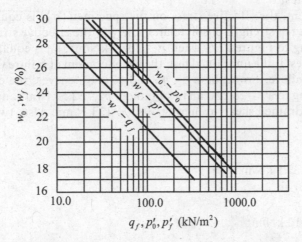

Figure 2.36 Relation between water content and state of stresses for Weald Clay (Lambe and Whitman, 1969).

Answer

The problem can be solved as follows:

a. The in situ stress before sampling

$$u_0 = 9.81 \times 10 = 98.1 \text{ kN/m}^2$$

$$\sigma_v = 17.66 \times 10 = 176.6 \text{ kN/m}^2$$

$$\sigma_v' = \sigma_v - u_0 = 176.6 - 98.1 = 78.5 \text{ kN/m}^2$$

$$\sigma_h = \sigma_h' + u_0 = K_0\sigma_v' + u_0 = 0.5 \times 78.5 + 98.1 = 137.35 \text{ kN/m}^2$$

b. The after-sampling stress
   Since the total stress is relieved, the pore water pressure is:

$$\begin{aligned} u &= u_0 + \Delta u = 98.1 + B\Delta\sigma_3 + AB(\Delta\sigma_1 - \Delta\sigma_3) \\ &= (1.0)(-137.34) + (0.2)(1.0)(-176.6 + 137.34) = 98.1 - 145.2 = -47.1 \text{ kN/m}^2 \end{aligned}$$

Considering the 15% of negative pore water pressure loss during sampling, thus

$$\sigma_v' = \sigma_v - 0.85u = 0 - (-40.04) = 40.04 \text{ kN/m}^2$$

$$\sigma_h' = \sigma_h - 0.85u = 0 - (-40.04) = 40.04 \text{ kN/m}^2$$

c.   Triaxial UU test

At the moment, the effective stress on the specimen (σ_3') is equal to $p_0' = 40.04$ kN/m². Since the strength of soil relates solely to the effective stress before shearing, according to Figure 2.36, when $p_0' = 40.04$ kN/m², w_0 is equal to 28.2%. Considering the test is the undrained test, the water content at failure would then equal that before failure. From the $w_f - p_f'$ and $w_f - q_f$ curves, the effective stresses corresponding to $w_f = 28.2\%$ can be obtained: $p_f' = 32.5$ kN/m² and $q_f = 12$ kN/m². Thus, the undrained shear strength $s_u = q_f = 12$ kN/m². Also, two equations can be obtained:

$$\frac{1}{2}(\sigma_1' + \sigma_3') = 32.5 \text{ kN/m}^2$$

$$\frac{1}{2}(\sigma_1' - \sigma_3') = 12 \text{ kN/m}^2$$

The stresses at failure are thus obtained as $\sigma_{1f}' = 44.5$ kN/m² and $\sigma_{3f}' = 20.5$ kN/m². The excess pore water pressure at failure $\Delta u_f = 78.5 - 20.5 = 58.0$ kN/m².

Example 2.5

With other conditions unchanged as in Example 2.4, suppose a loss of 25% of negative pore water pressure is estimated due to sampling disturbance. Estimate the undrained shear strength obtained by UU triaxial tests with a confining pressure of $\sigma_3 = 78.5$ kN/m².

Answer

Suppose 25% of the negative pore water pressure is lost during the process of sampling. Thus,

$$\sigma_v' = \sigma_v - 0.75u = 0 - (-35.3) = 35.3 \text{ kN/m}^2$$

$$\sigma_h' = \sigma_h - 0.75u = 0 - (-35.3) = 35.3 \text{ kN/m}^2$$

At the moment, the effective stress on the specimen $\sigma_c' = p_0' = 35.3$ kN/m². Refer to Figure 2.36, and the q_f can be found to be 10 kN/m². The undrained shear strength $s_u = q_f = 10$ kN/m².

As found by Examples 2.4 and 2.5, sampling disturbance will cause the undrained shear strength to decrease.

PROBLEMS

2.1   As shown in Figure P2.1, the depth of the groundwater level $H_1 = 2.0$ m, the moisture unit weight of clay $\gamma_m = 11.8$ kN / m³, and the saturated unit weight $\gamma_{sat} = 17.7$ kN / m³. The thickness of the clayey layer $H_2 = 40.0$ m. Under the clayey layer is a sandy layer where the piezometric level is the same as the groundwater level, i.e., $H_3 = 38.0\,\text{m}$. Compute and plot the total stress, the pore water pressure, and the effective stress with depth.

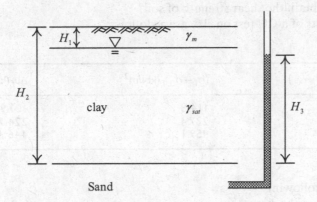

Figure P2.1

2.2   Same as above. Assume the piezometric pressure in the sandy soil is lowered, and render $H_3 = 30.0\,\text{m}$. Compute (1) the variation of the total stress, pore water pressure, and effective stress with depth for the short-term condition after lowering the water level and (2) the variation of the total stress, pore water pressure, and effective stress with depth for the long-term condition (when the seepage reaches the steady state condition).

2.3   Same as shown in Figure P2.1. Assume $\gamma_m = 13.2\,\text{kN/m}^3$, $\gamma_{sat} = 20.6\,\text{kN/m}^3$, $H_1 = 5.0\,\text{m}$, and $H_2 = 50\,\text{m}$. The piezometric level of the sandy layer is originally the same as the groundwater level ($H_3 = 45\,\text{m}$) and then is lowered by 5 m ($H_3 = 40.0\,\text{m}$) through dewatering. Compute the variation of the total stress, the pore water pressure, and effective stress in the clay layer for the short-term and long-term conditions.

2.4   Assume a homogeneous normally consolidated clayey ground where the groundwater level is on the ground surface and the properties of clay are as follows: $\gamma_{sat} = 20.6\,\text{kN/m}^3$, $K_0 = 0.5$, $c' = 0$, $\phi' = 27°$, $c_T = 0$, and $\phi_T = 15°$. Assume a horizontal failure surface in the soil at the depth of 10 m is generated under undrained condition, with the vertical total stress unchanged. The excess pore water pressure at failure is $49.1\,\text{kN/m}^2$. Compute the undrained shear strength of the soil (i.e., shear stress on the failure surface).

2.5   Same as the previous problem. Assume the major principal stress (total stress) occurs in the vertical direction at failure whereas the horizontal total stress is unchanged during the testing. The pore water pressure at failure is unknown, though the parameter of pore water pressure is $A_f = 0.6$. Compute the undrained shear strength (i.e., shear stress on the failure surface).

2.6   If we perform a direct shear test on sand, with the before-test vertical stress on the shear box $\sigma_v = 981\,\text{kN/m}^2$ and the shear stress on the failure surface $\tau_f = 63.77\,\text{kN/m}^2$, compute the effective angle of friction and the major principal stress at failure.

2.7   According to the principle of effective stress, the failure or deformation of soil relates solely to the effective stress, whereas it has nothing to do with the total stress. In the triaxial UU test, the pressure acting on the soil is the total stress rather than the effective stress. If so, why can the result of a triaxial UU test be used to obtain the shear strength of soil?

2.8   The results of a CU test on clay are as follows:

$\sigma_3(kN/m^2)$	$(\sigma_1 - \sigma_3)_f(kN/m^2)$	$\Delta u_f(kN/m^2)$
196.2	114.8	107.9
392.4	237.4	224.7
784.8	459.1	446.4

Find the following:
1. What are the strength parameters under the effective stress?
2. Which type of clay is it (NC or OC)?
3. If we take another specimen of the same type of soil, what will be the pore water pressure when the confining pressure reaches 147.2 kN/m² and the deviator stress is 172.8 kN/m²?

2.9   A triaxial CD test was performed on a saturated specimen of overconsolidated clay with the preconsolidation pressure equal to 147.2 kN/m². At failure, $\sigma'_{3f} = 294.3$ kN/m² and $\sigma'_{1f} = 882.9$ kN/m². A triaxial CU test was performed on another specimen of the same clay. At failure, the confining pressure (σ_{3f}) was 392.4 kN/m² and the axial stress at failure (σ_{1f}) was 784.8 kN/m². Find (1) the pore water pressure at failure for the CU test and (2) the angle between the failure surface and horizontal plane.

2.10 A ground is mainly composed of normally consolidated clay with the groundwater level at the ground surface and $\gamma_{sat} = 20.6$ kN / m³, $K_0 = 0.45$, $c' = 0$, and $\phi' = 27°$. A specimen was taken at the depth of 12 m, and 16% of the negative pore water pressure is estimated to be lost during the sampling process. A triaxial UU test is performed on the specimen with a confining pressure of 98.1 kN/m². Find (1) the undrained shear strength, (2) the pore water pressure at failure, and (3) the angle of the failure surface (assume that the pore water pressure parameter during sampling is the same as that at failure: $A = 0.3$).

2.11 Redo Problem 2.10. Assume 25% of the negative pore water pressure is lost during the sampling process.

2.12 Same as Problem 2.10. If the clay is Weald Clay, c' and ϕ' are unknown, and the relation between the water content and stress path is as shown in Figure 2.36. Find (1) the major and minor principal stresses at failure and (2) the pore water pressure at failure.

2.13 Same as Problem 2.11. If the clay is Weald Clay, c' and ϕ' are unknown, and the relation between the water content and stress path is as shown in Figure 2.36. Find (1) the major and minor principal stresses at failure and (2) the pore water pressure at failure.

2.14 The ratio of the undrained shear strength of clay, obtained from the triaxial CK_0U (axial compression) test, to the effective vertical overburden stress (s_u / σ_v') is 0.32. Is it feasible to use the ratio directly for stability analysis of an excavation or bearing capacity analysis of a foundation? If not, how must it be modified?

2.15 Redo Problem 2.14 if the effective vertical overburden stress (s_u / σ_v') from the triaxial CK_0U (axial compression) test is 0.28.

2.16 Assume a normally consolidated clayey ground with the groundwater level at the ground surface and the saturated unit weight $\gamma_{sat} = 20$ kN / m^3. Given that the ratio of the undrained shear strength of clay, obtained from the triaxial CK_0UC (axial compression) test, to the effective vertical overburden stress (s_u / σ_v') is 0.32 and $s_u / \sigma_v' = 0.21$ from the triaxial extension test (CK_0UE) for the ground. Assuming the embedment depth and width of a foundation are 2.0 and 3.0 m, respectively, estimate the undrained shear strength used in the calculation of the foundation bearing capacity.

2.17 Redo Problem 2.16, assuming the embedment depth and width of a foundation are 3.0 and 20 m, respectively.

2.18 Figure P2.18 is a profile of the undrained shear strength of a clayey ground. The groundwater level is 1.0 m below the ground surface. $\gamma_m = 11.77$ kN / m^3 above the groundwater level, whereas $\gamma_{sat} = 17.66$ kN / m^3 below it. Use Terzaghi's bearing capacity equation to compute the net ultimate bearing capacity of a 3-m-wide and 2-m-deep strip footing.

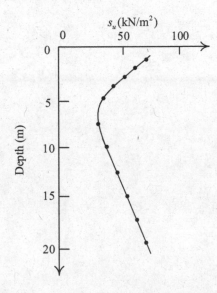

Figure P2.18

2.19 Redo Problem 2.18, assuming the embedment depth and width of a foundation are 3.0 and 10 m, respectively.

2.20 A 15-m-high embankment is to be constructed on a normally consolidated clayey ground with the groundwater table at the ground surface (see Figure 2.35). In fear of the shear failure of the soil below the embankment, a stability analysis has to be performed. How should one obtain the undrained shear strength (1) through the CU test and (2) through the UU test? (3) Compare the strengths and shortcomings of the two methods.

Chapter 3

Excavation methods and lateral supporting systems

3.1 INTRODUCTION

Deep excavations or the construction of basements include the construction of retaining walls, excavation, the installation of struts, and the construction of foundations and floor slabs. With the great variety of excavation methods and lateral supporting systems, to come to the most appropriate design, we have to consider, in combination, the local geological conditions, the environmental conditions, the allowable construction period, the budget, and the available construction equipment and make an overall plan accordingly. Considering that foundation construction involves many construction details and this book focuses on analysis and design, this chapter will only introduce some of the most commonly used excavation methods and lateral supporting systems.

This chapter will also introduce the construction process of the Taipei National Enterprise Center (TNEC) excavation. Its geological data are described in Appendix B. With the thorough documentation and studies of the excavation process, geological data, and monitoring results, we can get further understanding of excavation behaviors.

3.2 EXCAVATION METHODS

In the following section are introduced some commonly used excavation methods and their characteristics.

3.2.1 Full open cut method

The full open cut method can be further distinguished into the slope full open cut method and the cantilever full open cut method. As Figure 3.1 shows, in the slope full open cut method, the site is excavated with sloped sides and no struts and retaining walls are used. Since there is no strut to obstruct excavation, the cost is cheap if the excavation is not too deep. However, in deep excavations, or if the slopes are very gentle, the amount of excavated soil is tremendous and a great amount of soil will be needed to backfill after the construction is finished. Thus, the cost as a whole is not necessarily cheap.

Figure 3.2 illustrates the cantilever full open cut method, which is also called the strut-free excavation method. The stiffness of retaining walls is the sole source to keep them stable without temporary struts. The cantilever method, though requiring the construction of retaining walls, is not necessary to dig the slope and backfilling.

DOI: 10.1201/9780367853853-3

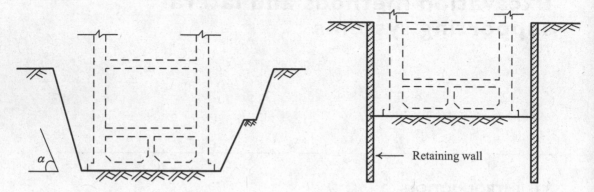

Figure 3.1 Slope open cut method.   Figure 3.2 Cantilever open cut method.

Therefore, the cost may not necessarily be higher than that of the slope method. Which is more economic and safer should be determined according to the results of analysis, design, and evaluation.

3.2.2 Braced excavation method

Installing horizontal struts in front of retaining walls to resist the earth pressure on the backs of walls is called the braced excavation method. Figure 3.3 shows the typical arrangement of horizontal struts. Figure 3.4 is the photo of a braced excavation. The bracing system of the braced excavation method includes struts, wales, end braces, corner braces, and center posts. The function of wales is to transfer the earth pressure on the back of retaining walls to horizontal struts. End or corner braces can help shorten the span of wales without increasing the number of struts. For example, some spacing between horizontal struts may slightly exceed allowable spacing out of construction expediency; adding one more strut, however, will make the spacing too small. End or corner braces can be used to adjust this spacing. Center posts are to support the struts, so that they will not fall out of their own weight.

In cases of deep excavations, construction is often carried out in stages. The following is the construction procedure of the braced excavation method (refer to Figure 3.3a):

1. Place center posts in the construction area.
2. Proceed to the first stage of excavation.
3. Install wales above the excavation surface, then install horizontal struts, and have them preloaded.
4. Repeat steps 2 and 3 till the designed depth.
5. Build the foundation of the building.
6. Demolish the struts above the foundation.
7. Construct floor slabs.
8. Repeat steps 6 and 7 till the construction of the floor slabs of the ground floor is completed.

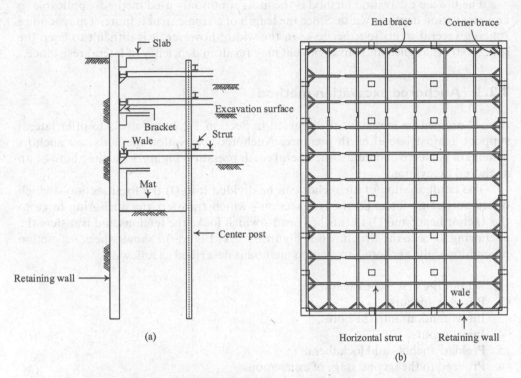

Figure 3.3 Braced excavation method. (a) Profile and (b) plan.

Figure 3.4 Photo of the braced excavation method.

The braced excavation method is the most commonly used method, applicable to any excavation depth or width. Since the length of a single strut is finite, it may require splicing several struts together to span the width; however, it is difficult to keep the spliced struts aligned and misalignment may result in deficiency of lateral resistance.

3.2.3   Anchored excavation method

Braced excavation methods, as discussed in Section 3.2.2, use struts to offer lateral support against lateral earth pressure. Anchored excavation methods use anchors instead of struts to counteract the lateral earth pressure. Figure 3.5 is the photo of an anchored excavation.

The configuration of an anchor can be divided into (1) the fixed section—which offers anchoring force; (2) the free section—which transfers the anchoring force to the anchor head; and (3) the anchor head—which locks the tendons and transfers the anchoring force to the structure (see Figure 3.6). As Figure 3.7 shows, the construction procedure of the anchored excavation method is described as follows:

1. Set out the first-stage excavation.
2. Bore for anchors.
3. Insert tendons into the bores.
4. Inject grouts.
5. Preload anchors and lock them.
6. Proceed to the second stage of excavation.
7. Repeat steps 1 to 6 till the designed depth.
8. Build the foundation of the building.
9. Construct the floor slabs from the foundation up to the ground sequentially.

Figure 3.5 Photo of the anchored excavation method.

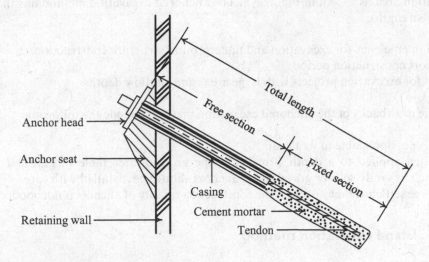

Figure 3.6 Basic configuration of an anchor.

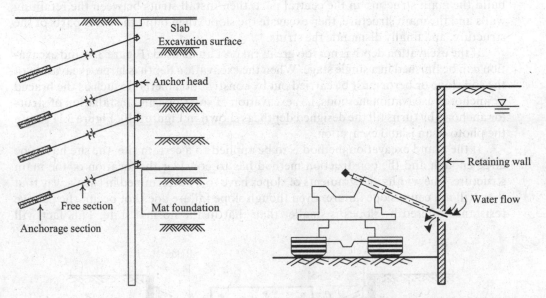

Figure 3.7 Profile of the anchored excavation method.

Figure 3.8 Problem of the anchored excavation
in the sandy soil with high groundwater level.

The anchored excavation counts solely on soil strength to offer the anchoring
force. The higher the soil strength, the stronger the anchoring force and vice versa.
Granular soils (such as sandy soils or gravel soils) have high strengths and thus offer
strong anchoring force, while clay has weak strength and creep will further decrease
the anchoring force. Therefore, the anchoring section should avoid being installed in
clay. Granular soils, however, usually have rather high permeability, while clayey soils
don't. Thus, an anchor installed in a granular soil with high groundwater level often
encounters difficulty in sealing bores because of the higher water pressure outside the

excavation area as shown in Figure 3.8. The anchored excavation method has the following strengths:

1. High efficiency for excavation and underground structure construction.
2. Short construction period.
3. Fit for excavation projects with large areas and shallow depths.

The drawbacks of the anchored excavation method include the following:

1. It is not applicable to weak soil.
2. When applied to a depth 10 m below the groundwater table in granular soils (such as sandy soils or gravel soils), anchors should be installed with care.
3. Large settlement may occur if the construction quality of anchors is not good.

3.2.4   Island excavation method

The conception of island excavation methods can be explained as follows: excavate the central part of the site first and keep the soil near the retaining walls to form slopes, build the main structure in the central part, then install struts between the retaining walls and the main structure, then excavate the slope, then build the other parts of the structure, and finally dismantle the struts.

If the excavation depth is not too great, rakers can be used (Figure 3.9) and excavation can be finished in a single stage. When the excavation depth is large, excavation of the soil slope or berm must be carried out by construction methods such as the braced or anchored excavation methods, i.e., excavation of soil slope and installation of struts (or anchors) by turns till the designed depth, as shown in Figure 3.10. Figure 3.11 shows the photo of an island excavation.

If the island excavation method is to be applied to a certain site, the site has to be large enough and the construction method has to consider the location of the main structure. The widths and gradients of slopes have to be determined in such a way that they will not cause slope failure. Even though slope failure does not occur, the passive resistance offered by slopes is smaller than that in the normal state. This fact will

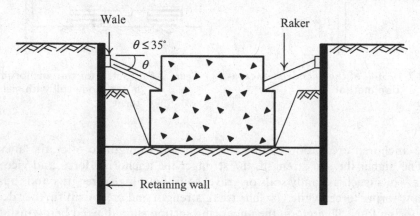

Figure 3.9 Island excavation method with a single level of struts.

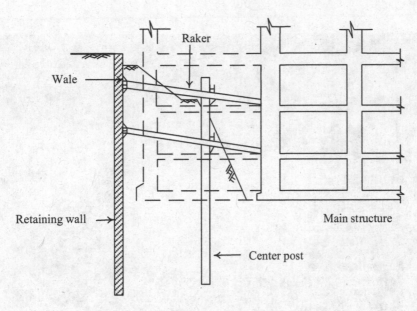

Figure 3.10 Island excavation method with multiple levels of struts.

bring about larger wall deflection or ground surface settlement. Therefore, analysis is required before excavation for the sake of adjacent building protection.

The most prominent strength of the island excavation method is the increase of efficiency and, accordingly, the shortened construction period. Compared to the braced excavation method, it requires fewer struts and reduces the cost of strut installation and dismantling. Used in large-area excavations, it can avoid the drawbacks of both the braced excavation method, where the larger spacing of struts may reduce the strut's resistance to lateral earth pressure, and the anchored method, where high water pressures obstruct the anchor installation. The island excavation method can resolve most of those problems. The major shortcoming is the possible water leakage or weak structural joints between the main structure in the center part and other structures in the surroundings. Besides, larger wall deflection and ground movement usually occur due to the lower passive resistance offered by slopes or berms smaller than those in the normal state, especially in soft soil.

3.2.5 Top-down construction method

Following the braced excavation method, excavation is carried out through to the designed depth, and then, raft foundations or foundation slabs are built; struts can then be removed level by level and floor slabs are built accordingly. Thus, the whole underground construction is finished. As such, the underground structure is constructed from bottom to top, which is the most conventional construction method and is generally called the bottom-up construction method. Precisely speaking, the anchored excavation method, the full open cut excavation method, and the braced excavation method all belong to the bottom-up construction method.

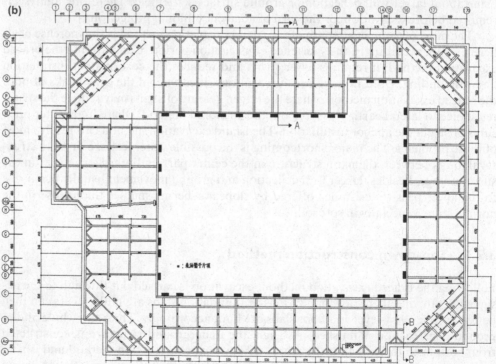

Figure 3.11 Photo and plan of the island excavation method.

Contrary to the bottom-up construction method, the top-down construction method is to erect molds and construct floor slabs right after each excavation. The floor slabs are permanent structures, which replace temporary steel struts in the braced excavation method to counteract the earth pressure on the back of the retaining wall. In this way, the underground structure construction is finished with the completion of excavation process. The construction of the underground structure is from the top to the bottom and is contrary to conventional foundation construction methods. The method is then called the top-down construction method.

The floor slabs used in the top-down construction method are heavier than the steel struts used in conventional excavation methods. In addition, the superstructure, which can be constructed simultaneously during excavation, puts more weight on the column. Thus, the bearing capacity of the column has to be considered. As a result, pile foundations are often chosen to be used for the top-down construction method. The typical construction procedure of the top-down construction method is as follows (see Figure 3.12):

1. Construct the retaining wall.
2. Construct piles. Place the steel columns where the piles are constructed.
3. Proceed to the first stage of excavation.
4. Cast the floor slab of the first basement level (B1 slab).
5. Begin to construct the superstructure.

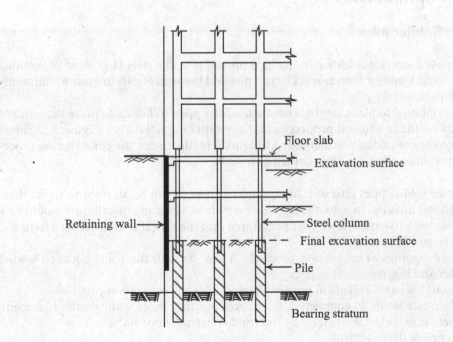

Figure 3.12 Top-down construction method.

6. Proceed to the second stage of excavation. Cast the floor slab of the second basement level (B2 slab).
7. Repeat the same procedures till the designed depth.
8. Construct foundation slabs, ground beams, etc. Complete the basement.
9. Keep constructing the superstructure till finishing it.

The strengths of the top-down construction method include the following:

1. The shortened construction period due to the simultaneous construction of the basement and the superstructure.
2. More operational space gained from the advanced construction of floor slabs.
3. The higher stiffness of floor slabs compared to steel struts which improves the safety of excavation.

The drawbacks of the top-down construction method are as follows:

1. It has a higher cost (due to the construction of pile foundations).
2. Since the construction period of the basement is quite long, the lateral displacement of retaining walls or ground settlement may possibly increase due to the influence of creep, if soft soil layers are encountered (refer to Section 6.6).
3. The construction quality may be influenced by worsened ventilation and illumination below floor slabs.

3.3 RETAINING WALLS

3.3.1 Soldier piles

The types of steel for soldier piles include the rail pile, the steel H-pile (or W-section), and the steel I-pile (or S-section). The rail pile and the steel H-pile are more commonly used than the steel I-pile.

It is optional to place laggings between soldier piles. Whether to place them or not depends on the in situ soil properties and strength characteristics. Figure 3.13 shows the photo of the soldier pile method. As Figure 3.14 illustrates, the construction procedure for soldier piles can be described as follows:

1. Strike soldier piles into soil. In non-urban areas, it will be all right to strike them into soil directly. In urban areas, however, static vibrating installation would be a better way to have soldier piles penetrated into the soil. If encountering a hard soil layer, pre-bore the soil.
2. Place laggings as excavation proceeds. Then, backfill the voids between soldier piles and laggings.
3. Install horizontal struts in proper places during the excavation process.
4. After excavation is completed, begin constructing inner walls of the basement. Then, remove the struts level by level and construct floor slabs.
5. Complete the basement.
6. Pull out the piles.

Figure 3.13 Soldier piles. (a) Front view and (b) section view.

The strengths of soldier piles include the following:

1. The construction is easier and faster with lower cost.
2. Piles can be easily pulled out.
3. Less ground disturbance is caused when pulling out the piles is necessary, compared to pulling out sheet piles.
4. The pile tip can be strengthened with special steel materials for use in gravel soils.
5. Soldier piles are reusable.

The drawbacks of soldier piles include the following:

1. Sealing is difficult. In sandy soils with high groundwater level, some dewatering measures may be necessary.
2. Installing soldier piles by striking will cause much noise and vibration. The latter will render sandy soils below the foundation denser in such a way that uneven settlement of adjacent buildings may occur.
3. Backfilling is necessary if soldier piles are driven using pre-boring. Deficiency in backfilling will cause bad effects in the vicinity.
4. The voids between retaining walls and surrounding soil need filling.
5. Removing piles will disturb the surrounding soil.

3.3.2   Sheet piles

Sheet piles can be driven into soil by striking or static vibrating and have them interlocked or connected with one another. Figure 3.15 shows the front view of sheet piles, and Figure 3.16 is a photo showing the sheet piles in an excavation. There are several shapes

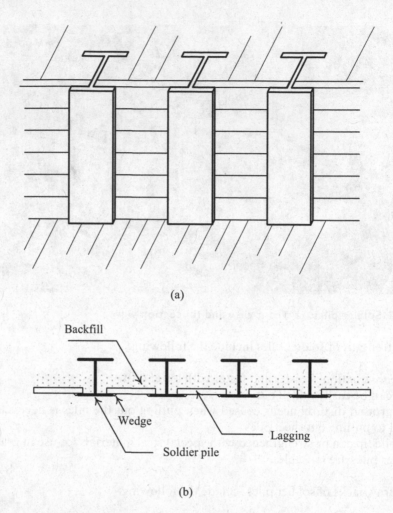

(a)

(b)

Figure 3.14 Photo of the soldier pile method.

of sheet pile sections, for example, the U-section and Z-section (Figure 3.17). Different shapes or types are used in different nations. If the interlocking is well done, sheet piles can be quite efficient in water sealing. If not, leaking may occur at joints. In clayey soils, having low permeability, sheet piles don't necessarily require perfect connection to prevent leaking. On the other hand, if they are used in sandy soils, with high permeability, any breach in sheet piles may well cause leaking. If leaks occur, sand in back of the retaining walls will very possibly flow out, which may cause settlement in turn. If leakages are too great, the excavation might be endangered. The construction method for sheet piles can be described as follows:

1. Drive sheet piles into soil by striking or static vibrating
2. Proceed to the first stage of excavation.
3. Place wales in proper places and install horizontal struts.
4. Proceed to the next stage of excavation.
5. Repeat steps 3 and 4 till the designed depth.
6. Complete excavation and begin to build the foundation of the building.

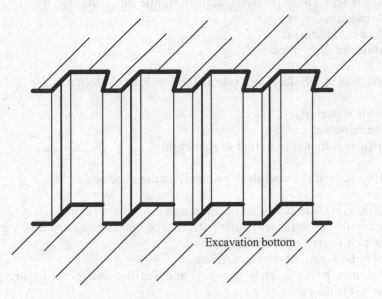

Excavation bottom

Figure 3.15 Steel sheet pile method.

Figure 3.16 Photo of the steel sheet pile method.

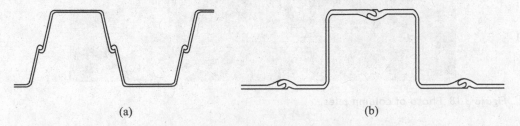

(a)   (b)

Figure 3.17 Sections of steel sheet piles. (a) U-pile and (b) Z-pile.

7. Build the inner walls of the basement. Dismantle the struts level by level and build floor slabs accordingly.
8. Complete the basement.
9. Dismantle the sheet piles.

The strengths of the steel sheet pile method include the following:

1. It is highly watertight.
2. It can be reusable.
3. The stiffness is higher than that of soldier piles.

The drawbacks of the steel sheet pile method are as follows:

1. The stiffness is lower than that of column piles or diaphragm walls.
2. It is susceptible to settlement during striking or dismantling in a sandy ground.
3. It is not easy to strike piles into hard soils.
4. A lot of noise is caused during striking.
5. Leaks cannot be completely avoided, and sealing and grouting are probably necessary if leaks occur.

3.3.3 Column piles

The column pile method is to construct rows of concrete piles as retaining walls by either the cast-in situ pile method or the precast pile method. Figure 3.18 is the photo of a column pile wall. The cast-in situ method can be divided into three subtypes according to their construction characteristics:

Figure 3.18 Photo of column piles.

1. Packed-in-place piles

    The packed-in-place (PIP) pile method can be described as follows: bore to the designed depth with a helical auger; while lifting the chopping bit gently, fill in prepacked mortar from the front end to press away the loosened soil to the ground surface; and after grouting finished, put steel cages or steel H-piles into the hole. The diameter of a PIP pile is around 300–600 mm. Since PIP piles are not capable of being installed completely vertically, gaps often exist between piles, causing leaks of ground water. PIP piles often have a problem of water tightness. Thus, if the PIP pile is adopted in sandy soils with high groundwater level, sealing and grouting are often required. Figure 3.19 illustrates the construction of a PIP pile.

2. Concrete piles

    The construction of concrete piles can be described as follows: drill a hole to the designed depth by machine, put the steel cages into it, and fill it with concrete using Tremie tubes. The reverse circulation drill method (also called the reverse method), which is to employ stabilizing fluid to stabilize the hole wall during drilling, is the most commonly used construction method for concrete piles. It is also feasible to build following the all-casing method, which is to drill with simultaneous casing installment to protect the hole wall. Since the wall is protected by casings, stabilizing fluid is not required. The cost of the all-casing method is rather high. Nevertheless, it can be easily applied to cobble-gravel layers or soils with seepage, whereas the reverse method cannot. The diameters of the concrete piles are around 0.6–2 m.

3. Mixed piles

    Mixed piles are also called mixed-in-place (MIP) piles or soil mixed wall (SMW). The method is to employ a special chopping bit to drill a hole with the concrete mortar sent out from the front of the bit to be mixed with soil. When the designed depth is reached, lift the bit gradually, keeping swirling and grouting simultaneously, and let mortar mix with soil thoroughly. After pulling out the drilling rod, put steel cages or H-piles into the hole if necessary. Figure 3.20 illustrates the construction process of a mixed pile. Figure 3.21 shows MIP piles with H-steels.

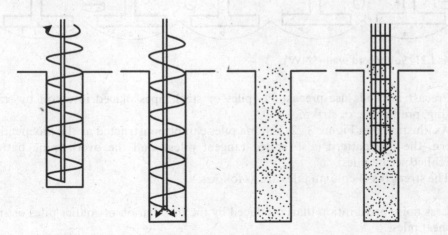

Figure 3.19 Construction procedure of a packed-in-place (PIP) pile.

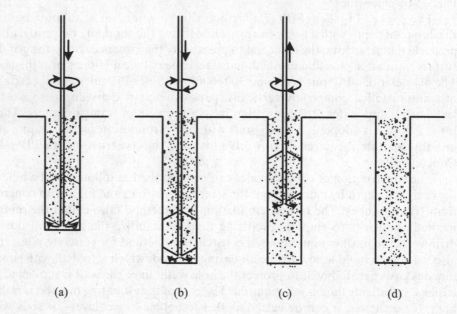

Figure 3.20 Construction procedure of a mixed-in-place (MIP) pile. (a) Swirl the drilling rod and inject mortar into the soil from the bottom of the drilling rod, (b) drill to the designed depth and treat the soil simultaneously while keeping swirling, (c) withdraw the drilling rod and inject the mortar simultaneously, and (d) finish the construction.

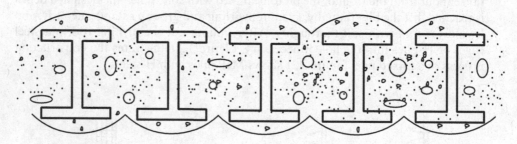

Figure 3.21 Soil mixed wall (SMW).

Precast methods use precast PC piles or steel pipes placed into soil by static pressing, pre-boring, or striking.

As illustrated in Figure 3.22, column piles can be constructed as the independent pattern, the line pattern (also called tangent piles), and the overlapping pattern (also called secant piles).

The strengths of column piles are as follows:

1. Less noise or vibration than produced by the installation of soldier piles or steel sheet piles.
2. Adjustable pile depth.

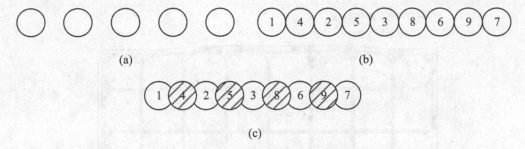

Figure 3.22 Layouts of column piles. (a) Independent pattern, (b) line pattern (tangent piles), and (c) overlapping pattern (secant piles).

3. Greater stiffness than soldier piles or steel sheet piles.
4. Being applicable to cobble-gravelly soils when equipped with a special bit.
5. Easier construction on sandy ground.

The drawbacks of column piles are as follows:

1. Nonexistence of an arching effect to decrease wall deformation and without lateral stiffness in the direction parallel to the excavation side.
2. Longer construction period than that for the soldier pile method or the steel sheet pile method.
3. Lower stiffness than diaphragm walls.
4. Being highly susceptible to construction deficiency.

3.3.4 Diaphragm walls

Diaphragm walls are also called slurry walls. Since they were first adopted in Italy in the 1950s, they have been widely used in the world. With technological advances, more and more new methods and construction equipment have been developed.

A diaphragm wall is formed by many individual panels. Before construction, the diaphragm wall should be segmented into many individual panels on a layout plan. Figure 3.23 shows a typical panel segmentation of a diaphragm wall in an excavation project. Diaphragm wall panels include primary panels and secondary panels that are connected by joints. Figure 3.24 shows different types of joints with the primary panel and secondary panel schematically shown in the figure. Among those joints, rigid joints are formed by overlapping the horizontal steel bars of two adjacent diaphragm wall panels, and transverse bending moment can be transmitted from one panel to adjacent panels. Since horizontal steel bars of two adjacent diaphragm walls are not overlapped in flat joints, H-steel joints, and connection-pipe joints, transverse bending moment cannot be transmitted, but it is easier to construct and to meet water tightness requirements.

Basically, the longer length of the panel, the more economical the construction cost, but the greater instability of the panel trench, which may induce a larger ground settlement. Besides, a heavier reinforcement steel cage and higher capacity of cranes

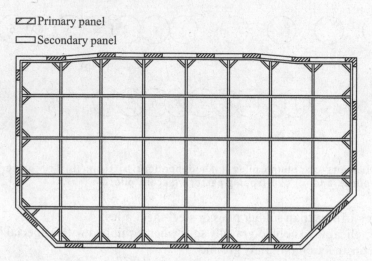

Figure 3.23 A typical panel segmentation of diaphragm wall in deep excavations.

are required. On the other hand, the shorter the diaphragm wall panel, the more stable the panel trench, so the higher the construction cost. Generally speaking, the length of the primary panel is around 2.5–4.5 m, while that of the secondary panel is around 4.7–7.5 m.

The procedure for each panel construction includes the stages of construction of guided ditch, trench excavation, reinforcement cage placement, and concrete

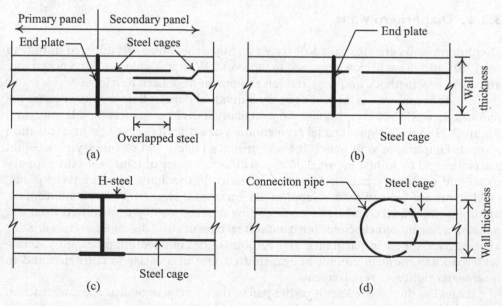

Figure 3.24 Different types of joints between panels. (a) Rigid joint, (b) flat joint, (c) H-steel joint, and (d) connection-pipe joint.

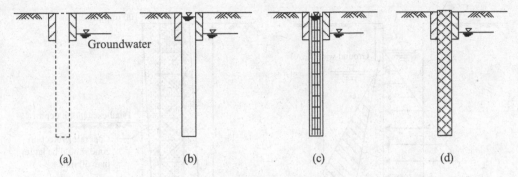

(a)   (b)   (c)   (d)

Figure 3.25 Construction procedure of a diaphragm wall panel. (a) Construction of the guided wall, (b) excavation of the trench, (c) placement of reinforcements, and (d) concrete casting.

casting, as shown in Figure 3.25. Each stage of construction is explained in the following items.

3.3.4.1 Construction of guided ditch

The guided ditch is a part of the trench where its side should be protected with guided wall to maintain the stability of the ditch. The objective of the guided ditch is to direct excavators to proceed trench excavation and ensure the accuracy of excavation and reinforcement steel cage placement. Figure 3.26 shows the photo of a guided ditch and its wooden supports.

Figure 3.26 Photo of a guided ditch and its wooden supports.

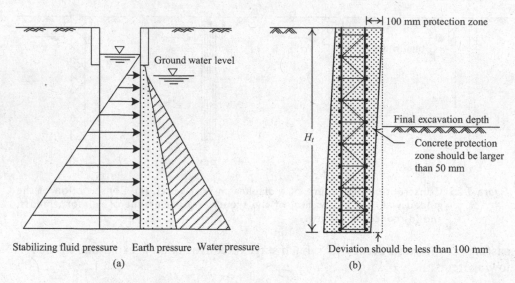

Figure 3.27 Necessity for a trench. (a) Maintain trench stability and (b) maintain a good vertical accuracy of trench.

3.3.4.2 Excavation of trench

Trench excavation should satisfy the stability of the trench, designed depth and thickness of the diaphragm wall, and vertical accuracy of diaphragm wall. Therefore, an appropriate excavator should be adopted and the stabilizing fluid should be rigorously managed. During trench excavation, the stabilizing fluid in the trench should be able to provide a sufficient pressure to maintain the trench stability and avoid the collapse of the side soil into the trench. Figure 3.27a shows the mechanism of trench stability. Therefore, the stabilizing fluid should be with a higher specific gravity than water and a higher level than the groundwater level. The larger the specific gravity, the more stable the trench, but the worse the quality of the concrete. A very good vertical accuracy of the trench should be achieved. Otherwise, it will affect the placement of the reinforcement steel cage and thickness of protection zone of the diaphragm wall (Figure 3.27b). In general, the vertical accuracy of the trench is highly related to the construction quality and selection of an excavator. The vertical angle of deviation is normally less than $10/H_t$ (radian), where H_t is the depth of the trench (in centimeter). Besides, the deviation of the bottom of the base should be less than 50 mm.

3.3.4.3 Placement of reinforcement steel cage

Detailing of reinforcement steel cages is determined by moment and shear diagrams (Figure 10.1). The reinforcement steel can be categorized into vertical main steel, horizontal steel, and shear steel (Figure 10.2). The vertical main steel is for the resistance against bending moment due to excavation of soil, and the horizontal steel is for temperature effect. The shear steel is subject to shear force and also provides a sufficient

Figure 3.28 Photo of reinforcement steel cage placement.

rigidity to avoid distortion of the steel cage during the operation of hanging the cage into the trench. Figure 3.28 is a photo of placing a steel cage into the trench with a crane.

3.3.4.4   Casting of concrete

Concreting the trench is conducted by transporting concrete from the ground surface to the trench using Tremie pipes. The concrete flows out of the bottom of the Tremie pipe and rises gradually from the bottom of the Tremie pipe. The rising concrete would expel the stabilizing fluid in the trench. The rising concrete will squeeze the poor concrete that is deteriorated by the stabilizing fluid locating on the interface between the concrete and stabilizing fluid, and spread it out to the side (Figure 3.29). That poor concrete would reduce the quality of the concrete near the joint of the panel. Part of the poor concrete would be mixed into the fresh concrete, which certainly would lower the quality of the concrete. There are many ways to reduce the amount of poor concrete. Interested readers are advised to refer to related references.

Beginning with trench excavation to the completion of concrete casting, a stabilizing fluid must be used to ensure the stability of the trench and exclude soil crumbs or particles generated during trench excavation to maintain the quality of concrete. The stabilizing fluid in the trench can produce fluid pressure against the water and/or earth pressure outside the trench. Considering that the unit weight of stabilizing fluid is less than that of saturated soil, the level of the stabilizing fluid in the trench should be higher than the water level in soil so that a sufficient fluid pressure

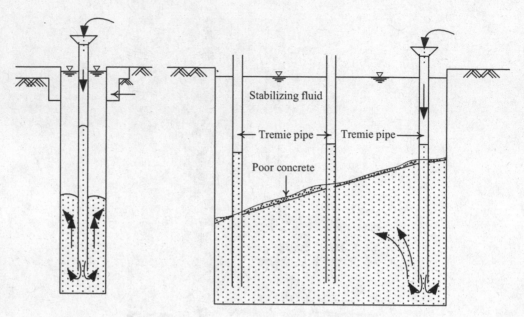

Figure 3.29 Influence of the stabilizer on the quality of Tremie concrete.

can be provided (Figure 3.27a). Stabilizing fluid can be prepared by mixing bentonite or polymer (e.g., polyacrylamide) with water.

Bentonite stabilizing fluid can yield an impermeable muddy cake on the surface of the trench (Figure 3.30), in which fluid pressure can effectively act on the cake and also avoid the collapse of the side soil into the trench. Bentonite stabilizing fluid also has a function of suspending soil crumbs or particles, which would be taken out of the trench along with the stabilizing fluid by a positive circulation method or a reverse circulation method.

The polymer in the polymer stabilizing fluid has a large helical molecular structure that are connected by molecular bonds. The helical structure is opened when it is in water, forming a flexible long-chain structure that will hold the water molecules, so that it has swelling properties, and the viscosity of the fluid is thus increased. The flexible molecular long chains in the solution contain many polar groups, similar to the sucker on an octopus claw. If there are suspended soil crumbs or particles in the fluid, the molecular chain can adsorb these suspended substances, so that the suspended substances are wrapped up or intertwined, causing the suspended soil crumbs or particles to aggregate and precipitate. The settled soil crumbs or particles can be taken out of the trench by excavators or grab buckets, so it will not affect the quality of the concrete casting afterward. Figure 3.31 shows the mechanism of aggregation of soil crumbs or particles. In addition, polymer stabilizing fluid will not form a water-impermeable muddy cake on the side soil of the trench, but its high viscosity will make it difficult to run out or permeate into the soil layers. Nevertheless, some stabilizing fluid still can permeate into the soil layer, which can increase the strength of the soil layer and prevent the side soil from collapsing.

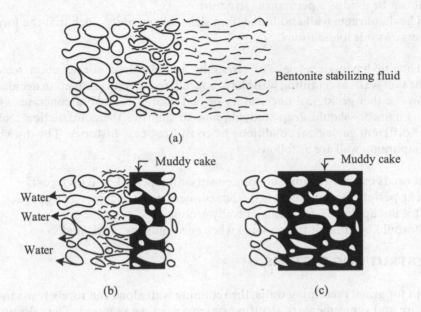

Bentonite stabilizing fluid

(a)

Muddy cake   Muddy cake

Water
Water

Water

(b)   (c)

Figure 3.30 Formation of a muddy cake in bentonite stabilizing fluid.

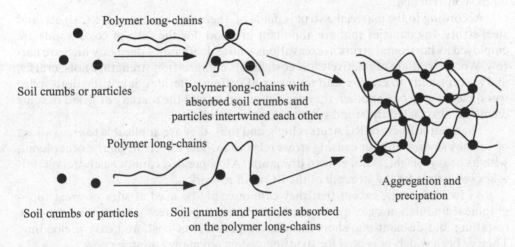

Polymer long-chains

Soil crumbs or particles

Polymer long-chains with
absorbed soil crumbs and
particles intertwined each other

Polymer long-chains

Soil crumbs or particles

Soil crumbs and particles absorbed
on the polymer long-chains

Aggregation and
precipitation

Figure 3.31 Mechanism of precipitation of soil crumbs and particles in polymer stabilizing
fluid.

In general, the diaphragm wall has the following strengths:

1. It has low vibration, low noise, high rigidity, and relatively small deformation.
2. It can be adjustable in thickness and depth.
3. It has good sealing capability.

4. It can be used as a permanent structure.
5. The diaphragm wall and foundation slabs form a unity, such that the former can serve as pile foundations.

Though having a variety of strengths and matured construction technology, diaphragm walls as retaining walls have engendered several excavation accidents. The reason is either geological uncertainty or bad quality control of concrete casting, or both. Engineers should keep studying how to improve the construction technology under different geological conditions by consulting case histories. The drawbacks of the diaphragm wall are as follows:

1. It needs massive equipment, long construction period, and high cost.
2. The peripheral equipment (e.g., the sediment pool) occupies a large space.
3. It is not applicable to cobble-gravelly grounds.
4. It would be difficult to construct when encountering quicksand.

3.4 STRUTTING SYSTEMS

Except for gravity retaining walls, the retaining wall alone can rarely resist the lateral pressure and supplementary strutting systems are thus required. The selection of the strutting system depends on not only the magnitude of lateral pressure, but also the period it will take to install the strutting system and the obstruction it may bring about on the construction.

According to the material a strut is made of, there are wood struts, RC struts, and steel struts. In countries that are abundant in wood, for the sake of cost, woods are employed as horizontal struts in excavations whose depth is not deep or which are narrow. Wood struts are of relatively low cost, but its compression strength, knots, cracks, and susceptibility to erosion lead to low axial stiffness. Besides, it is difficult to splice wood pieces with one another, they are not reusable, and the scarcity of wood in some countries also makes the wood strut an unwelcome choice.

The axial stiffness of RC struts is high, and thus, they are applicable to excavations of various shapes without causing stress relaxation. The RC strut, on the other hand, with its heavy weight, is not easy to dismantle. Also, preload cannot easily act on it. It takes some time for the strength of the RC strut to work.

As to steel struts, except that they cannot easily be used at sites of great topographical undulation or of great width, the strengths of steel struts are many: easy installing and dismantling, short installation period, low cost, and easy preloading. They've been widely accepted for strutting systems in many countries now.

According to the function of a strut, it is classified as an earth berm, a horizontal strut, a raker, an anchor, or a top-down floor slab. Figure 3.32 shows an earth berm, which is made by removing the soil in the central area while retaining an earth berm with a certain width for the lateral support of retaining walls. The earth berm is usually supplementary to island excavation methods. With the limitation of the width, the earth berm has accordingly limited lateral resistance and is useful only on grounds with high strength, and is rendered useless on soft ground.

Horizontal struts can be made of wood, RC, or steel, whose strengths and drawbacks are as mentioned above.

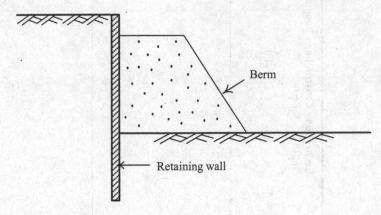

Figure 3.32 Earth berm as lateral strut.

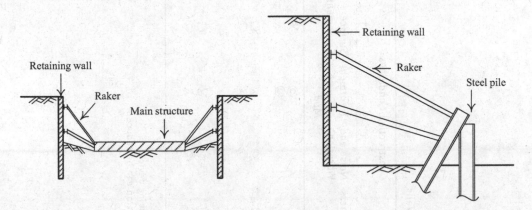

Figure 3.33 Rakers.

A raker is a type of strut and can also be made of wood, RC, or steel. Known from the systematic characteristics of a structure, the lateral support from the raker is smaller than that from the horizontal strut. Rakers are mostly employed in the island excavation method (Figure 3.33). Anchors and top-down floor slabs are two other types of struts.

3.5   SELECTION OF A RETAINING STRUT SYSTEM

The retaining strut system consists of retaining walls and strutting systems. The selection of retaining walls has to consider the excavation depth, geological conditions, groundwater conditions, adjacent building conditions, the site size, the construction period, the budget, etc. The characteristics of retaining walls are as described in Section 3.3. Table 3.1 lists the application ranges of various retaining walls for design or construction reference.

Table 3.1 Application Conditions for Retaining Walls

Wall Type	Soil Type			Sealing and Stiffness		Construction Conditions				Excavation Depth	Construction Period	Budget
	Soft Clay	Sand	Gravel Soil	Sealing	Stiffness	Noise and Vibration	Treatment of Dump Mud	Surface Settlement	Underground Obstruction			
Soldier pile	×	○	○[a]	×	×	×[b]	◎	×	○	×	◎	◎
Steel sheet pile	○	◎	×	○	×	×[b]	◎	×	○	×	◎	◎
PIP pile	◎	○	×	○	◎	◎	×	○	×	○	×	○
Reinforced concrete column pile	◎	◎	×	○	◎	◎	×	○	×	◎	×	×
MIP pile	○	○	×	○	○	◎	○	○	×	○	○	○
Diaphragm wall	◎	◎	○	◎	◎	◎	×	◎	×	◎	×	×

◎: good; ○: acceptable; ×: not good.

a Should be applied along with a special drill and a striking device.

b If driven into soil by static vibrating, noise and vibration can be reduced.

Table 3.2 Nominal Stiffness (Before Reduction)

Retaining Wall		E (MPa)	I (cm⁴/m)	E I (MN-m²/m)	Stiffness Ratio
Method	Type and Dimension				
Soldier pile[a]	H300×300×10×15	2.04×10^5	20,400	41.6	1.0
	H350×350×12×19	2.04×10^5	40,300	82.2	2.0
Steel sheet pile	SP-III	2.04×10^5	16,400	33.5	0.8
	SP-IV	2.04×10^5	31,900	65	1.6
Column pile	30 cm (diameter)	2.1×10^4	132,500	27.8	0.7
	80 cm (diameter)	2.1×10^4	2,513,300	527.8	12.7
MIP pile	SMW method H400×200×8×13	2.04×10^5	59,250	120.9	2.9
Diaphragm wall	0.5 m thick	2.1×10^4	1,041,700	219	5.3
	1.0 m thick	2.1×10^4	8,333,300	1750	42.0

[a] The distance between H-steels is assumed to be 1.00 m.

Table 3.2 illustrates the nominal stiffness per unit length where the value of moment of inertia (I) is not reduced. A stress or deformation analysis, however, has to consider the reduction of stiffness of a soldier pile or a steel sheet pile because of repetitive use. A discount of 80% of the nominal stiffness is often suggested.

The f_c'-value of a PIP pile is about 17 MPa, while that of a concrete pile, a reverse circulation drill pile, or a diaphragm wall is about 21 MPa or more. Their Young modulus can be derived from $E = 4,700\sqrt{f_c'}$ MPa. Considering the cracks in the retaining wall due to bending moment, the moment of inertia may be reduced by 30%–50%. Since the compressive strength of a mixed pile is very small, about 0.5 MPa, its stiffness can be estimated solely on the basis of the stiffness of the H-steel. Listed in the sixth column of Table 3.2 are the stiffness ratios where the inverse values represent the deformation ratios under the same construction condition (i.e., the same excavation depth, strut location, and strut stiffness). For example, the deformation of a SP-III steel sheet pile is 6.63 (5.3/0.8) times as great as a 500-m-thick diaphragm wall.

The selection of the strutting system is highly related to the excavation method, and thus, there aren't many choices for selection. This section won't go into further discussion of the topics.

3.6 CONSTRUCTION OF THE TNEC PROJECT

The author and his group (1998, 2000a, 2000b) have conducted thorough investigations of the excavation of the TNEC. The objects of the investigations include the soil properties of the site, the stress behavior of excavation, and the strain and displacement behaviors of soil. The excavation of TNEC is a typical case of the top-down construction method. Since the process has been thoroughly documented, the TNEC excavation is a very good example for case study. The soil properties and related test data are described in Appendix B. The monitoring results of stress and deformation caused by excavation are shown in Chapter 6, which can be used as calibration for numerical analysis (Chapters 7 and 8).

Figure 3.34a shows the plan of the construction site of the TNEC. As shown in the figure, the site was basically a trapezoid 60–105 m long and 43 m wide. There were six

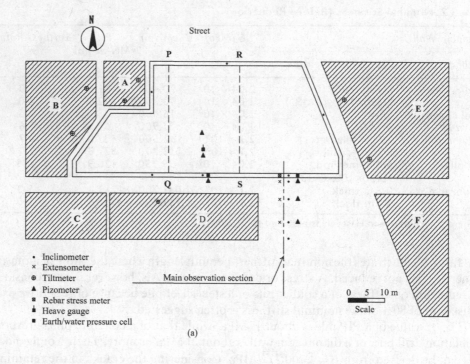

- • Inclinometer
- × Extensometer
- ⊕ Tiltmeter
- ▲ Pizometer
- ▨ Rebar stress meter
- ⬚ Heave gauge
- ▪ Earth/water pressure cell

Main observation section

0 5 10 m
Scale

(a)

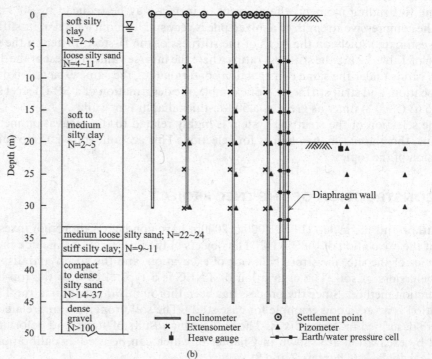

(b)

Figure 3.34 Excavation of the Taipei National Enterprise Center (TNEC). (a) Plan and (b) profile.

buildings, A, B, C, D, E, and F, in the neighborhood of the site. C and D were old four-story RC apartments with individual footings. A, B, E, and F were high-rise buildings with mat foundations. On A, B, D, and E were installed tiltmeters. Three in-wall incli-nometer casings were installed on the southern diaphragm of the site, the eastern one of which was also equipped with strain gauges on steel and earth/water pressure cells on this section. To the south of the site were placed many ground settlement marks, serial extensometers, and in-soil inclinometer casings. Figure 3.34b shows the profile of those instruments along with soil profile. For the details of the allocation of the instruments, please refer to the author's previous studies (Ou et al., 1998, 2000a, 2000b).

The construction of the diaphragm wall of the TNEC started on August 13, 1991, and was completed on November 10, 1991. The monitoring devices, including earth/water pressure cells, strain gauges on steel, and inclinometer casings, were installed as the diaphragm wall was being built. For the convenience of description, take the first day that diaphragm wall construction started as the base, and the completion day of the diaphragm wall construction would be the 89th day. Figure 3.35 and Table 3.3 show the top-down construction process and the schedule of the construction operation, respectively, and Figure 3.36 shows the excavation profile along with the subsoil depos-its. Known from Table 3.3, the excavation is divided into 16 construction stages where the 1st, 3rd, 5th, 7th, 9th, 11th, and 13th are excavation stages and the 2nd, 4th, 6th, 8th, 10th, 12th, 14th, and 15th are strut installation or floor slab construction stages. The table also records the detailed start and end time of each stage.

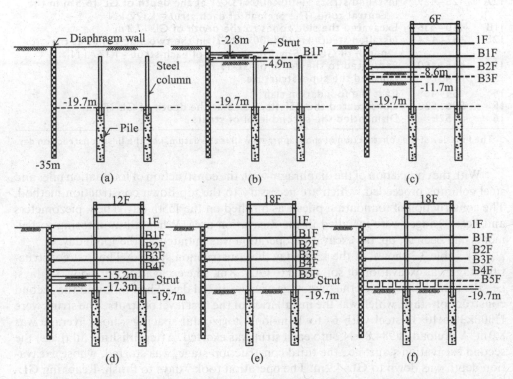

Figure 3.35 Construction procedure of the TNEC (see Table 3.3 for the description of the construction procedure).

Table 3.3 Excavation Process of the TNEC

Stage	Day	Excavation Activities
	−29	Installed devices outside of the excavation zone, including in-soil inclinometers, extensometers, observation wells, and electronic piezometers
	1–89	Constructed the diaphragm wall, including installation of the earth/water pressure cells, in-wall rebar strain meters, and in-wall inclinometers
	89–147	Constructed piles and the steel columns
	147–155	Installed devices inside of the excavation zone, including the piezometers and heave gauges
1	156–162	Excavated to the depth of GL-2.80 m
2	164–169	Installed struts H300×300×10×15 at the depth of GL-2.0 m. The preload of each strut = 784.8 kN
3	181–188	Excavated to the depth of GL-4.9 m
4A	217	Constructed B1F floor slab at the depth of GL-3.5 m
4B	222–238	Dismantled the first level of strut and constructed the 1F floor slab. Started the construction of the superstructure
5	233–255	Excavated to the depth of GL-8.6 m
6	279	Constructed the B2F floor slab at the depth of GL-7.1 m
7	318–337	Excavated to the depth of GL-11.8 m
8	352	Constructed the B3F floor slab at the depth of GL-10.3 m
9	363–378	Excavated to the depth of GL-15.2 m
10	400	Constructed the B4F floor slab at the depth of GL-13.7 m
11A	419–423	Excavated the central zone to the depth of GL-17.3 m
12A	425–429	Installed struts H400×400×13×21 at the depth of GL-16.5 m in the central zone. The preload of each strut = 1,177 kN
11B	430–436	Excavated the side zones to the depth of GL-17.3 m
12B	437–444	Installed struts H400×400×13×21 in the two side zones at the depth of GL-16.5 m. The preload of each strut = 1,177 kN
13	445–460	Excavated to the depth of GL-19.7 m
	457	Finished the superstructure
14	464–468	Cast the foundation slab
15	506–520	Constructed the B5F floor slab at the depth of GL-17.1 m
16	528	Dismantled the second level of struts

The first day of the construction of the diaphragm wall is the datum, i.e., the first construction day.

With the completion of the diaphragm wall, the construction of foundation piles and steel columns proceeded, which are necessary to the top-down construction method. The construction of foundation piles was finished on the 155th day. Then, piezometers and heave gauges were installed in the excavation area. When all monitoring instruments had been set up, the excavation operation was initiated on the 156th day.

As Table 3.3 shows, at the 156th day the construction started. The first construction stage was excavation of soil down to GL-2.8 m. The excavation period was 6 days; i.e., the completion day of the first stage was the 162nd day. Then followed the second construction stage, which was the installation of the first level of struts. The struts were H300×300×10×15 steel with 6- to 11-m-long horizontal spacing whose average was 8.0 m. A preload of 784.8 kN onto each strut was exerted. After finishing all these, the second excavation stage, i.e., the third construction stage, was started, whose excavation depth was down to GL-4.9 m. The operation took 7 days to finish. Reaching GL-4.9 m, a 100-mm-thick layer of poor concrete was placed onto the excavation surface.

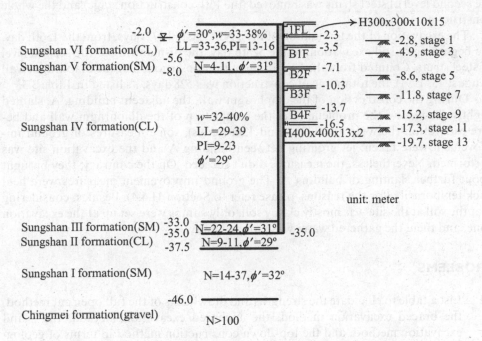

Figure 3.36 Excavation profile of the TNEC along with the subsoil deposits.

Formworks were set up for the construction of the B1F floor slab at GL-3.5 m. As the B1F floor slab was finished and had reached a certain degree of strength, the construction began to proceed upward (i.e., superstructure construction) and downward (i.e., basement construction) simultaneously. The 1F floor slab and the superstructure were built, while at the same time the third excavation stage (the fifth construction stage) was started, whose depth was down to GL-8.6 m. After that, the excavation and the construction of floor slabs were repeated till the completion of the construction of the B4F floor slabs (the tenth stage).

After the 10th stage, considering that the deformation of the retaining walls had accumulated up to 8.0 cm (see Figure 6.10 in Chapter 6), to reduce the wall deformation, the zoned excavation method was adopted. That is, the central area was excavated (the area PQRS in Figure 3.34) first. The central area was excavated to the depth of GL-17.3 m, while at the depth of GL-16.5 m the second level of steel struts was set up. The east and west parts were excavated following the installation of the struts in the central area. Then, the struts of the side areas were also installed. Thus, the 11th and 12th construction stages were further divided into the 11A, 11B, 12A, and 12B stages. The second level of steel struts was H400×400×13×21 steel whose spacing ranged from 2.5 to 6.0 m with an average of 3.4 m. Each strut was preloaded to 1,177 kN. The seventh excavation stage (the 13th construction stage) followed, reaching the final excavation bottom, which was GL-19.7 m deep. On the final excavation bottom was built the raft foundation. The construction of the raft foundation was divided into two stages: first, the building of the raft foundation, and then, the building of the B5F floor slab (the 15th construction stage). When the concrete of the B5F floor got enough strength, the second level of steel struts was removed (the 16th construction stage) and the whole construction was finished.

The excavation of the foundation of the TNEC lasted 372 days, from the 156th day, the beginning of soil excavation, to the 528th day, the dismantling of the second level of steel struts. Counted from the beginning of the construction of the diaphragm wall (August 13, 1991), the total time of construction was 528 days, as listed in Table 3.3.

During the construction of the diaphragm wall, the adjacent building, A, slanted slightly. To resolve the problem, after the completion of the diaphragm wall and before excavation (between the 102nd and 124th days), some measures of ground improvement were taken: jet grouting between building A and the excavation site was performed. Nevertheless, the measures didn't succeed. On the contrary, they brought about further slanting of building A. The ground improvement measures were held back temporarily (for the reasons, please refer to Section 11.4.4). Besides, considering that the soil at the site was mostly clayey soil, only sumps were set up at the excavation zone, and then, the gathered water was pumped away.

PROBLEMS

3.1   List a table to elucidate the strengths and drawbacks of the full open cut method, the braced excavation method, the anchored excavation method, the island excavation method, and the top-down construction method in terms of geological conditions (sandy soils, clayey soils, and gravel soils), construction periods, excavation depths, and excavation widths.

3.2   List a table to elucidate the strengths and drawbacks of the soldier pile, the sheet pile, the column pile, and the diaphragm wall from the viewpoint of geological conditions (sandy soils, clayey soils, and gravel soils), construction period, excavation depth, and sealing capability.

3.3   According to the construction characteristics of solider piles and sheet piles, what kinds of excavation calamites and building damage might be produced if those retaining walls are used?

3.4   According to the construction characteristics of PIP column piles, what kinds of excavation calamites and building damages might be produced if those retaining walls are used?

3.5   What kinds of excavation calamites and building damages might be produced if the diaphragm wall is used as a retaining wall in sandy soils with high groundwater pressure?

3.6   From the literature, references, or excavations in progress, find an excavation case which adopts a different excavation method other than the five methods introduced in this chapter and elucidate its excavation process.

3.7   Elucidate the commonly used construction methods of diaphragm walls from the literature or references and explain their applicability.

3.8   Elucidate the process of the diaphragm wall construction.

3.11   Explain the difference in the mechanism of maintaining trench stability between bentonite and polymer stabilizing fluid in diaphragm wall construction.

3.12   Explain the difference in the mechanism of maintaining concrete quality between bentonite and polymer stabilizing fluid in diaphragm wall construction.

3.13   Explicate the construction process of the reverse circulation drill method.

Chapter 4

Lateral earth pressure

4.1 INTRODUCTION

Problems of deep excavation, no matter if it is a stability analysis (Chapter 5), stress analysis, or deformation analysis (Chapters 6–8), entail the distribution of earth pressures. Though introductory books on soil mechanics or foundation engineering have discussed quite a few earth pressure theories along with many examples, a systematic organization is lacking. In actual analyses, a wrong choice of earth pressure theory may lead to an uneconomical or even unsafe design. This chapter is going to do systematic organization and to simplify the complicated calculations for excavation analyses and design. Most of the methods introduced here have been frequently used in engineering practice though they haven't been introduced in general textbooks. Common examples of the calculation of earth pressures, since many exist in books on soil mechanics and foundation engineering, are not to be discussed in the chapter but will be left in the exercise problems at the end of the chapter.

Earthquakes will engender lateral and vertical acceleration, which will in turn increase the active earth pressure on the retaining wall and decrease the passive earth pressure in front of the wall. Since excavation support systems are basically temporary structures, earthquakes are usually not considered in the design of temporary structures. Therefore, this chapter skips the discussion of the earth pressure induced by earthquakes. Interested readers are advised to refer to related literatures.

4.2 LATERAL EARTH PRESSURE AT REST

Figure 4.1 shows a vertical retaining wall with a height of H. Assume friction doesn't exist between the retaining wall and the soil. When the wall is restrained from movement, the stresses at depth, z, below the ground surface are under elastic equilibrium with no shear stress. Supposing σ'_v represents the effective vertical overburden pressure, the effective lateral pressure σ'_h is:

$$\sigma'_h = K_0 \sigma'_v \tag{4.1}$$

where K_0 = coefficient of lateral earth pressure at rest.
The total lateral pressure is:

$$\sigma_h = \sigma'_h + u \tag{4.2}$$

DOI: 10.1201/9780367853853-4

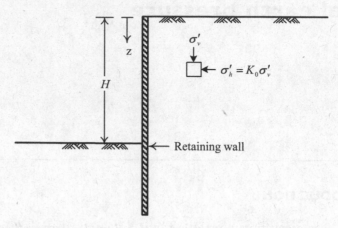

Figure 4.1 Lateral earth pressure at rest.

where u is the pore water pressure, which is the summation of the hydrostatic water pressure and excess pore water pressure.

For both cohesive soil and cohesionless soil, K_0 can be estimated by Jaky's (1944) equation:

$$K_0 = 1 - \sin\phi' \tag{4.3}$$

where ϕ' = effective internal angle of friction or drained friction angle.

When soil is in the preconsolidated or overconsolidated state, K_0 can be estimated by the following equation (Schmidt 1967, Alpan 1967, Ladd et al. 1977):

$$K_{0,\mathrm{OC}} = K_{0,\mathrm{NC}}(\mathrm{OCR})^\alpha \tag{4.4}$$

where

$K_{0,\mathrm{OC}}$ = coefficient of lateral earth pressure at rest for overconsolidated soil with overconsolidation ratio, OCR;

$K_{0,\mathrm{NC}}$ = coefficient of lateral earth pressure at rest for a normally consolidated soil;

α = empirical coefficient; $\alpha \approx \sin\phi'$.

Generally speaking, for normally consolidated cohesive or cohesionless soil, Eq. 4.3 can produce quite satisfactory results. For overconsolidated soil, nevertheless, cohesive and cohesionless alike, the results from Eq. 4.4 are relatively unreliable. One of the reasons is that the formation process of overconsolidated soil is complicated. To obtain the most appropriate K_0-value, the best way is to carry out an in situ test.

4.3 RANKINE'S EARTH PRESSURE THEORY

Rankine (1857) developed a theory of lateral earth pressure in conditions of failure in front of and in back of a retaining wall on the basis of the concept of plastic equilibrium.

As Figure 4.2a shows, the parameters of the Mohr–Coulomb failure line both in front of and in back of the retaining wall are c and ϕ. Suppose there exists no friction between the retaining wall and the soil and the earth pressure both in front of and in

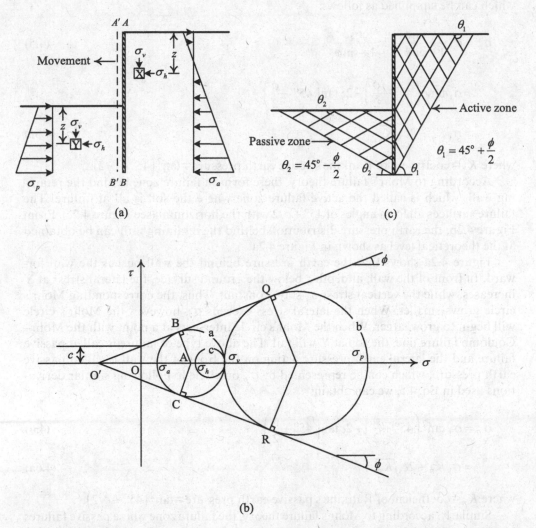

Figure 4.2 Rankine's earth pressure theory. (a) Earth pressure distributions, (b) failure zones, and (c) Mohr's circles.

back of the retaining wall is at K_0 before the retaining wall moves. The vertical stress, σ_v, at X and Y near the wall and at depth, z, below the ground surface are the same, and the stress conditions can be represented by the Mohr's circle (circle c) in Figure 4.2b. Due to the earth pressure on the back of the retaining wall, the wall is pushed toward A'B'. The horizontal stress decreases while the vertical stress condition is unchanged. The Mohr's circle grows larger and will intersect at a point with the Mohr–Coulomb failure line when the soil at X is at failure. The type of failure is called active failure and the lateral earth pressure on the retaining wall is called active earth pressure, as is represented by σ_a on circle a. Therefore,

$$\sin\phi = \frac{AB}{AO'} = \frac{AB}{OO' + OA} = \frac{(\sigma_v - \sigma_a)/2}{c\cot\phi + (\sigma_v + \sigma_a)/2}$$

which can be simplified as follows:

$$\sigma_a = \sigma_v \frac{1-\sin\phi}{1+\sin\phi} - 2c\frac{\cos\phi}{1+\sin\phi} \tag{4.5}$$

$$= \sigma_v \tan^2\left(45^\circ - \frac{\phi}{2}\right) - 2c\tan\left(45^\circ - \frac{\phi}{2}\right)$$

$$= \sigma_v K_a - 2c\sqrt{K_a} \tag{4.5a}$$

where K_a = coefficient of Rankine's active earth pressure = $\tan^2\left(45^\circ - \phi/2\right)$.

According to Mohr's failure theory, there forms a failure zone behind the retaining wall, which is called the active failure zone where the soil is all at failure. The failure surfaces all form angles of $45^\circ + \phi/2$ with the horizontal (see Figure 4.2c). From Figure 4.2b, the earth pressure distribution behind the retaining wall can be obtained at the theoretical level as shown in Figure 4.2a.

Figure 4.2a shows that the earth pressure behind the wall pushes the wall forward. In front of the wall, at depth z below the ground surface, the lateral stress at Y increases, while the vertical stress σ_v stays constant. Thus, the corresponding Mohr's circle grows smaller. When the lateral stress exceeds σ_v, however, the Mohr's circle will begin to grow larger. When the Mohr's circle intersects at a point with the Mohr–Coulomb failure line, the soil at Y will fail. The above type of failure is called passive failure and the lateral earth pressure acting on the front of the wall is called passive earth pressure, which can be represented by σ_p on circle **b**. Following similar derivations used in Eq. 4.5, we can obtain:

$$\sigma_p = \sigma_v \tan^2\left(45^\circ + \frac{\phi}{2}\right) + 2c\tan\left(45^\circ + \frac{\phi}{2}\right) \tag{4.6}$$

$$= \sigma_v K_p + 2c\sqrt{K_p} \tag{4.6a}$$

where K_p = coefficient of Rankine's passive earth pressure = $\tan^2\left(45^\circ + \phi/2\right)$.

Similarly, according to Mohr's failure theory, the failure zone where passive failures happen is called the passive failure zone. The soil in the area is all at failure, whose failure surfaces form angles of $45^\circ - \phi/2$ with the horizontal plane, as shown in Figure 4.2c.

Suppose the wall moves forward by such a great distance that the soil before the wall is completely at passive failure (see Section 4.5.1). The distribution diagram of the passive earth pressures before the wall is also as shown in Figure 4.2a.

For effective stress analysis or drained analysis, the parameters of the Mohr–Coulomb failure line should be expressed in terms of the effective cohesion (c') and the effective internal angle of friction (ϕ'). Eqs. 4.5 and 4.6 should be rewritten in terms of the effective stress as follows:

$$\sigma_a' = \sigma_v' \tan^2\left(45^\circ - \frac{\phi'}{2}\right) - 2c'\tan\left(45^\circ - \frac{\phi'}{2}\right) \tag{4.7}$$

$$= \sigma_v' K_a - 2c'\sqrt{K_a} \tag{4.7a}$$

$$\sigma_p' = \sigma_v' \tan^2\left(45^\circ + \frac{\phi'}{2}\right) + 2c'\tan\left(45^\circ + \frac{\phi'}{2}\right) \tag{4.8}$$

$$= \sigma'_v K_p + 2c'\sqrt{K_p} \qquad\qquad (4.8a)$$

where the coefficients of Rankine's active and passive earth pressures are $K_a = \tan^2\left(45^\circ - \phi'/2\right)$ and $K_p = \tan^2\left(45^\circ + \phi'/2\right)$, respectively.

The angle between the active failure surface and the horizontal plane is $45^\circ + \phi'/2$, and that between the failure passive surface and the horizontal plane is $45^\circ - \phi'/2$.

For total stress undrained analysis of clay, the "$\phi = 0$" concept is applied. As shown in Figure 4.3, assume the σ_v and the undrained shear strength (s_u) of clay at location X and Y are the same (note: as a matter of fact, they may not be the same at X and Y).

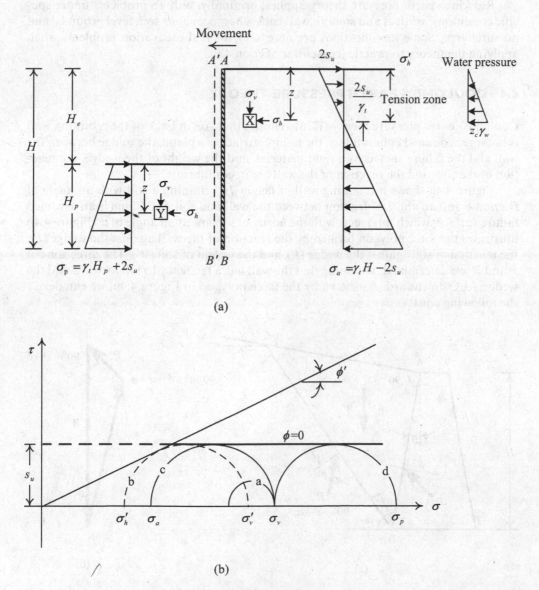

(a)

(b)

Figure 4.3 Active and passive earth pressures under undrained conditions. (a) Earth pressure distributions and (b) Mohr's circles.

Substitute $\phi=0$ into Eqs. 4.5 and 4.6, and we can obtain the active earth pressure (σ_a) and passive earth pressure (σ_p) on the wall as follows:

$$\sigma_a = \sigma_v - 2s_u \tag{4.9}$$

$$\sigma_p = \sigma_v + 2s_u \tag{4.10}$$

According to Eqs. 4.9 and 4.10, we can obtain the theoretical distribution of earth pressure as shown in Figure 4.3a. The total stress Mohr's circles at the active and passive states, along with their effective Mohr' circle, are shown in Figure 4.3b.

Rankine's earth pressure theory applies, originally, only to problems under specific conditions: vertical and smooth wall backs, homogeneous soil, level grounds, and no surcharge. Some modifications are necessary for real excavation problems when applying the theory to practical cases (see Section 4.6).

4.4   COULOMB'S EARTH PRESSURE THEORY

Coulomb's earth pressure theory (1776) assumes that soil in back of the retaining wall is homogeneous and cohesionless, the failure surface is a plane, the wedge between the wall and the failure surface is a rigid material, and the weight of the wedge, the reaction of the soil, and the reaction of the wall are in equilibrium.

Figure 4.4a shows a retaining wall of height H, retaining a soil with an angle of friction ϕ and an angle of friction between the wall and soil δ. BC line is an assumed failure surface, which intersects with the horizontal plane at an angle of α. Figure 4.4b illustrates the force polygon formed by the reaction of the wall against the wedge (P), the reaction of soil against the wedge (R), and the weight of soil (W). The directions of P and R are determined, assuming that the wall has a tendency to fall forward and the wedge ABC downward. As shown by the force polygon in Figure 4.4b, we can derive the following equation:

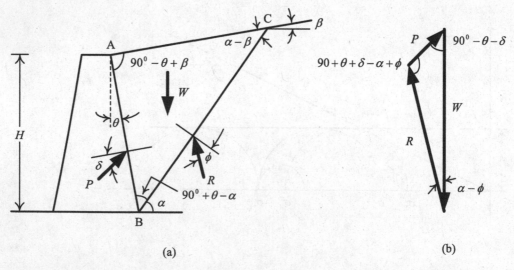

(a)   (b)

Figure 4.4 Coulomb's active earth pressure. (a) All the forces acting on the failure wedge and (b) force polygon.

$$\frac{W}{\sin\left(90^\circ+\theta+\delta-\alpha+\phi\right)}=\frac{P}{\sin\left(\alpha-\phi\right)} \tag{4.11}$$

$$P=\frac{W\sin\left(\alpha-\phi\right)}{\sin\left(90^\circ+\theta+\delta-\alpha+\phi\right)} \tag{4.12}$$

$$=\frac{1}{2}\gamma H^2\left[\frac{\cos(\theta-\alpha)\cos(\theta-\beta)\sin(\alpha-\phi)}{\cos^2\theta\sin(\alpha-\beta)\sin\left(90^\circ+\theta+\delta-\alpha+\phi\right)}\right] \tag{4.12a}$$

where γ is the unit weight of soil and parameters γ, ϕ, δ, θ, β, and H are constants. α is a variable because BC is an assumed failure surface. P varies with α; the minimum of P represents the active earth pressure P_a, that is,

$$\frac{dP}{d\alpha}=0 \tag{4.13}$$

Substitute the critical α-value derived from Eq. 4.13 into Eq. 4.12, and we will obtain the active earth pressure $\left(P_a\right)$ as follows:

$$P_a=\frac{1}{2}\gamma H^2 K_a \tag{4.14}$$

$$K_a=\frac{\cos^2\left(\phi-\theta\right)}{\cos^2\theta\cos\left(\delta+\theta\right)\left[1+\sqrt{\dfrac{\sin\left(\delta+\phi\right)\sin\left(\phi-\beta\right)}{\cos\left(\delta+\theta\right)\cos\left(\theta-\beta\right)}}\right]^2} \tag{4.14a}$$

where K_a is the coefficient of Coulomb's active earth pressure. When $\theta=0$, $\beta=0$, and $\delta=0$, $K_a=\tan^2\left(45^\circ-\phi/2\right)$, which is identical with Rankine's.

Figure 4.5 illustrates the passive soil failure in back of the retaining wall, which is pushed outward by an external force. BC line is an assumed failure surface. The directions of the soil reaction force (R) and the reaction of the wall (P) are determined on the basis that the wall is pushed against soil and the wedge (ABC) moves upward. Following the similar method used in obtaining the active earth pressure, the passive earth pressure $\left(P_p\right)$ can be derived and expressed by the following equation:

$$P_p=\frac{1}{2}\gamma H^2 K_p \tag{4.15}$$

$$K_p=\frac{\cos^2\left(\phi+\theta\right)}{\cos^2\theta\cos\left(\delta-\theta\right)\left[1-\sqrt{\dfrac{\sin\left(\phi+\delta\right)\sin\left(\phi+\beta\right)}{\cos\left(\delta-\theta\right)\cos\left(\beta-\theta\right)}}\right]^2} \tag{4.15a}$$

where K_p is the coefficient of Coulomb's passive earth pressure.

When $\theta=0$, $\beta=0$, and $\delta=0$, Eq. 4.15a can be rewritten as $K_p=\tan^2\left(45^\circ+\phi/2\right)$, which is identical with that of Rankine.

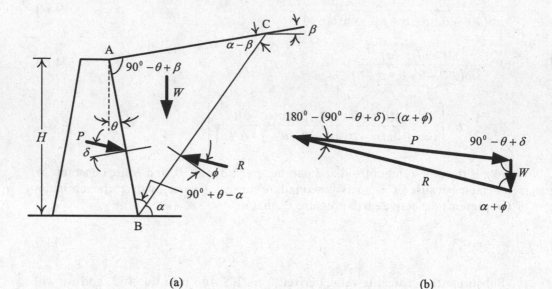

Figure 4.5 Coulomb's passive earth pressure. (a) All the forces acting on the failure wedge and (b) force polygon.

4.5 GENERAL DISCUSSION OF VARIOUS EARTH PRESSURE THEORIES

4.5.1 Displacement and earth pressure

According to the loading conditions of soil, when the soil's strained state changes from K_0 state to the active state, the direction of its principal stresses will remain unchanged. However, if it is from K_0 state to the passive state, its direction will change by a rotation of 90°. That is, σ_1 is originally vertical $(\sigma_1 = \sigma_v)$ and then changes to be horizontal $(\sigma_1 = \sigma_h)$. Both the major and minor principal stresses rotate by 90°.

Thus, the strain for soil to come to passive failure should be greater than that to reach active failure. As to problems of retaining walls, there have been many experiments that explore the relationships between wall displacement and earth pressure. Figure 4.6 shows the results from experiments on rigid gravity walls provided by NAVFAC DM7.2 (1982). As shown in Figure 4.6, the necessary wall displacement inducing the passive condition for cohesionless soil is about four times larger than that inducing active conditions. For cohesive soil, the relationship is about two times.

Concerning problems of deep excavation, since the rigidity of the excavation wall is much smaller than that of the gravity retaining wall, under the working load the displacement (see Figure 4.7) at the bottom of the excavation wall may be too small to induce passive failure. Thus, it is reasonable to infer that the actual earth pressure near the wall bottom is smaller than the passive earth pressure. However, the displacement near the excavation bottom is large enough to cause the passive failure. On the other hand, with a smaller displacement necessary to produce the active failure, the active failure may well occur in the soil behind the retaining wall (Ou et al., 1998). Thus, the actual distribution of earth pressures in front of and behind the wall should be as illustrated in Figure 4.7b.

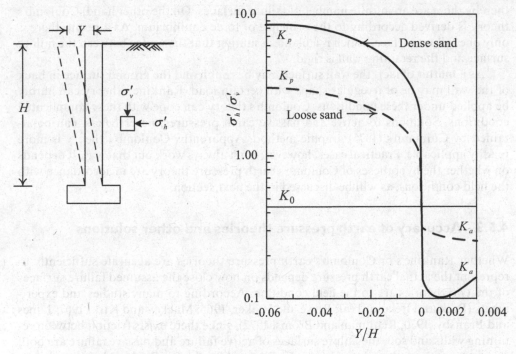

Figure 4.6 Effect of wall movement on earth pressure (Y is the lateral movement at the top of the wall, and H is the wall height) (NAVFAC DM7.2, 1982).

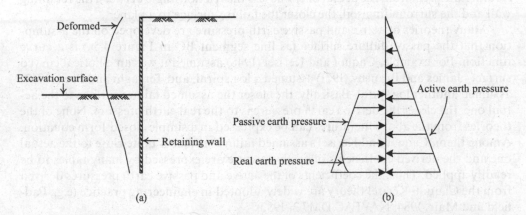

Figure 4.7 Excavation-induced deformation of a retaining wall and the distribution of lateral earth pressure. (a) Wall deformation and (b) distribution of earth pressure.

4.5.2 Comparison of Rankine's and Coulomb's earth pressure theories

As discussed in Section 4.4, the results obtained from Rankine's and Coulomb's earth pressure theories are identical under the same conditions (smooth wall surfaces, level grounds, and homogeneous cohesionless soil) though the two theories are quite differently based. As described in Section 4.3, Rankine's earth pressure theory is based on the principle of the plasticity equilibrium of the strained soil. That is to say, soil at any point within the failure zone (also called the wedge) is indiscriminately at failure and thereby there are an infinite number of failure surfaces. On the other hand, Coulomb's theory is derived according to the principle of force equilibrium. As a result, there is only one failure surface, which is a plane, assuming that the wedge between the failure surface and the retaining wall is rigid.

As a matter of fact, the wall surface may be rough and the ground surface in back of the wall may be of irregular shape with certain load. Rankine's theory can hardly be applied under these conditions. Coulomb's theory can cope with these complicated conditions. Solutions to active and passive earth pressures, nevertheless, can be attained by Culmann's (1875) graphic method. Apparently, Coulomb's theory is more readily applicable. Practical cases, however, don't always work out that way. It depends on whether the hypotheses of Coulomb's earth pressure theory are in accordance with the field conditions, as will be discussed in the next section.

4.5.3 Accuracy of earth pressure theories and other solutions

Whether Rankine's or Coulomb's earth pressure theories are accurate sufficiently to represent the actual earth pressure depends on how close the assumed failure surfaces of the two theories are to the field condition. According to many studies and experiments (Peck and Ireland, 1961; Rowe and Peaker, 1965; Mackey and Kirk, 1967; James and Bransby, 1970; Rehnman and Broms, 1972), since there exists friction between retaining walls and soil, the failure surfaces of active failure and passive failure are both curved surfaces rather than planes. As Figure 4.8a shows, Coulomb's active failure surface is close to the actual one. On the other hand, Coulomb's passive failure surface is rather different from the actual one. The less the friction angle between the retaining wall and the surrounding soil, the closer the failure surface to a plane.

Many theories of active and passive earth pressure are developed on the assumption that the passive failure surface (as line segment BC in Figure 4.5a) is a curve function. For example, Caquot and Kerisel (1948) assumed it was an elliptical curved surface, James and Bransby (1971) assumed a log spiral, and Terzaghi and Peck (1967) assumed another log spiral. Basically, the closer the assumed failure surface to the actual one, the closer the derived earth pressure is to the real earth pressure. None of the theories from the above literatures can be expressed in a simple closed-form equation. Among them, Caquot and Kerisel's assumed failure surface is quite close to the actual one and the derived coefficients of earth pressure are expressed by many tables to be readily applied. Thus, the coefficients of the active and passive earth pressures derived from the Caquot–Kerisel theory are widely adopted in engineering practice (e.g., Padfield and Mair, 1984; NAVFAC DM7.2, 1982).

For active earth pressure, when $\delta = 0$ (δ is the friction angle between the wall and soil), the coefficients of active earth pressure obtained from Rankine's, Coulomb's, and Caquot and Kerisel's theories are identical. When $\delta > 0$, the coefficients of Rankine's active earth pressure come out the largest, while those of Coulomb's and Caquot and Kerisel's are still very close. In the extreme condition, that is, $\delta = \phi'$, the differences in the coefficients of Coulomb's and Caquot and Kerisel's active earth pressures should be largest, and however, as shown in Figure 4.9 (level ground and vertical wall), the differences are slight for all ϕ'. In practical application, K_a can be obtained directly from Coulomb's equation (Eq. 4.14). When $\delta \neq 0$, the direction of P_a can refer to Figure 4.11. The horizontal component of the earth pressure $\sigma_{a,h} = \gamma H K_{a,h} = \gamma H K_a \cos\delta$ and $P_{a,h} = \left(\gamma H^2 K_a \cos\delta\right)/2$ can then be computed accordingly.

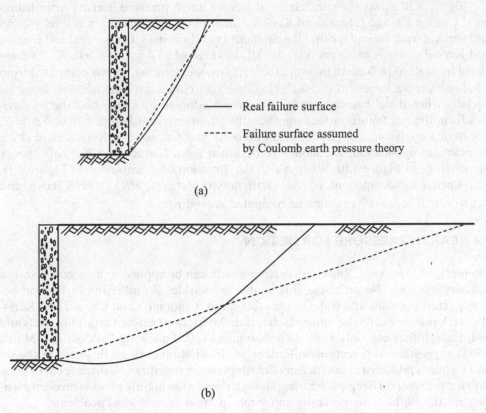

(a)

(b)

Figure 4.8 Real failure surfaces and failure surfaces assumed by Coulomb's earth pressure theory. (a) Active condition and (b) passive condition.

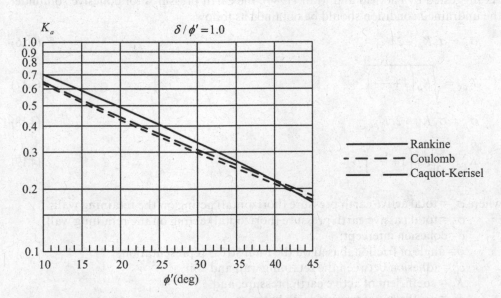

Figure 4.9 Coefficients of Rankine's, Coulomb's, and Caquot and Kerisel's active earth pressure for $\delta = \phi$.

Figure 4.10 shows the coefficients of passive earth pressure derived from Rankine's, Coulomb's, and Caquot and Kerisel's theories on the conditions of level ground and vertical wall. We can see that the coefficients of Rankine's, Coulomb's, and Caquot and Kerisel's passive earth pressure are all the same when $\delta = 0$. If $\delta > 0$, K_p, as computed from Rankine's earth pressure theory, is the smallest, while that computed from Coulomb's is the largest. If $\phi' > 40°$, Coulomb's K_p comes out unreasonably large especially when $\delta = \phi'$ because the failure surface adopted by Coulomb's theory deviates from the real failure surface significantly. Observed from Figure 4.10, if $\delta \leq 0.5\phi'$, Coulomb's coefficients of passive earth pressure and Caquot and Kerisel's are close. In practical application, K_p should be obtained from Caquot and Kerisel's theory by referring to Figure 4.10. When $\delta \neq 0$, the direction of P_p can refer to Figure 4.11. The horizontal component of the earth pressure $\sigma_{p,h} = \gamma H K_{p,h} = \gamma H K_p \cos\delta$ and $P_{p,h} = (\gamma H^2 K_p \cos\delta) / 2$ can then be computed accordingly.

4.6   EARTH PRESSURE FOR DESIGN

Theoretically speaking, Rankine's earth pressure can be applied to both cohesive and cohesionless soils. Nevertheless, it is unable to consider the adhesion or friction between retaining walls and soil. On the other hand, Coulomb's and Caquot and Kerisel's earth pressure can take into consideration the friction between retaining walls and soil, though they can only apply to cohesionless soil. Following Padfield and Mair's (1984) suggestion, this section will adopt modified Rankine's earth pressure theory and Caquot and Kerisel's coefficient of earth pressure together to calculate earth pressure for problems of deep excavation. Though the section mainly focuses on deep excavations, the methods are generally applicable to other geotechnical problems.

4.6.1   Cohesive soil

As suggested by Padfield and Mair (1984), the earth pressures for cohesive soil under the undrained condition should be obtained as follows:

$$\sigma_a = \sigma_v K_a - 2c K_{ac} \tag{4.16}$$

$$K_{ac} = \sqrt{K_a \left(1 + \frac{c_w}{c}\right)} \tag{4.17}$$

$$\sigma_p = \sigma_v K_p + 2c K_{pc} \tag{4.18}$$

$$K_{pc} = \sqrt{K_p \left(1 + \frac{c_w}{c}\right)} \tag{4.19}$$

where σ_a = total active earth pressure (horizontal) acting on the retaining wall;
$\quad \sigma_p$ = total passive earth pressure (horizontal) acting on the retaining wall;
$\quad c$ = cohesion intercept;
$\quad \phi$ = angle of friction, based on the total stress representation;
$\quad c_w$ = adhesion between the retaining wall and soil;
$\quad K_a$ = coefficient of active earth pressure; and
$\quad K_p$ = coefficient of passive earth pressure.

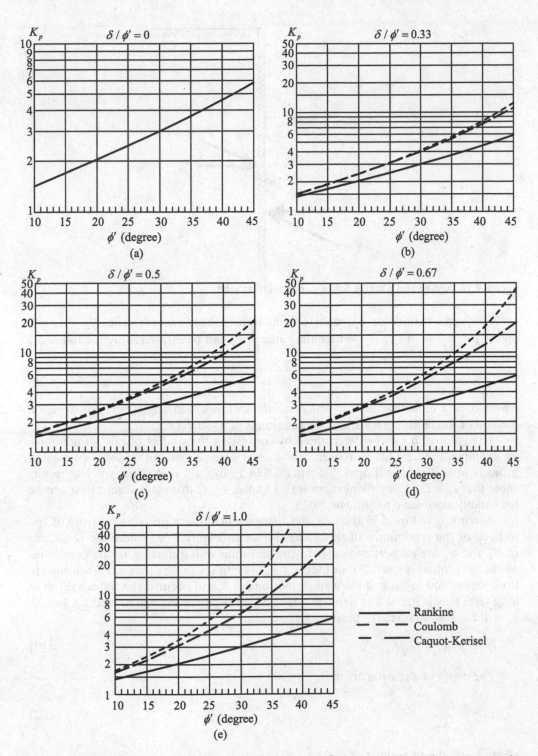

Figure 4.10 Coefficients of Rankine's, Coulomb's, and Caquot and Kerisel's passive earth pressure.

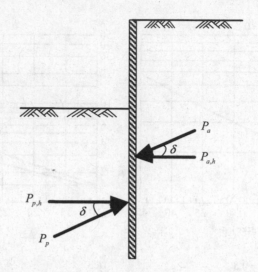

Figure 4.11 Active and passive forces on a vertical wall.

For the $\phi = 0$ condition, c is equal to the undrained shear strength s_u; i.e., $K_a = K_p = 1$, and $K_{ac} = K_{pc} = \sqrt{1 + c_w/s_u}$, where the value of c_w can be estimated by the following equation:

$$c_w = \alpha s_u \tag{4.20}$$

where α is a reduction factor, which relates to the soil strength, the construction method of the retaining wall, and the roughness of its surface.

The value of α can be determined by consulting the studies of pile foundations, for example, Figure 4.12. Moreover, case studies were carried out on some excavation projects in soft clay in Taipei, Singapore, San Francisco, and Chicago. The results show that $c_w = 2s_u/3$ (for diaphragm walls) and $c_w = s_u/2$ (for steel sheet piles) can be reasonably assumed (Ou and Hu, 1998).

According to Eqs. 4.16 and 4.18, the distribution of earth pressures in front of and in back of the retaining wall should be similar to Figure 4.3a, where the resultants of σ_a and σ_p are all perpendicular to the retaining wall (referring to the conditions where the ground is level). From Figure 4.3a, we can see there exists a tension zone in the cohesive soil in back of the wall. A tension crack will occur if the soil is subject to long-term tensile stress. The depth of a tension crack can be estimated by Eq. 4.16 with $c_w = 0$. Let the lateral earth pressure be 0; then,

$$\sigma_a = \gamma z K_a - 2c\sqrt{K_a} = 0 \tag{4.21}$$

The depth of a tension crack is:

$$z_c = \frac{2c}{\gamma\sqrt{K_a}} \tag{4.22}$$

where z_c = depth of a tension crack.

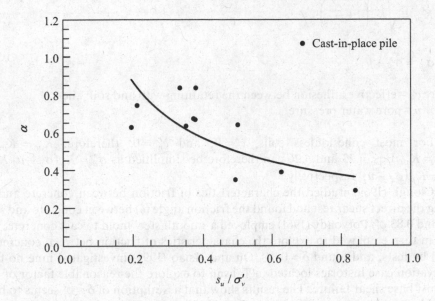

Figure 4.12 Relation between adhesion and undrained shear strength of clay (Lin and Lin, 1999).

Once tension cracks are produced, the soil is no longer able to bear tensile stress. Thus, to be conservative, we often suppose there already exist tension cracks in the design. Rain and environmental factors will further make cracks full of water, so that they will push the retaining wall as shown in Figure 4.3a.

The long-term behavior of cohesive soil should be analyzed on the basis of the complete dissipation of the excess pore water pressure. The distribution of earth pressure is similar to that of cohesionless soil, as will be discussed in the following section.

4.6.2   Cohesionless soil

The excess pore water pressure of cohesionless soil dissipates quickly as soon as shearing occurs. As a result, the analysis should follow the effective stress method. Supposing there exists friction between the retaining wall and the surrounding soil, the earth pressure for design can be represented as follows (Padfield and Mair, 1984):

$$\sigma_a' = K_a(\sigma_v - u) - 2c'K_{ac} \tag{4.23}$$

$$K_{ac} = \sqrt{K_a\left(1 + \frac{c_w'}{c'}\right)} \tag{4.24}$$

$$\sigma_a = \sigma_a' + u \tag{4.25}$$

$$\sigma_p' = K_p(\sigma_v - u) + 2c'K_{pc} \tag{4.26}$$

$$K_{pc} = \sqrt{K_p\left(1 + \frac{c'_w}{c'}\right)} \qquad\qquad (4.27)$$

$$\sigma_p = \sigma'_p + u \qquad\qquad (4.28)$$

where c'_w = effective adhesion between the retaining wall and soil; and
u = pore water pressure.

For most cohesionless soils, $c' = 0$, and $c'_w = 0$; therefore, $K_{ac} = K_a$, and $K_{pc} = K_p$. Eqs. 4.23 and 4.26 can therefore be simplified as $\sigma'_a = K_a(\sigma_v - u)K_a$ and $\sigma'_p = K_p(\sigma_v - u)$, respectively.

Clough (1969) studied the characteristics of friction between concrete and sand using the direct shear test and found the friction angle (δ) between concrete and sand is around $0.83\ \phi'$. Potyondy (1961) employed a smooth steel mold to cast concrete, which was in turn employed to explore the characteristics of friction between concrete and sand by tests, and found $\delta = 0.8\phi'$. Ou and Hsiao (1999) investigated nine no-failure excavation case histories located in Taiwan to explore the reasonable factor of safety against base shear failure. The results show that assumption of $\delta = \phi'$ seems to be reasonable for diaphragm wall with sandy soils. However, to be conservative in analysis, $\delta = 2\phi'/3$ and $\delta = \phi'/2$ can be assumed for diaphragm wall and steel sheet piles with sandy soils, respectively.

A tension zone or a tension crack zone will not occur in cohesionless soil because $c' = 0$. Sometimes a value of c' may be found, which is other than 0 in design, the reason for which might be that c' is an apparent earth pressure or a deviation due to regression analysis of test results. Whichever is the case, there cannot exist a tension zone or a tension crack zone in cohesionless soil and some necessary modifications are required.

4.6.3 Alternated layers

To compute the distribution of earth pressures for alternated layers composed of both cohesive and cohesionless soils, the total stress analysis should be adopted for (short-term behavior of) cohesive soil, the pore water pressure should not be considered in the analysis while the effective stress analysis is used for cohesionless soil, and the pore water pressure should be computed. Note that pore water pressure in cohesionless soil is not necessary in the hydrostatic condition because of the influences of geology formation, the environment change, and past dewatering history. If working with the long-term behavior, the effective stress analysis should be applied to both cohesive and cohesionless soils and the pore water pressure should be estimated separately. Figure 4.13 shows a typical alternated soil deposit with a "possible" distribution of pore water pressure.

4.6.4 Sloping ground

According to Rankine's earth pressure theory, the coefficients of the earth pressure for a sloping ground (Figure 4.14) can be computed as follows (Bowles, 1988):

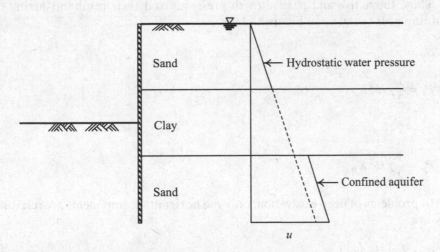

Figure 4.13 Distribution of water pressure for alternated layers.

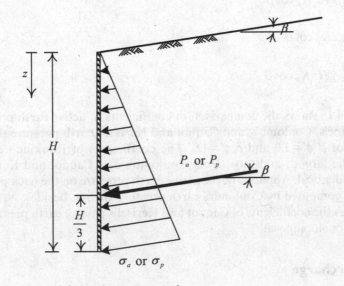

Figure 4.14 Earth pressure for a sloping ground.

$$K_a = \cos\beta \frac{\cos\beta - \sqrt{\cos^2\beta - \cos^2\phi}}{\cos\beta + \sqrt{\cos^2\beta - \cos^2\phi}} \qquad (4.29)$$

$$K_p = \cos\beta \frac{\cos\beta + \sqrt{\cos^2\beta - \cos^2\phi}}{\cos\beta - \sqrt{\cos^2\beta - \cos^2\phi}} \qquad (4.30)$$

Thus, the active and passive earth pressures and their resultants acting on the retaining wall as shown in Figure 4.14 are:

$$\sigma_a = \gamma z K_a \tag{4.31}$$

$$P_a = \frac{1}{2}\gamma H^2 K_a \tag{4.32}$$

$$\sigma_p = \gamma z K_p \tag{4.33}$$

$$P_p = \frac{1}{2}\gamma H^2 K_p \tag{4.34}$$

In problems of deep excavation, only the horizontal components are relevant, and they are:

$$\sigma_{a,h} = \gamma z K_a \cos\beta \tag{4.35}$$

$$P_{a,h} = \frac{1}{2}\gamma H^2 K_a \cos\beta \tag{4.36}$$

$$\sigma_{p,h} = \gamma z K_p \cos\beta \tag{4.37}$$

$$P_{p,h} = \frac{1}{2}\gamma H^2 K_p \cos\beta \tag{4.38}$$

Figure 4.15 shows the comparison of coefficients of active earth pressure derived from Rankine's, Coulomb's, and Caquot and Kerisel's earth pressure theories for the conditions of $\beta/\phi' = 1.0$ and $\delta/\phi' = 1.0$. The coefficients of Rankine's earth pressure come out the largest, and those from Coulomb's and Caquot and Kerisel's theories are almost identical. Apparently, the active earth pressure on the back of the retaining wall can be computed by Coulomb's earth pressure theory. Besides, NAVFAC DM7.2 also provides the coefficients of Caquot and Kerisel's passive earth pressure for a sloping ground or sloping wall.

4.6.5 Surcharge

As shown in Figure 4.16, the uniformly distributed load, q, acts on the entire ground surface, while p acts on the entire excavation bottom. The load-induced active and passive earth pressures on the retaining wall are separately:

$$\sigma_a = qK_a \tag{4.39}$$

$$\sigma_p = pK_p \tag{4.40}$$

If the load is neither uniformly nor extensively distributed over the entire area, the thrust of load against the wall will then be a problem of the theory of elasticity.

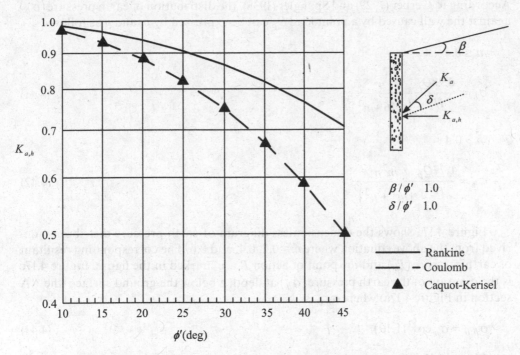

Figure 4.15 Coefficients of Rankine's, Coulomb's, and Caquot and Kerisel's horizontal active earth pressure.

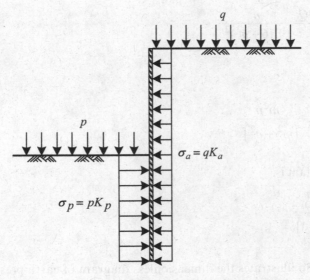

Figure 4.16 Lateral pressure produced by a uniformly distributed load.

According to Gerber (1929) and Spangler (1938), the distribution of earth pressure (σ_h) against the wall caused by a point load Q_p can be expressed by m and n as follows:

$m \leq 0.4$

$$\sigma_h = \frac{0.28Q_p}{H^2} \frac{n^2}{\left(0.16+n^2\right)^3} \tag{4.41}$$

$m > 0.4$

$$\sigma_h = \frac{1.77Q_p}{H^2} \frac{m^2n^2}{\left(m^2+n^2\right)^3} \tag{4.42}$$

Figure 4.17a shows the dimensionless diagram of earth pressure distribution derived from the above equation where $m = 0.2, 0.4$, and 0.6. The corresponding resultant of earth pressure (P_h) and the point of action R are marked in the figure. Figure 4.17c is the diagram of the earth pressure (σ_h) at depth z below the ground surface (the AA section in Figure 4.17a) where $\sigma_{h,\theta}$ is:

$$\sigma_{h,\theta} = \sigma_h \cos^2\left(1.1\theta\right) \tag{4.43}$$

Figure 4.18a illustrates the earth pressure distribution caused by a line load (Q_ℓ) parallel to the retaining wall, whose earth pressure and earth pressure resultant can be computed as follows:

$m \leq 0.4$

$$\sigma_h = \frac{0.203Q_\ell}{H} \frac{n}{(0.16+n^2)^2} \tag{4.44}$$

$m > 0.4$

$$\sigma_h = \frac{1.28Q_\ell}{H} \frac{m^2n}{\left(m^2+n^2\right)^2} \tag{4.45}$$

and their resultant is:

$$P_h = \frac{0.64Q_\ell}{\left(m^2+1\right)} \tag{4.46}$$

Figure 4.18b illustrates the dimensionless diagram of earth pressure distribution derived from the above equation where $m = 0.1, 0.3, 0.5$, and 0.7, respectively. The corresponding point of action R is also marked in the figure.

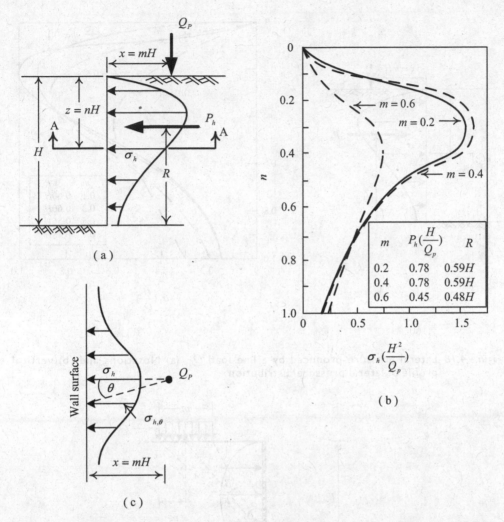

Figure 4.17 Lateral pressure produced by a point load Q_p. (a) Notations, (b) vertical profile of lateral pressure distribution, and (c) horizontal profile of lateral pressure distribution.

Figure 4.19 illustrates a strip load (Q_s) paralleling the retaining wall. According to the theory of elasticity, the lateral earth pressure (σ_h) at depth z is:

$$\sigma_h = \frac{Q_s}{H}(\beta - \sin\beta\cos 2\alpha) \tag{4.47}$$

Considering the restraining effect of the retaining wall, the earth pressure acting on the wall should be modified as follows:

$$\sigma_h = \frac{2Q_s}{H}(\beta - \sin\beta\cos 2\alpha) \tag{4.48}$$

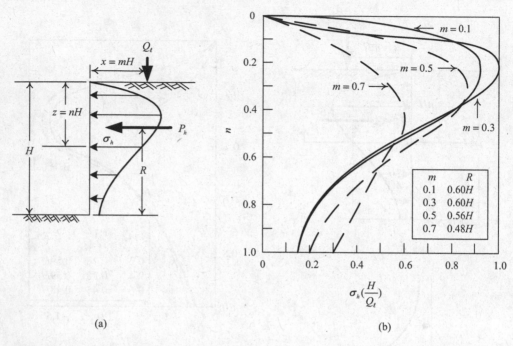

(a)

(b)

Figure 4.18 Lateral pressure produced by a line load Q_ℓ. (a) Notations and (b) vertical profile of lateral pressure distribution.

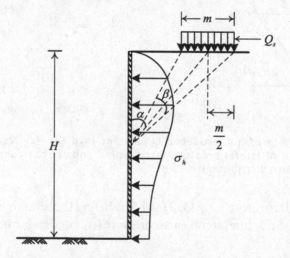

Figure 4.19 Lateral pressure produced by a strip load Q_s.

4.6.6 Seepage

Cohesionless soil has a high degree of permeability. As a result, the difference of the water levels in front of and behind the wall will cause seepage. Cohesive soil exhibits undrained behavior in the short-term condition. That is to say, no seepage is to be considered. In the long run, cohesive soil may also produce seepage and can be analyzed in the same way as cohesionless soil.

Seepage will affect the pore water pressures inside and outside the excavation zone and also the effective stresses. The variation of pore water pressure can be estimated by flow net or using the finite element method. Figure 4.20 is a typical diagram of the flow net. For simplicity, the behavior of seepage can be assumed to be one-dimensional; that is, the head loss per unit length of a flow path is the same.

As shown in Figure 4.21a, the difference of the total heads between the upstream water level (outside the excavation zone) and the downstream water level is $H_e + d_i - d_j$. The length of the path of the water flowing from the upstream water level along the retaining wall down to the downstream water level is $2H_p + H_e - d_i - d_j$. Assuming that the datum of the elevation head is set at the upstream water level, the total head (h) at a distance of x from the upstream water level would be:

$$h = 0 - \frac{x\left(H_e + d_i - d_j\right)}{2H_p + H_e - d_i - d_j} = -\frac{x\left(H_e + d_i - d_j\right)}{2H_p + H_e - d_i - d_j} \tag{4.49}$$

Let h_e be the elevation head and h_p the pressure head. The pressure head at a distance of x from the upstream water level would be:

$$h_p = h - h_e = \frac{x\left(H_e + d_i - d_j\right)}{2H_p + H_e - d_i - d_j} - (-x) = \frac{2x\left(H_p - d_i\right)}{2H_p + H_e - d_i - d_j} \tag{4.50}$$

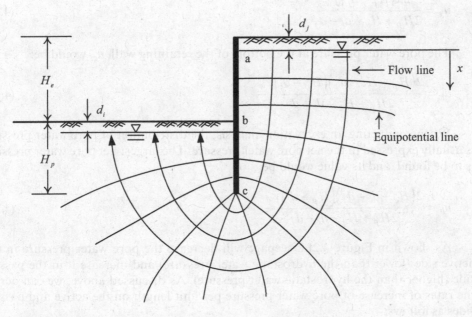

Figure 4.20 Seepage in an excavation zone.

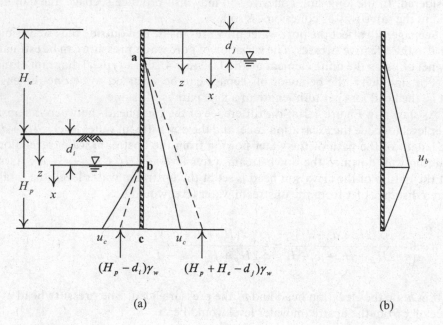

$(H_p - d_i)\gamma_w \qquad (H_p + H_e - d_j)\gamma_w$

(a)

(b)

Figure 4.21 Simplified analysis method for seepage. (a) Distribution of water pressure and (b) net water pressure.

Thus, the pore water pressure at a distance of x from the upstream water level, u_x, would be:

$$u_x = \frac{2x\left(H_p - d_i\right)\gamma_w}{2H_p + H_e - d_i - d_j} \tag{4.51}$$

The pore water pressure at the bottom of the retaining wall, u_c, would be:

$$u_c = \frac{2\left(H_p + H_e - d_j\right)\left(H_p - d_i\right)\gamma_w}{2H_p + H_e - d_i - d_j} \tag{4.52}$$

When conducting an excavation analysis, the distribution of pore water pressure is usually expressed in the net pore water pressure. The largest net pore water pressure is to be found, and its value would be:

$$u_b = \frac{2\left(H_e + d_i - d_j\right)\left(H_p - d_i\right)\gamma_w}{2H_p + H_e - d_i - d_j} \tag{4.53}$$

As shown in Figure 4.21a, seepage will decrease the pore water pressure in the active side (lower than the hydrostatic water pressure) and increase it in the passive side (higher than the hydrostatic water pressure). As discussed above, we can derive the rates of increase of pore water pressure per unit length on the active and passive sides as follows:

$$\mu_a = \frac{u_c}{H_p + H_e - d_j} \tag{4.54}$$

$$\mu_p = \frac{u_c}{H_p - d_i} \tag{4.55}$$

where μ_a and μ_p represent the rates of increase of the pore water pressure per unit length in the active and passive sides, respectively.

Thus, the pore water pressures x below the water level at the active and passive sides are separately $u_x = \mu_a x$ (the active side) and $u_x = \mu_p x$ (the passive side).

Therefore, μ_a and μ_p can be seen as the modified unit weights of water in the active side and passive side, respectively. As a result, the total lateral earth pressure at a depth of z below the ground surface on the back of the wall would be:

$$\sigma_h = \sigma_v' K_a + u \tag{4.56}$$

$$= \left[\sigma_v - \frac{(z - d_j)}{H_p + H_e - d_j} u_c \right] K_a + \frac{(z - d_j)}{H_p + H_e - d_j} u_c \tag{4.57}$$

Similarly, the total lateral earth pressure at a distance of z below the ground surface on the front of the wall would be:

$$\sigma_h = \sigma_v' K_p + u \tag{4.58}$$

$$= \left[\sigma_v - \frac{(z - d_i)}{H_p - d_i} u_c \right] K_p + \frac{(z - d_i)}{H_p - d_i} u_c \tag{4.59}$$

PROBLEMS

4.1 Figure P4.1 shows a two-story basement. The soil is silty sand, and the groundwa-
ter level $H = 3.0$ m. $\gamma = 18$ kN/m³, $\gamma_{sat} = 22$ kN/m³, $c' = 0$, and $\phi' = 34°$. Compute the
total lateral force on the outer wall of the basement (including the resultant and
the location of line of action).

4.2 Same as above. Assume that the soil is normally consolidated clay. $H = 2.5$ m,
$\gamma = 16$ kN/m³, $\gamma_{sat} = 20$ kN/m³, $c' = 0$, $\phi' = 32°$, and $s_u = 55$ kN/m². Compute the
total lateral force on the outer wall of the basement (including the resultant and
the location of line of action).

4.3 Assume the soil as shown in Figure P4.3 is sand and the groundwater level is very
deep. $H_e = H_p = 13.0$ m, $c' = 0$, $\phi' = 34°$, and $\gamma = 18$ kN/m³. Compute the active
earth pressure using the following earth pressure theories, respectively (including
the resultant and the location of line of action):
   a. Rankine's earth pressure theory
   b. Coulomb's earth pressure theory $(\delta = \phi')$
   c. Caquot and Kerisel's earth pressure theory $(\delta = \phi')$

4.4 Same as above. If $\delta = \phi'/2$, compute the active earth pressure using Rankine's,
Coulomb's, and Caquot and Kerisel's earth pressure theories again.

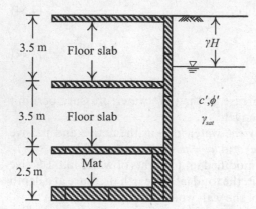

Figure P4.1

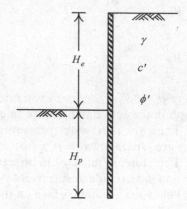

Figure P4.3

4.5 Same as Problem 4.3. Compute the passive earth pressure.

4.6 Same as Problem 4.4. Compute the passive earth pressure.

4.7 Assume the ground as shown in Figure P4.7 consists of clayey soil. $H_1 = 4.0$ m, $s_u = 100$ kN/m², $\gamma_{sat1} = 18$ kN/m³, $\gamma_{sat2} = 18$ kN/m³, and $s_u/\sigma'_v = 0.3$. (a) Use Rankine's earth pressure theory and (b) use Eq. 4.16 with $c_w = 0.5 s_u$ to compute the total active earth pressure (including the resultant and the location of line of action).

4.8 Redo Problem 4.7 assuming $c_w = s_u$.

4.9 Same as Problem 4.7. (a) Use Rankine's earth pressure theory and (b) use Eq. 4.18 with $c_w = 0.5 s_u$ to compute the total passive earth pressure (including the resultant and the location of line of action).

4.10 Figure P4.10 shows a retaining wall and the soil profile. $H_e = H_p = 9.5$ m, $H_1 = 4.0$ m, $c' = 0$, $\phi' = 32°$, $\gamma = 16$ kN/m³, $\gamma_{sat} = 18$ kN/m³, and $s_u/\sigma'_v = 0.35$. The pore water pressure in the sand layer $u = 300$ kN/m². Compute the active and passive earth pressures and their locations of line of action using Rankine's earth pressure theory.

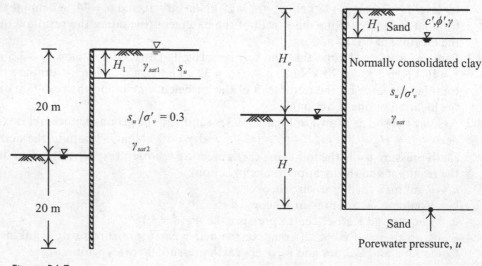

Figure P4.7

Figure P4.10

4.11 Same as above. Assume the earth pressure in the sandy layer is computed according to Caquot and Kerisel's theory $(\delta = 0.5\phi')$. The active and passive earth pressures for the clayey layer are computed following Eqs. 4.16 and 4.18 $(c_w = 0.5s_u)$. Compute the lateral forces on the active and passive sides.

4.12 Redo Problem 4.10. Assume $H_e = H_p = 15$ m, $H_1 = 5.0$ m, $c' = 0$, $\phi' = 35°$, $\gamma = 14$ kN/m^3, $\gamma_{sat} = 20$ kN/m^3, and $s_u / \sigma'_v = 0.30$. The pore water pressure in the sandy layer $u = 300$ kN/m^2.

4.13 Figure P4.13 shows a clayey layer with a line load of Q_ℓ. The soil above the groundwater level is saturated. $H_e = H_p = 10$ m, $d_j = 1.0$ m, $\gamma_{sat} = 18$ kN/m^3, $s_u = 20$ kN/m^2, $Q_\ell = 500$ kN/m, and $d = 3.0$ m. Compute the earth pressure induced by the line load, and compare the result with Rankine's active earth pressure.

4.14 Same as the previous problem. Assume the clayey layer is acted on by a point load $Q_p = 500$ kN. Compute the earth pressure on the section nearest to the point load Q_p. Compare the result with Rankine's active earth pressure, too.

4.15 Redo Problem 4.13. Assume $H_e = H_p = 15$ m, $d_j = 1.0$ m, $\gamma_{sat} = 20$ kN/m^3, $s_u = 20$ kN/m^2, $Q_\ell = 900$ kN/m, and $d = 2.0$ m.

4.16 Figure P4.16 shows a retaining wall and the soil profile. Assume the excavation width $B = 30$ m, $D = 5$ m, $H_e = H_p = 10$ m, $d_i = 0.5$ m, $d_j = 1.0$ m, $\gamma = 18$ kN/m^3, $\gamma_{sat} = 22$ kN/m^3, $c' = 0$, and $\phi' = 34°$. According to Caquot and Kerisel's theory, compute the total lateral force, including the resultant and the location of the line of action under the following conditions (assuming $\delta = \phi'$):

a. If no seepage occurs

b. If seepage occurs, use the simplified method (assuming the seepage is one-dimensional)

c. If seepage occurs, draw the flow net to compute the pore water pressure

4.17 Redo Problem 4.16. Assume $H_e = H_p = 15$ m, $d_i = 0.5$ m, $d_j = 3.0$ m, $\gamma = 12$ kN/m^3, $\gamma_{sat} = 20$ kN/m^3, $c' = 0$, and $\phi' = 34°$.

4.18 Redo Problem 4.16. Use Rankine's theory.

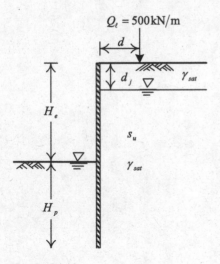

Figure P4.13

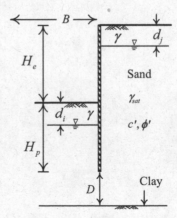

Figure P4.16

Chapter 5

Stability analysis

5.1 INTRODUCTION

Failure or collapse of excavations is disastrous. At worst, it endangers construction workers and adjacent properties. Its influence range is usually large: Much ground settlement may arise, and adjacent properties within the influence range of settlement may be damaged significantly. Since they are very influential, to avoid failure or collapse is of the first importance and stability analyses are therefore required.

Failure of an excavation may arise from the stress on the support system exceeding the strength of its materials, for example, when the strut load exceeds the buckling load of struts or the bending moment of the retaining wall exceeds the limiting bending moment. Please see Chapter 10 for more detailed discussions. Failure can also arise from the shear stress in the base soil exceeding the shear strength. The methods of analyzing whether the soils at the excavation site are able to bear the stress generated by excavation are called stability analysis and are the main subject of this chapter.

Stability analysis includes analysis of base shear failure, sand boiling, and upheaval. This chapter will introduce those analysis methods in detail.

5.2 TYPES OF FACTORS OF SAFETY

There are basically three methods to determine the factor of safety for stability analysis: the strength factor method, the load factor method, and the dimension factor method, which are explained as follows:

5.2.1 Strength factor method

The method considers the soil strength involving much uncertainty and has the strength reduced by a factor of safety. If the factor of safety for the strength factor method is represented as FS_s, the soil parameters for the effective stress analysis are as follows:

$$\tan\phi'_m = \frac{\tan\phi'}{FS_s} \tag{5.1a}$$

$$c'_m = \frac{c'}{FS_s} \tag{5.1b}$$

DOI: 10.1201/9780367853853-5

The parameter for the undrained analysis is

$$s_{u,m} = \frac{s_u}{FS_s} \tag{5.2}$$

After conducting a force equilibrium or a moment equilibrium analysis with the parameters c'_m, ϕ'_m, or $s_{u,m}$ derived as above, we can design the penetration depth. The method locates the factor of safety at the source where the largest uncertainty arises and is therefore quite a reasonable method. Since the reduced parameters will lead to a smaller K_p and a larger K_a, the distribution of earth pressures on the retaining wall will be skewed. As a result, the method is applicable only to stability analysis and cannot be applied to deformation analysis or stress analysis (see Chapters 6–8).

5.2.2   Load factor method

The factor of safety for the load method, FS_ℓ, can be defined as follows:

$$FS_\ell = \frac{R}{D} \tag{5.3}$$

where R represents the resistant force and D is the driving force. R and D can be either the resistant moment and the driving moment, or the bearing force and the external force.

FS_ℓ considers uncertainty arising from the soil strength, the analysis method, and external forces synthetically.

5.2.3   Dimension factor method

Suppose retaining walls are in the ultimate state and the soil strengths are fully mobilized. With the force equilibrium (the horizontal force equilibrium, the moment equilibrium, or other type of force equilibrium), the penetration depth of retaining walls in the ultimate state can be found. The penetration depth for design is

$$H_{p,d} = FS_d H_{p,\text{cal}} \tag{5.4}$$

where

FS_d = factor of safety for the dimension factor method
$H_{p,\text{cal}}$ = penetration depth computed from the limit equilibrium.

The factor of safety is usually defined as a ratio of the resistant force to driving force or as a factor to reduce the strength. Eq. 5.4 is empirically oriented and cannot properly express the meaning of the factor of safety, leading to unreasonable results sometimes, and is not recommended (Burland and Potts, 1981). If applied, cross-checking by other methods is necessary.

5.3   BASE SHEAR FAILURE

When the shear stress generated by the unbalanced force in most of the soil below the excavation bottom reaches the shear strength, the soil exhibits a large amount of

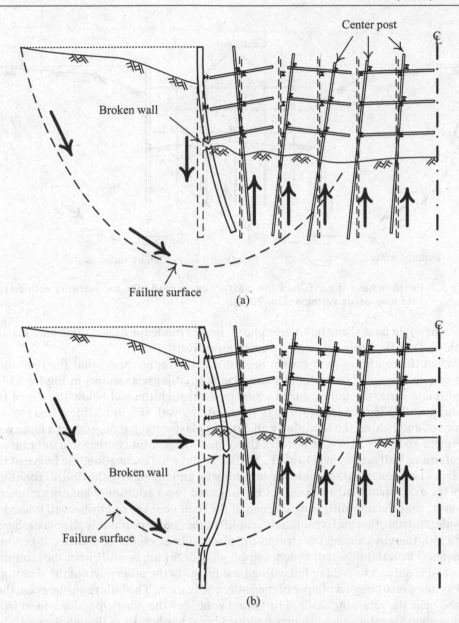

Figure 5.1 Possible base shear failure (or plastic heave) mode for excavations with elastoplastic retaining-strut systems (Do, 2015).

displacement or heave, leading to the failure of the excavation system. This phenomenon is called the base shear failure. When the shear stress is equal to the shear strength, the name of "soil failure" is used in the conventional soil mechanics, while "plastic state" is called in the finite element method. Therefore, the base shear failure is also called the plastic basal heave, simply referred to as "basal heave" or "plastic heave." The base shear failure may occur in the excavation of clay or sand. Figures 5.1 and 5.2

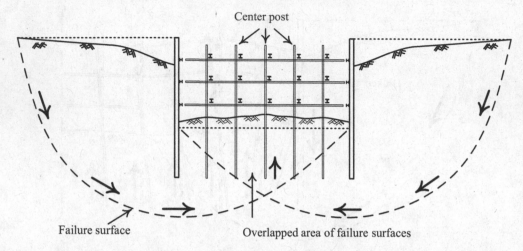

Figure 5.2 Possible base shear failure (or plastic heave) mode for excavations with elastic retaining-strut systems (Do, 2015).

show a possible base shear failure (or plastic heave) mode for excavations with elasto-plastic and elastic retaining-strut systems, respectively.

When the retaining-strut system is elastoplastic (representing that the retaining-strut system may yield or fail), the excavation may collapse as shown in Figure 5.1. In the ultimate state, the failure surface may pass through the soil below the toe of the retaining wall (Figure 5.1a) or through the retaining wall (Figure 5.1b). When the soil in front of and behind the retaining wall enters a plastic state, it may cause a large wall movement and/or an excessive heave of the center post, leading to the yield or failure of the retaining wall and strut (Do et al., 2016). For the wide excavation, the heave of the center post in the central zone is relatively uniform, and the strut in the central zone only bears the axial strut load and may not be subjected to an additional bending moment. However, near the retaining wall, one end of the strut near the retaining wall is fixed to the wale, and the other end is subjected to uplift of the center post due to the plastic heave of the soil, thereby causing the strut at the place to bear excessive additional bending moments. The retaining-strut system can get yield or failure easily. Either the retaining wall or strut subject to yield or fail will cause damage to the other parts of the retaining-strut system, resulting in collapse of the entire excavation. The failure surfaces are thus formed near the retaining walls. The failure zones (or the plastic/passive zones) from the two opposite retaining walls may not overlap each other. Even though the retaining wall penetrates deeper into the soil (Figure 5.1b), due to a large displacement of the retaining wall, the soil below the excavation bottom and behind the wall will still enter a plastic state. The failure surface will pass through the retaining wall, causing the failure of the retaining-strut system and the collapse of the entire excavation.

The failure mechanism for the narrow excavation is similar to that of the wide excavation, but the failure surfaces in the both side of the wall do overlap each other.

If the retaining-strut system is elastic or strong sufficiently, in theory, in the ultimate state, yield or failure of the retaining-strut system will not occur, and the failure surface or plastic zone will occur in the soil behind the retaining wall and below the excavation bottom. A large amount of soil would settle behind the wall, and a large amount of soil flows into the excavation area. The soil displacement was much larger than that shown in Figure 5.1. Since the retaining-strut system is elastic, in theory, the retaining-strut system can still be maintained, and collapse of the excavation does not necessarily happen. In fact, the retaining-strut system is rarely elastic.

Due to the $\phi = 0$ of clay, the failure surface in Figures 5.1 and 5.2 is a circular arc in shape. The failure mechanism of the excavation in sand is similar to Figures 5.1 and 5.2 but with non-circular arc in shape because of $\phi \neq 0$.

5.4   FREE EARTH SUPPORT METHOD AND FIXED EARTH SUPPORT METHOD

There are two analysis methods for the base shear failure: free earth support method and fixed earth support method. As shown in Figure 5.3a, the free earth support method assumes that the embedment of the retaining wall is not large so that the failure surface passes through the soil below the bottom of the retaining wall (Figure 5.1a). In the ultimate state, a significant amount of movement occurs at the bottom of the retaining wall. Therefore, the earth pressure on the retaining wall in the ultimate state can be assumed as shown in Figure 5.3b.

The fixed earth support method assumes that the embedment of the retaining wall is very large so that the failure surface passes through the retaining wall. The retaining wall seems to be fixed at a point below the excavation surface. The embedded part may rotate about the fixed point, as shown in Figure 5.4a. Thus, when the retaining wall is in the ultimate state, the lateral earth pressure around the fixed point on the two sides of the retaining wall does not necessarily reach the active or passive pressures, as shown in Figure 5.4b.

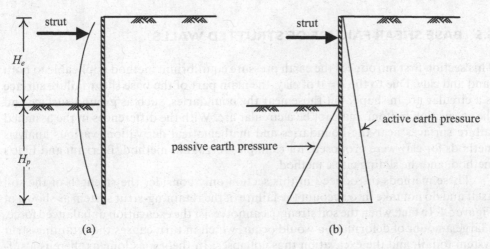

Figure 5.3 Free earth support method: (a) deformation of the retaining wall and (b) earth pressure distribution.

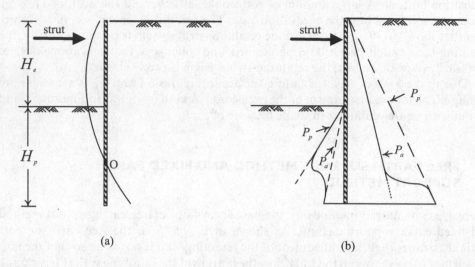

Figure 5.4 Fixed earth support method: (a) deformation of the retaining wall and (b) earth pressure distribution.

The cantilever wall can be designed with the fixed earth support method. If a cantilever wall is designed based on the free earth support method, no fixed point exists in the embedded part of the wall. The external forces, only passive and active forces, on the retaining wall are not able to come to equilibrium. Therefore, the free earth support method is not applicable to cantilever walls. On the other hand, if the free earth support method is applied to a strutted wall, the forces acting on the wall will include both the passive and active forces and the strut load. With external forces on the wall coming to equilibrium, the method is applicable to a strutted wall. On the other hand, if the fixed earth support method is applied to a strutted wall, the penetration depth of the wall will be too large to be economical.

5.5 BASE SHEAR FAILURE OF STRUTTED WALLS

This section first introduces the earth pressure equilibrium method applicable to both sand and clay. Due to the $\phi = 0$ of clay, the main part of the base shear failure surface is a circular arc in shape, but those near the boundaries, such as ground surface and the excavation bottom, may not be a circular arc. With the differences in the assumed failure surfaces near the boundaries and mathematical derivation, various analysis methods for clay were proposed, for example, Terzaghi's method, Bjerrum and Eide's method, and the sliding circle method.

These methods introduced in this section only consider the strength of the soil itself and do not take into account the failure of the retaining-strut system as shown in Figure 5.1. In fact, when the soil strength cannot resist the excavation unbalanced force, a large amount of deformation would occur, which in turn causes the retaining-strut system to fail, and the excavation thus collapses. In theory, as long as there is a sufficient safety factor to make the soil strength resistant to the excavation unbalanced

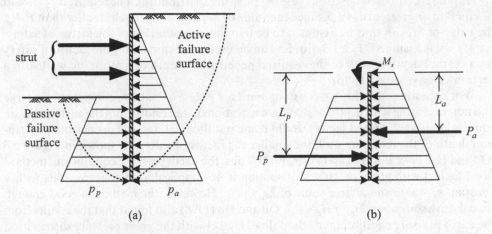

Figure 5.5 Earth pressure equilibrium method (load factor): (a) earth pressure distribution and (b) free body diagram.

force, the deformation is not too great, and the safety of the retaining-strut system can be guaranteed.

5.5.1 Earth pressure equilibrium method (load factor)

Using the concepts of the load safety factor and the force equilibrium, the method developed is called the load factor earth pressure equilibrium method. As discussed in Section 5.4, for a strutted wall, the free earth support method is commonly used. As shown in Figure 5.5a, the earth pressures acting on the front and back of the retaining wall in excavations will reach the active and the passive earth pressures in the ultimate state, respectively. Take the retaining wall below the lowest level of strut as a free body and conduct a force equilibrium analysis (Figure 5.5b), and we can then find the factor of safety (F_b) against base shear failure as follows:

$$F_b = \frac{M_r}{M_d} = \frac{P_p L_p + M_s}{P_a L_a} \tag{5.5}$$

where
M_r = resisting moment
M_d = driving moment
P_a = resultant of the active earth pressure on the front of the wall below the lowest level of strut
L_a = length from the lowest level of strut to the point of action P_a
M_s = allowable bending moment of the retaining wall
P_p = resultant of the passive earth pressure on the back of the retaining wall below the excavation surface
L_p = length from the lowest level of strut to the point of action P_p

Eq. 5.5 is computed based on the gross pressure distribution. The required F_b should be equal to or greater than 1.5. Since the value of M_s is usually much smaller than P_pL_p, the value of M_s can thus be assumed to be 0 in the computation of the factor of safety. As $M_s = 0$ is assumed, $F_b \geq 1.2$. Eq. 5.5 can be used to obtain either the factor of safety for a certain depth of wall or the required penetration depth of a retaining wall with a certain value of safety factor.

For cohesive soils with $s_u = $ constant and $2s_u / \gamma H_e \geq 0.7$ (i.e., s_u comparatively large and not increasing with depth, or shallow excavations), the results from the earth pressure equilibrium method (load factor) would come out illogical, i.e., the deeper the penetration depth of the retaining wall, the smaller the factor of safety, as shown in Figure 5.6 (Ou and Hu, 1998; Burland and Potts, 1981). Thus, the earth pressure equilibrium method (load factor) with gross pressure distribution is not applicable to cohesive soils with a constant s_u-value and a large value of $2s_u / \gamma H_e$. However, the method is good enough for cohesive soils when $2s_u / \gamma H_e < 0.7$. Ou and Hu (1998) also found that the results from the earth pressure equilibrium method (load factor) with the gross pressure distribution will not come out illogical when applied to cohesive soils with $s_u / \sigma_v' = $ constant or s_u increasing with the increase of depth. For cohesionless soils with ϕ' smaller than 22°, the penetration depth obtained from the earth pressure equilibrium method (load factor) would come out too large, according to Burland and Potts (1981).

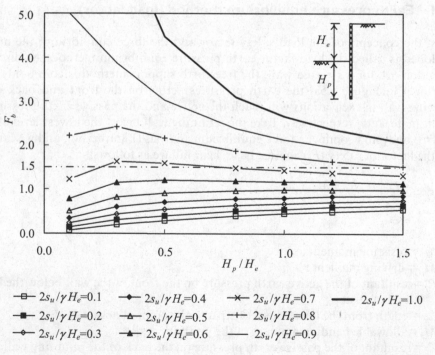

Figure 5.6 Relationship of factors of safety against base shear failure from the earth pressure equilibrium method (load factor) and wall penetration depth for clay with constant su.

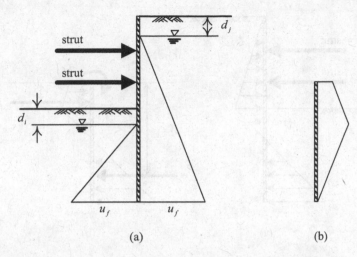

(a)   (b)

Figure 5.7 Distribution of water pressure due to seepage: (a) water pressure distribution and (b) net water pressure. (Source: u_f = water pressure due to seepage.)

As stated in Section 4.6, Eqs. 4.16 and 4.18 can be adopted for the earth pressures on a retaining wall in cohesive soils. Both equations have to consider the adhesion (c_w) between the wall and soil, which can be estimated by the following equation:

$$c_w = \alpha s_u \tag{4.20}$$

where α can be called a strength reduction factor.

Section 4.6.1 also discusses the selection of the α-value. Basically, for concrete walls such as diaphragm walls, $c_w \approx 0.67\ s_u$; for steel walls such as sheet pile walls, $c_w \approx 0.5\ s_u$.

To estimate the factor of safety in cohesionless soils (sandy soils) with a high groundwater level, though the gross pressure distribution can be used along with either the gross water pressure distribution or the net water pressure distribution, the latter is more reasonable according to Padfield and Mair's study (1984), as shown in Figure 5.7. Moreover, as stated in Section 4.5.3, for cohesionless soils, Caquot–Kerisel's or Coulomb's active earth pressure should be adopted for the active earth pressure acting on the retaining wall. For the passive earth pressure, Caquot–Kerisel's is preferable. For more detailed explanation, refer to Section 4.5.3. Section 4.6.2 discusses some findings on values of δ and recommends that for concrete walls such as diaphragm walls, $\delta = 0.67\phi'$; for steel walls such as sheet pile walls, $\delta = 0.5\phi'$.

As shown in Figure 5.8a, the active and passive earth pressures on the two sides of the retaining wall are expressed in terms of net values. After conducting a force equilibrium analysis of the net forces on the retaining wall below the lowest level of strut, viewed as a free body (Figure 5.8b), we will thus derive an equation similar to Eq. 5.5 to compute the factor of safety. Nevertheless, according to Burland and Potts (1981) and Ou and Hu (1998), any slight difference in the penetration depth will lead to a large change in the factor of safety computed with the net pressure distribution. That is to say, the factor of safety is too sensitive to the penetration depth and the net pressure distribution is not good for stability analysis of base shear failure.

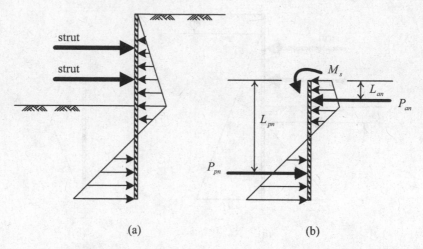

(a)   (b)

Figure 5.8 Analysis of base shear failure by the net pressure method: (a) net earth pressure distribution and (b) free body diagram.

5.5.2   Earth pressure equilibrium method (strength factor)

Using the concepts of the strength factor method, and the force equilibrium, the method developed is called the strength factor earth pressure equilibrium method. As described in Section 5.2, after dividing the soil strength parameters by a safety factor specified in the codes, compute the active earth pressure and the passive earth pressure distribution using the reduced strength parameters (Eqs. 5.1 or 5.2) (Figure 5.5), and then, compute the driving moment and resistant moment with respect to the lowest level of struts, respectively. When the resistant moment is greater than or equal to the driving moment, the required penetration depth of the retaining wall can be obtained, that is,

$$P_p L_p + M_s \geq P_a L_a \tag{5.6}$$

where definitions of P_p, P_a, L_p, L_a, and M_s are exactly the same as those in Section 5.5.1.

Similarly, M_s can be assumed to be equal to 0 in the analysis. In sand, the net pore water pressure distribution is suggested to use.

If the safety factor is known, the required penetration depth can be found with trial and error; that is, you can first assume a relatively small penetration depth, then reduced the soil strength parameters and then compute the resistant moment and the driving moment, respectively. Then, check if the resistant moment and driving moment satisfy Eq. 5.6. If not satisfied, increase the penetration depth and repeat the above procedure until Eq. 5.6 is satisfied. On the other hand, if the penetration depth is known, the corresponding safety factor can be computed following a step similar to the above procedure, or refer to the computation method of the case study in Section 5.7.

Results from the strength factor earth pressure equilibrium method are reasonable, which does not have a phenomenon as shown in Figure 5.6. It is recommended by Eurocode 7 and CIRIA report C580 (Gaba et al., 2003). The required factor of safety $F_b = 1.2$.

Some prefer to reduce the passive earth pressure only by a factor of safety, but the active earth pressure is not reduced. For example, $K_{p,\text{design}} = K_p/F_p$. Then, the penetration depth is computed following Eq. 5.6. As a matter of fact, this is exactly the same as the load factor earth pressure equilibrium method.

5.5.3 Earth pressure equilibrium method (dimension factor)

Some engineers prefer to assume $F_b = 1.0$ first and then use Eq. 5.5 to determine the preliminary penetration depth and then added 20%–40% to be the penetration depth for design. The method is referred to as the dimension factor earth pressure equilibrium method, as discussed in Section 5.2. According to Burland and Potts' study (1981), the factor of safety obtained following the dimension factor method does not conform to the definition of the factor of safety and may lead to unreasonable results.

5.5.4 Terzaghi's method

Terzaghi's method is only applicable to clay. Terzaghi's method assumes that the failure surface of base shear failure, i.e., plastic heave, covers the entire excavation area (Figure 5.9). Therefore, the failure surface in the excavation base initiates with a circular arc and then develops outward to the excavation surface level **ab** and then extends upward to the ground surface. The soil weight within the width of B_1 that acts on plane **ab** is treated as a driving force that causes the excavation to fail.

According to Terzaghi's bearing capacity theory, the bearing capacity of clay below plane **ab** can be denoted as $P_{\max} = 5.7 s_u$. When the soil weight above plane **ab** is greater than the soil bearing capacity, the excavation will fail. Besides, the failure surface will be restrained by stiff soils. Let D represent the distance between the excavation surface and the stiff soil. Terzaghi's method can then be discussed in two parts:

When $D \geq B/\sqrt{2}$, as shown in Figure 5.9a, the formation of a failure surface is not restrained by the stiff soil. Suppose the unit weight of the soil is γ. The soil weight (containing the surcharge q_s) that ranges B_1 on plane **ab** will be:

$$W = (\gamma H_e + q_s)(B_1 \times 1) = (\gamma H_e + q_s)\frac{B}{\sqrt{2}} \tag{5.7}$$

The ultimate load, Q_u, of clay below plane **ab** will be:

$$Q_u = 5.7 s_{u2}(B_1 \times 1) = (5.7 s_{u2})\frac{B}{\sqrt{2}} \tag{5.8}$$

When a basal heave failure occurs, vertical failure plane **bc** can offer shear resistance $(s_{u1}H_e)$ and the factor of safety against basal heave (F_b) will be:

$$F_b = \frac{Q_u}{W - s_{u1}H_e} = \frac{5.7 s_{u2}\, B/\sqrt{2}}{(\gamma H_e + q_s)\, B/\sqrt{2} - s_{u1}H_e} \tag{5.9}$$

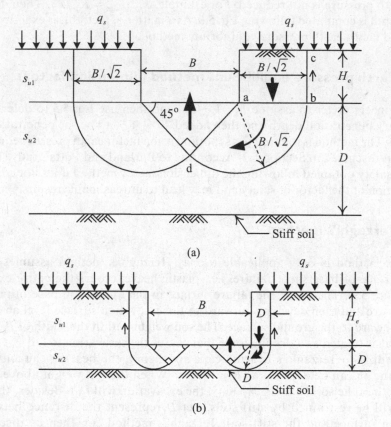

Figure 5.9 Terzaghi's method: (a) $D \geq B/\sqrt{2}$ and (b) $D < B/\sqrt{2}$.

where s_{u1} and s_{u2} represent, respectively, the undrained shear strengths of the soils above and below the excavation surface in the failure zone; q_s denotes surcharge on the ground surface.

When $D < B/\sqrt{2}$, under such a condition, the failure surface will be restrained by stiff soil, as shown in Figure 5.9b, and its factor of safety (F_b) will be:

$$F_b = \frac{Q_u}{W - s_{u1}H_e} = \frac{5.7 s_{u2}D}{(\gamma H_e + q_s)D - s_{u1}H_e} \tag{5.10}$$

Terzaghi's method does not take the influence of the penetration depth and stiffness of the retaining wall into account. That is to say, it is assumed that the retaining wall does not exist in the analysis. It only considers whether the bearing capacity of the soil below the bottom of the excavation can withstand the weight of the soil outside the excavation. Therefore, it is also called the bearing capacity method. Such a consideration is reasonable because, for strutted excavations, the free earth support method is normally designed, resulting in a shorter penetration of the retaining wall. In the ultimate condition, the failure surface should pass through the soil below the toe of the retaining wall and cover the entire excavation depth. Therefore, the depth and the stiffness of the penetrated part of the retaining wall have no influence on the factor of safety against base shear failure.

Terzaghi's method is more suitable for wide excavations. Detailed discussion can be seen in Section 5.6.2. For most excavation cases, Terzaghi's factor of safety (F_b) should be greater than or equal to 1.5 (Mana and Clough, 1981; JSA, 1988).

5.5.5   Bjerrum and Eide's method

Bjerrum and Eide's method is applicable to clay only. Bjerrum and Eide's method assumes that the unloading behavior caused by excavation is analogous to the building foundation being subject to an upward loading and that the shape of the failure surface is similar to the failure mode of the deep (pile) foundation. Then, using the bearing capacity equation for the deep (pile) foundation, we can obtain the ultimate unloading pressure. The factor of safety is the ratio of the ultimate unloading pressure to the unloading pressure. Bjerrum and Eide's method is also called the negative bearing capacity method. Bjerrum and Eide's method fixes the disadvantages of Terzaghi's method, and it can be applied to both wide and narrow excavations.

As shown in Figure 5.10, Bjerrum and Eide's method assumes that the failure surface with a radius $B/\sqrt{2}$ covers the entire excavation width. The factor of safety against the base shear failure can be computed as follows:

$$F_b = \frac{N_c \cdot s_u}{\gamma \cdot H_e + q_s} \tag{5.11}$$

where N_c is the Skempton's bearing capacity factor (1951).

In Eq. 5.11, s_u is the average undrained shear strength below the excavation surface in the failure zone. N_c can be determined from Figure 5.11 or be computed by the following equation:

$$N_{c(\text{rectangular})} = N_{c(\text{square})}\left(0.84 + 0.16\frac{B}{L}\right) \tag{5.12}$$

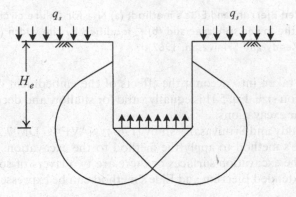

Figure 5.10 Bjerrum and Eide's method.

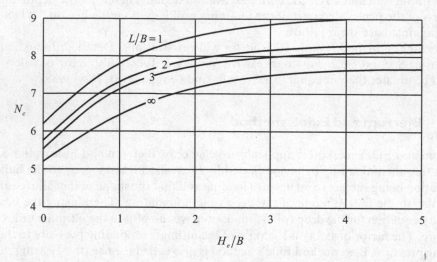

Figure 5.11 Skempton's bearing capacity factor (Skempton, 1951).

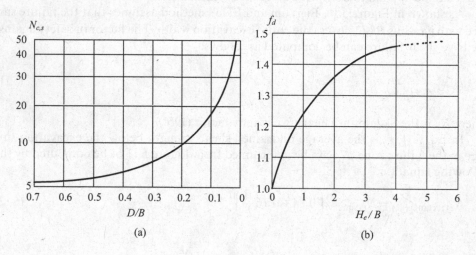

Figure 5.12 Extended Bjerrum and Eide's method: (a) $N_{c,s}$ for failure circles tangent to the top of the lower soil layer and (b) f_d modified by the width (NAVFAC DM7.2, 1982; Reddy and Srinivasan, 1967).

Since N_c has taken into account the effects of the embedment depth of foundations and excavation size, Eq. 5.11 is equally valid for shallow and deep excavations, as well as rectangular excavations.

Following Reddy and Srinivasan's study (1967), NAVFAC DM 7.2 (1982) modified Bjerrum and Eide's method to apply the method to the excavations where there are stiff soils below the excavation surfaces or there are two layers of soils. As shown in Figure 5.12, the extended Bjerrum and Eide's method can be expressed as follows:

$$F_b = \frac{s_u N_{c,s} f_d f_s}{\gamma H_e} \tag{5.13}$$

where

$N_{c,s}$ = bearing capacity factor that considers the stiff soil, which can be determined from Figure 5.12a.

f_d = depth correction factor, which can be found in Figure 5.12b.

f_s = shape correction factor, which can be estimated by the following equation:

$$f_s = 1 + 0.2 \frac{B}{L} \tag{5.14}$$

where B is the excavation width and L the excavation length.

Similar to Terzaghi's method, Bjerrum and Eide's method does not consider the influence of the penetration depth and stiffness of the retaining wall. That is to say, it is assumed that the retaining wall does not exist in the analysis. Bjerrum and Eide's method takes into account the effects of excavation shape, width, and depth. Therefore, the methods are applicable to various shapes of excavations, shallow excavations as well as deep excavations.

For most excavations, the factor of safety obtained according to Bjerrum and Eide's method (F_b) should be larger than or equal to 1.5 (NAVFAC DM 7.2, 1982).

On the other hand, as the average s_u is adopted between the excavation bottom and depth of the failure surface, exactly the same as that used in Terzaghi's method, the obtained F_b would be close to Terzaghi's.

5.5.6 Slip circle method

The slip circle method is also applicable to clay only. As shown in Figure 5.13a, the method assumes the main part of the trial failure surface below the excavation bottom to be a circular arc, with its center at the lowest level of strut. The failure surface then grows outward to the excavation surface level and extends vertically

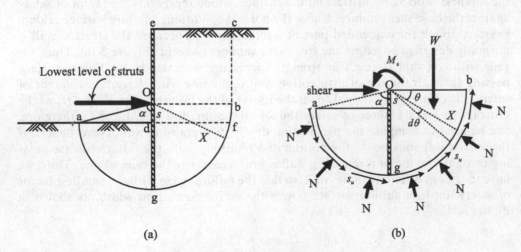

Figure 5.13 Slip circle method: (a) the failure surface and (b) free body diagram.

to the ground surface. The shear strength on the vertical failure plane (line **bc** in Figure 5.13a) is ignored. Take the retaining wall and soil below the lowest level of struts as well as above the circular arc as a free body (Figure 5.13b). The soil weight above the excavation surface behind the wall can be treated as the driving force and the shear strength along the failure surface as the resistant force. The ratio of the resistant moment (M_r) to the driving moment (M_d) with respect to the lowest level of strut will be:

$$F_b = \frac{M_r}{M_d} = \frac{X \int_0^{\frac{\pi}{2}+\alpha} s_u (X d\theta) + M_s}{W \cdot \frac{X}{2}} \tag{5.15}$$

where

M_s = allowable bending moment of the retaining wall
s = the distance between the lowest level of strut and the excavation surface
X = radius of the failure circle
W = total weight of the soil in front of the vertical failure plane and above the excavation surface, including the surcharge on the ground surface

Then, compute the ratio of the resistant moment to the driving moment for different trial failure surfaces. The smallest ratio among them is thus the factor of safety (F_b) against base shear failure for the excavation.

The original source of the slip circle method is untraceable. Nevertheless, TGS (2001) and JSA (1988) adopted the method in their building codes. Similar to the earth pressure equilibrium method, $M_s \approx 0$ and the required factor of safety (F_b) should be greater than or equal to 1.2.

In analysis, we should try out different failure surfaces and find the one with the smallest ratio as the critical failure surface, which represents the factor of safety against the base shear failure. Under the normal condition, a failure surface seldom passes through the embedded part of a retaining wall because the strutted wall is normally designed based on the free earth support concept (Figure 5.1a). Thus, for soils with constant strength or strength increasing with depth, the failure surface passing through the toe of a retaining wall is the one with the smallest factor of safety. Therefore, it is rational to set the radius of the failure circle $X = s + H_e$ as the critical one and its factor of safety should be the smallest one. Eq. 5.15, therefore, can be used to compute the penetration depth of a retaining wall. Nevertheless, if there exist soft soils below the bottom of a retaining wall, the failure surface passing through the toe of a retaining wall is not necessarily the critical one. Thus, we have to try out different values of X to find the failure circle with the smallest factor of safety until the failure surface covers the entire excavation width, as shown in Figure 5.14.

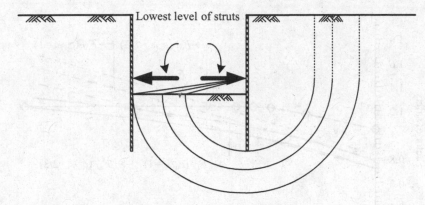

Figure 5.14 Various trial failure surfaces for the slip circle method.

5.6   GENERAL DISCUSSION OF ANALYSIS METHODS OF BASE SHEAR FAILURE

5.6.1   Effect of the stiffness of the retaining-strut system

The above-mentioned methods do not take the stiffness of the retaining-strut system into account. If the retaining wall and the strut are elastoplastic, when a large deformation occurs, the retaining wall or strut may yield or fail first, causing the base shear failure of soil, or even the collapse of the excavation (Figure 5.1). If the retaining wall and the strut are elastic, the retaining wall or the strut never yields or fails, and the base shear failure of soil will occur first, leading to the collapse of the entire excavation (Figure 5.2). The latter is certainly of a higher factor of safety than the former.

Both Terzaghi's method and Bjerrum and Eide's method assume that the retaining wall penetration depth is zero; that is, the presence of the retaining wall is not considered. The slip circle method assumes that the failure surface passes through, or below, the toe of the retaining wall. However, as shown in the failure mechanism shown in Figure 5.1b, the failure surface passing through the retaining wall is also possible. Under such a condition, the failure of the excavation is also related to the stiffness of the retaining-strut system.

Figure 5.15 shows the relationship between the safety factor and H_p/H_e for $s_u/\sigma'_v = 0.22$, which are obtained from the finite element method (F_{fem}, see Section 8.10), the earth pressure equilibrium method with load factor (F_{LF}) and strength factor (F_{SF}), and the slip circle method (F_{SC}). For the finite element method, the wall and struts are assumed to be elastoplastic. As $H_p/H_e < 1.3$, F_{fem} increases with H_p/H_e where excavation failure at the ultimate condition is due to soil instability and the failure mode is close to Figure 5.1a. As $H_p/H_e > 1.3$, even if the H_p continues to increase, F_{fem} no longer increases, but remains constant because the yielding of wall and/or struts initiates excavation failure, and the failure mode is close to Figure 5.1b. This figure also shows that both F_{SC} and F_{SF} increase, almost linearly, with H_p/H_e.

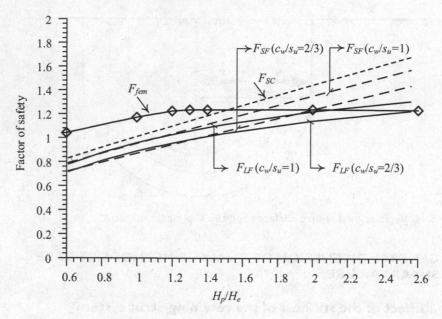

Figure 5.15 Relationship between the safety factors from various methods and the wall penetration depth for $s_u/\sigma'_v = 0.22$.

Both F_{SF} and F_{LF} increase with c_w/s_u. Moreover, as H_p/H_e is greater than a certain value, the variation of the tendency of F_{LF} deviates from F_{SF} and F_{SC} significantly. For a specific F_{LF}, a relatively large H_p is normally obtained, which is conservative, as compared with other methods. The F_{SF} ($c_w/s_u = 1.0$) is comparable with the F_{SC} for all the values of H_p/H_e.

Figure 5.16 further shows that increasing the thickness of the retaining wall can increase the safety factor for a large penetration depth, for example, $H_p/H_e = 2.6$. Because the thickness of the retaining wall increases, the retaining wall is relatively less to yield. But when the penetration depth is not large (for example, $H_p/H_e = 0.6$), the safety factor does not increase with the thickness of the retaining wall. This is because the failure surface or the plastic zone passes below the retaining wall, causing the center post to heave, leading to the yield or failure of the strut.

5.6.2 Difference in various analysis methods

In the ultimate or near-failure state, the soils in front of and the retaining wall should reach passive and active states, respectively. The failure surface may be circular ($\phi = 0$) or non-circular ($\phi \neq 0$), depending on the type of soil. In earth pressure equilibrium methods, passive earth pressure and active earth pressure are used to express the interaction between the wall and the surrounding soils, and the safety factor is then obtained. Terzaghi's method, Bjerrum and Eide's method, and the slip circle method, respectively, assume different failure circles and obtain safety factors. The basic principles of these methods for obtaining the safety factor are similar, but only differences in the assumed failure surfaces near the boundaries and their computation methods.

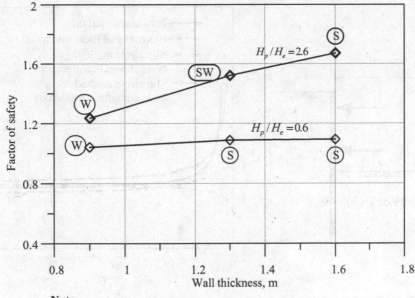

Figure 5.16 Effect of wall thickness and penetration depth on the factor of safety for $s_u/\sigma'_v = 0.22$ (Do, 2015).

Note:
S: first yielding of support system occurs at strut
W: first yielding of support system occurs at wall
SW: first yielding of support system occurs at both strut and wall

Assuming a hypothetical excavation as shown in Figure 5.17a and Figure 5.17b shows the relationship between the F_b and the excavation width (B) for the constant undrained shear strength of the soil (s_u =constant) as computed using Terzaghi's method and Bjerrum and Eide's method. As shown in this figure, Terzaghi's F_b is unreasonably high for a narrow excavation because the undrained shear strength on the vertical failure surface (**bc** in Figure 5.9) has a big influence on the computed F_b when the excavation width (B) is small, according to Eq. 5.9. The F_b then decreases with the increase of the excavation width until B/H_e is close to 1.0. The values of F_b then change little as B/H_e increases. Obviously, Terzaghi's method is appropriate for B/H_e equal to or greater than 1.0, which is often treated as wide excavations, but it is not suitable for narrow excavations.

For Bjerrum and Eide's method, the F_b decreases steadily with the increase of B for s_u=constant. The tendency of the variation of Bjerrum and Eide's F_b seems more reasonable than that from Terzaghi's. This is because Bjerrum and Eide's method takes the depth factor into account in Figure 5.12b.

Assuming the failure surface of the slip circle method passes through, or below, the toe of the retaining wall and covers the entire excavation width, similar to the assumption by Terzaghi, and Bjerrum and Eide's methods, the variation of the F_b with the excavation width is also shown in Figure 5.17. Basically, the F_b increases with the excavation width for s_u =constant. The changing tendency of the F_b seems to be opposite to those from Terzaghi's and Bjerrum and Eide's methods for narrow excavations. The F_b is unreasonably low for narrow excavations because the side strength (**bc** in Figure 5.13a) is not taken into account in Eq. 5.15. Figure 5.17b also provides the results

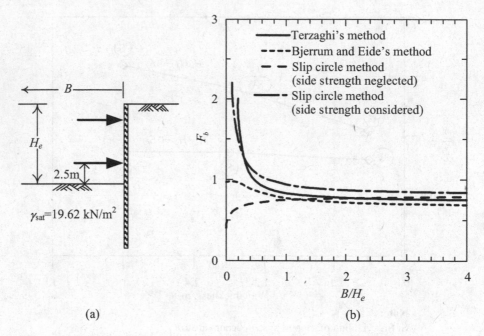

Figure 5.17 Difference in various analysis methods: (a) profile of an assumed excavation and (b) relationship between the safety factors from various methods and the excavation width for constant undrained shear strength $\left(s_u = 25 \, \text{kN/m}^2 \right)$.

of the slip circle method with the side shear strength considered. We can see from the figure that the changing tendencies of the F_b from the slip circle method with the side strength considered and those from Terzaghi's method are similar, though the values are different. The reason for the difference in the values is that the assumed shape of the failure surface near the boundaries with the slip circle method is different from that from Terzaghi's method. When B/H_e is greater than 1.0 for wide excavations, the values of F_b computed from those three methods are similar.

Similar conclusions can be drawn for Terzaghi's, Bjerrum and Eide's, and the slip circle methods with the undrained shear strength increasing with the depth or $s_u / \sigma'_v =$ constant, but the values of F_b computed from those three methods vary to a certain extent.

5.6.3   Improvement of factors of safety

When the computed safety factors for an excavation in clay are insufficient, under the condition that the s_u increases with the depth, the penetration depth of the wall should be increased to make the calculated safety factor meet the requirements. Of course, conducting ground improvement is another effective way to increase the safety factor. However, if the shear strength or s_u is constant throughout the entire soil deposits, only ground improvement is an effective method to improve the safety factor.

5.7   CASE STUDY OF BASE SHEAR FAILURE

The excavation case was located in Taipei. The width of the excavation was 17.6–25.8 m; the length was 100.1 m; and the depth was 13.45 m. A 70-cm-thick, 24-m-deep diaphragm wall was used as the earth retaining wall. There were four levels of struts, and the excavation was carried out in five stages. The excavation profile and the related soil test data are as illustrated in Figure 5.18a. The construction collapsed one and half hour after the completion of excavation and caused a serious damage to not only the construction site and the adjacent properties but also the public facilities (Do et al., 2016).

c_u and ϕ_u as shown in Figure 5.18a, the total stress strength parameters of the clayey soils, obtained from the triaxial CU test, were adopted by the original designer, who assumed that no adhesion or friction exists between the retaining wall and the soil. Following Eq. 5.5, the designer computed the factor of safety using the earth equilibrium method (load factor) to be 1.5, and the factor of safety is 2.3 using Eq. 5.15.

Based on the author's studies, the normalized undrained shear strength at the construction site was about $s_u / \sigma'_v = 0.22$. To simplify the analyses and be conservative, we assume the soil below the lowest level of struts (GL-10.15 m) to be a clayey layer, and the adhesion between the retaining wall and the soil $c_w = 2s_u/3$. The active earth pressure and the passive earth pressure of the clayey soil on the diaphragm wall can be estimated following Eqs. 4.16 and 4.18, respectively. Then, we can find the total stress, the pore water pressure, the undrained shear strength, and the active (or passive) earth pressure at each depth.

At GL-10.15 m

$$\sigma'_v = \sigma_v - u = 185.5 - 72.1 = 113.3 \text{ kN/m}^2$$

$$s_u = 0.22\sigma'_v = 0.22 \times 113.3 = 24.9 \text{ kN/m}^2$$

$$\sigma_{h,a} = \sigma_v K_a - 2s_u \sqrt{K_a \left(1 + \frac{c_w}{s_u}\right)} = 185.5 - 2(24.9)\sqrt{1 + \frac{2}{3}} = 121.2 \text{ kN/m}^2$$

At GL-13.45 m
Before excavation

$$\sigma'_v = \sigma_v - u = 248 - 104.5 = 143.5 \text{ kN/m}^2$$

$$s_u = 0.22\sigma'_v = 0.22 \times 143.5 = 31.6 \text{ kN/m}^2$$

After excavation, $\sigma_v = 0$ on the passive side, but s_u-value stayed unchanged. Thus,

$$\sigma_{h,p} = \sigma_v K_p + 2s_u \sqrt{K_p \left(1 + \frac{c_w}{s_u}\right)} = 0 + 2(31.6)\sqrt{1 + \frac{2}{3}} = 81.5 \text{ kN/m}^2$$

At GL-24.0 m

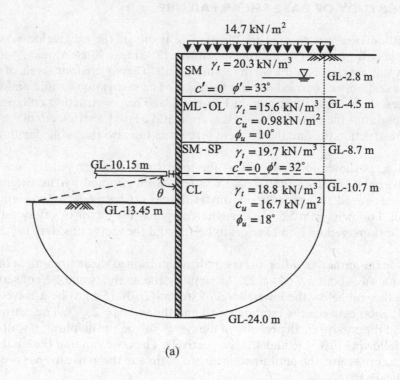

(a)

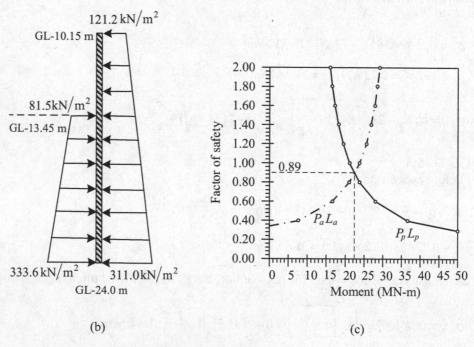

(b)

(c)

Figure 5.18 Stability analysis of a case study: (a) excavation profile, (b) earth pressure distribution, and (c) strength factor analysis.

The active side

$$\sigma'_v = \sigma_v - u = 446.3 - 208 = 238.3 \text{ kN/m}^2$$

$$s_u = 0.22\sigma'_v = 0.22 \times 238.3 = 52.4 \text{ kN/m}^2$$

$$\sigma_{h,a} = \sigma_v K_a - 2s_u\sqrt{K_a\left(1 + \frac{c_w}{s_u}\right)} = 446.3 - 2(52.4)\sqrt{1 + \frac{2}{3}} = 311.0 \text{ kN/m}^2$$

The passive side
s_u stayed constant after excavation,

$$\sigma_v = 18.8 \times (24.0 - 13.45) = 198.3 \text{ kN/m}^2$$

$$\sigma_{h,p} = \sigma_v K_p + 2s_u\sqrt{K_p\left(1 + \frac{c_w}{s_u}\right)} = 198.3 + 2(52.4)\sqrt{1 + \frac{2}{3}} = 333.6 \text{ kN/m}^2$$

Figure 5.18b shows the earth pressures on both sides of the retaining wall. According to Eq. 5.5, we obtain the factor of safety as

$$F_b = \frac{81.5 \times 10.55 \times (10.55/2 + 3.3) + (333.6 - 81.5) \times 10.55 \times 0.5 \times (10.55 \times 2/3 + 3.3)}{121.2 \times 13.85 \times 13.85/2 + (311.0 - 121.2) \times 13.85 \times 0.5 \times 13.85 \times 2/3}$$
$$= 0.89$$

The factor of safety can also be computed following the strength safety factor method. With this method, the soil strength parameter is first divided by a reduction factor (assumed to be F), and then, the resistant moment $\left(P_p L_p\right)$ and the driving moment $\left(P_a L_a\right)$ are computed, respectively. Then, increase the value of the reduction factor F, and then, use the same method to compute $P_a L_a$ and $P_p L_p$. The computed results are plotted as shown in Figure 5.18c. At the point of $P_a L_a = P_p L_p$, the corresponding reduction factor is the factor of safety. As shown in this figure, the factor of safety obtained by the strength safety factor method is 0.89. The result is exactly the same as that from the earth pressure equilibrium method (load factor). According to the study by Do et al. (2016), the factors of safety by the load safety factor method and the strength safety factor method are quite close to each other for the clay with $s_u / \sigma'_v = 0.22$. When the $s_u / \sigma'_v \neq 0.22$, the values by the load safety factor and the strength safety factor methods are not necessarily the same.

Similarly, we can compute the factor of safety against base shear or basal heave using Eq. 5.15. The average undrained shear strength (active side) of the soil between GL-10.15 m and GL-24.0 m is

$$s_{u,a} = \frac{24.9 + 52.4}{2} = 38.7 \text{ kN/m}^2$$

The average undrained shear strength (passive side) of the soil between GL-13.45 m and GL-24.0 m is

$$s_{u,p} = \frac{31.6 + 52.4}{2} = 42.0 \text{ kN/m}^2$$

The central angle of the failure circular arc on the passive side is

$$\theta = \cos^{-1}\left(\frac{3.3}{24 - 10.15}\right) = 1.33$$

The factor of safety against circular arc failure would be

$$F_b = \frac{13.85 \times 1.33 \times 42.0 \times 13.85 + 13.85 \times 1.57 \times 38.7 \times 13.85}{\sigma_{v(GL-13.45)} \times 13.85 \times 13.85 / 2} = \frac{22,370}{23,786} = 0.94$$

Assuming $B = 17.6$ m, and according to Terzaghi's method, the average undrained shear strength of soil within the range of the failure surface can be computed as follows: At GL-$(13.45 + B/\sqrt{2} = 25.9$ m)

$$s_u = 0.22\sigma_v' = 0.22 \times (\sigma_v - u) = 0.22 \times (482.1 - 226.6) = 56.2 \text{ kN/m}^2$$

The average undrained shear strength within the range of the failure surface can be evaluated as $s_u = (31.6 + 56.2)/2 = 43.9$ kN/m^2. As computed earlier, the total vertical pressure outside the excavation zone acting at the excavation surface level is equal to 248.0 kN/m^2. To simplify the analysis and be conservative, we assume that the soil above the excavation surface is clay and has undrained shear strength expressed as $s_u / \sigma_v' = 0.22$. The average undrained shear strength of the soil outside the excavation zone and above the excavation surface level would be

$$s_u = \frac{0.22\sigma_{v(GL-13.45)}'}{2} = \frac{31.6}{2} = 15.8 \text{ kN/m}^2$$

The factor of safety according to Terzaghi's method can be computed as

$$F_b = \frac{Q_u}{W - s_{u1}H_e} = \frac{5.7 \times 43.9 \times 17.6 / \sqrt{2}}{248.0 \times 17.6 / \sqrt{2} - 15.8 \times 13.45} = \frac{3,118}{2,874} = 1.08$$

With $H_e/B = 0.76$ and $L/B = 5.7$ given, we would have $N_c = 6.2$, according to Figure 5.11. Therefore, the factor of safety following Bjerrum and Eide's method is computed as

$$F_b = \frac{s_u N_c}{\gamma H_e + q_s} = \frac{43.9 \times 6.2}{248.0} = 1.10$$

Assuming $B = 25.8$ m, in a similar way to the above procedure, the F_b values computed with Terzaghi's method and Bjerrum and Eide's method are 1.20 and 1.30, respectively. Therefore, for Terzaghi's method, the safety of the excavation should be evaluated based on $B = 17.6$ m.

In the above computation, the average s_u is adopted between the excavation bottom (GL-13.45 m) and depth of the failure surface (GL-25.9 m), exactly the same as that used in Terzaghi's method, resulting in Bjerrum and Eide's F_b very close to Terzaghi's. However, if the average s_u is taken between 13.45- $B/\sqrt{2} = 1.0$ m and $13.45 + B/\sqrt{2} = 25.9$ m, the obtained F_b is even smaller than that from the earth pressure equilibrium method.

The factors of safety for this case are, as above computed, always smaller than 1.0 or slightly larger than 1.0. The above computation did not include the surcharge on ground surface 14.7 kN/m² (please refer to Figure 5.18a) yet. If it is considered, the factors of safety will be even smaller. We can thus see why the excavation failed.

Example 5.1

Assume a 9.0-m-deep excavation in a sandy ground and the lowest level of struts is 2.5 m above the excavation surface. The level of groundwater outside the excavation zone is ground surface high while that within the excavation zone is as high as the excavation surface. The unit weight of saturated sandy soils $\gamma_{sat} = 20$ kN/m², the effective cohesion $c' = 0$, and the effective angle of friction $\phi' = 30°$. Because of the difference between the levels of groundwater, seepage will occur. Assume that the friction angles (δ) between the retaining wall and soil on both the active and passive sides are $0.5\phi'$ and the factor of safety against base shear failure, $F_b = 1.5$. Compute the required penetration depth (H_e) using the load factor and strength factor earth pressure equilibrium methods.

Solution

Let z represent the depth from the ground surface and x the depth from the groundwater level (see Figure 4.21).

1. Determine the coefficient of the earth pressure

    Compute both the active and passive earth pressures following Caquot–Kerisel's earth pressure theory. When $\delta = 0.5\phi'$, the coefficient of active is computed from Coulomb's earth pressure theory and that of passive earth pressure can be found from Figure 4.10 to be 0.3 and 4.6 separately. Thus, the coefficients of the horizontal active and passive earth pressure would be

$$K_{a,h} = 0.3\cos\delta = 0.3\cos0.5\phi' = 0.29$$

$$K_{p,h} = 4.6\cos\delta = 4.6\cos0.5\phi' = 4.4$$

2. Compute the effective active earth pressure on the wall

    At the lowest level of strut ($z = 6.5$ m, $x = 6.5$ m)

$$\sigma'_{a,h} = \sigma'_v K_{a,h}$$

$$\sigma_v = 20 \times 6.5 = 130 \text{ kN/m}^2$$

According to Eq. 4.51, the pore water pressure at x away from upstream water level would be

$$u = \frac{2x(H_p - d_i)\gamma_w}{2H_p + H_e - d_i - d_j} = \frac{2 \times 6.5 \times H_p \times 9.81}{2H_p + 9} = \frac{63.77 H_p}{H_p + 4.5}$$

$$\sigma'_{a,h} = (\sigma - u)K_{a,h} = \left(130 - \frac{63.77 H_p}{H_p + 4.5}\right) \times 0.29 = 37.7 - \frac{18.49 H_p}{H_p + 4.5}$$

At the toe of the retaining wall ($z = 9 + H_p$, $x = 9 + H_p$)

$$u = \frac{2x(H_p - d_i)\gamma_w}{2H_p + H_e - d_i - d_j} = \frac{2 \times (9 + H_p) \times H_p \times 9.81}{2H_p + 9} = \frac{9.81 H_p^2 + 88.29 H_p}{H_p + 4.5}$$

$$\sigma'_{a,h} = \left(180 + 20H_p - \frac{9.81 H_p^2 + 88.29 H_p}{H_p + 4.5}\right) \times 0.29$$

$$= 52.2 + 5.8H_p - \frac{2.84 H_p^2 + 25.60 H_p}{H_p + 4.5}$$

3. Compute the lateral effective passive earth pressure on the wall
    At the toe of the retaining wall

$$\sigma_v = 20 \times H_p = 20H_p$$

$$u = \frac{9.81 H_p^2 + 88.29 H_p}{H_p + 4.5}$$

$$\sigma'_{p,h} = \left(20H_p - \frac{9.81 H_p^2 + 88.29 H_p}{H_p + 4.5}\right) \times 4.4 = 88H_p - \frac{43.16 H_p^2 + 388.48 H_p}{H_p + 4.5}$$

4. Compute the maximum net pore water pressure (at the excavation surface)
    According to Eq. 4.53, the maximum net pore water pressure would be

$$u_b = \frac{2(H_e + d_i - d_j)(H_p - d_i)\gamma_w}{2H_p + H_e - d_i - d_j} = \frac{2 \times 9 \times H_p \times 9.81}{2H_p + 9} = \frac{88.29 H_p}{H_p + 4.5}$$

5. The effective earth pressure on both sides of the wall and the distribution of the net pore water pressure are as shown in Figure 5.19a
6. Compute the driving moment (M_d) and the resistant moment (M_r) for the free body below the lowest level of struts

$$M_d = P_{a,h} L_a$$

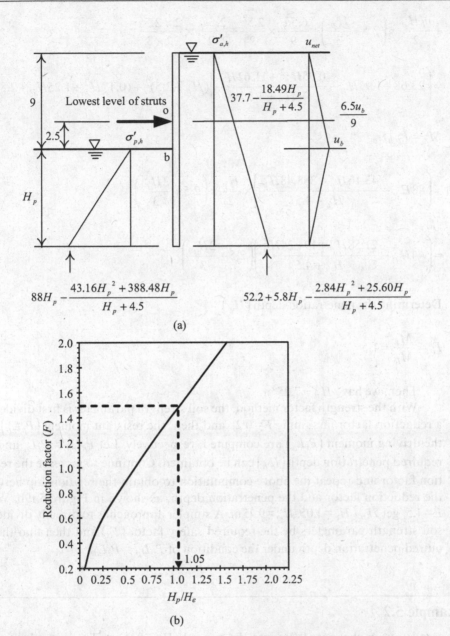

$$88H_p - \frac{43.16H_p^{\,2} + 388.48H_p}{H_p + 4.5}$$

$$52.2 + 5.8H_p - \frac{2.84H_p^{\,2} + 25.60H_p}{H_p + 4.5}$$

(a)

(b)

Figure 5.19 (a) Earth pressure distribution in sand and (b) strength factor analysis.

$$= \left(37.7 - \frac{18.49H_p}{H_p + 4.5}\right) \times \frac{(H_p + 2.5)^2}{2} + \left(14.5 + 5.8H_p - \frac{2.84H_p^2 + 7.11H_p}{H_p + 4.5}\right)$$
$$\times \frac{2(H_p + 2.5)^2}{2 \times 3}$$

$$+\frac{u_b H_p}{2} \times \left(2.5 + \frac{H_p}{3}\right) + \frac{6.5 u_b}{9} \times \frac{2.5^2}{2} + \frac{2.5 u_b}{9} \times \frac{2 \times 2.5^2}{2 \times 3}$$

$$= \left(23.68 + 1.93 H_p - \frac{0.95 H_p{}^2 + 11.62 H_p}{H_p + 4.5}\right)(H_p + 2.5)^2 + (0.17 H_p^2 + 1.25 H_p + 2.84) u_b$$

$$M_r = P_{p,h} L_p$$

$$= \left(88 H_p - \frac{43.16 H_p^2 + 388.48 H_p}{H_p + 4.5}\right) \times \frac{H_p}{2} \times \left(2.5 + \frac{2 H_p}{3}\right)$$

$$= \left(44 H_p^2 - \frac{21.58 H_p^3 + 194.24 H_p^2}{H_p + 4.5}\right)\left(2.5 + \frac{2 H_p}{3}\right)$$

7. Determine the penetration depth $\left(H_p\right)$

$$F_b = \frac{M_r}{M_d} = 1.5$$

Then, we have $H_p = 7.25$ m.

With the strength factor method, the soil strength parameter is first divided by a reduction factor (assuming $F = 0.2$), and then, the resistant moment $\left(P_p L_p\right)$ and the driving moment $\left(P_a L_a\right)$ are computed, respectively. Let $P_p L_p = P_a L_a$ and the required penetration depth $\left(H_p\right)$ can be obtained. Continue to increase the reduction factor and repeat the above computation to obtain the relationship between the reduction factor and the penetration depth, as shown in Figure 5.19b. When $F = 1.5$, get $H_p / H_e = 1.05$, $H_p = 9.45$ m. A simpler approach is to directly divide the soil strength parameters by the required safety factor $\left(F_b\right)$ and then find the required penetration depth under the condition of $P_p L_p = P_a L_a$.

Example 5.2

An excavation in clay goes 9.0 m into the ground ($H_e = 9.0$ m). The groundwater behind the wall is at the ground surface level while that in the front of the wall is at the level of the excavation surface. $\gamma_{sat} = 17.0$ kN/m^3. The undrained shear strength $s_u = 45$ kN/m^2. Suppose the excavation width $B = 10$ m and the excavation length $L = 30$ m. Compute the factor of safety against basal heave according to Terzaghi's method and Bjerrum and Eide's method, respectively.

Solution

In this example, the surcharge $q_s = 0$
According to Terzaghi's method,

$$F_b = \frac{5.7 s_{u2} B/\sqrt{2}}{(\gamma H_e + q_s) B/\sqrt{2} - s_{u1} H_e} = \frac{5.7 \times 45 \times 10/1.414}{17 \times 9 \times 10/1.414 - 45 \times 9} = 2.68$$

According to Bjerrum and Eide's method,

$$\frac{L}{B} = \frac{30}{10} = 3.0$$

$$\frac{H_e}{B} = \frac{9}{10} = 0.9$$

According to Figure 5.11, we have $N_c = 7.1$

$$F_b = \frac{s_u N_c}{\gamma_t H_e} = \frac{45 \times 7.1}{17 \times 9} = 2.09$$

Example 5.3

Same as Example 5.2 except assume there exists a stiff clayey layer 3.0 m below the excavation surface and the undrained shear strength of the clay is 90 kN/m^2. Compute the factor of safety against base shear failure (or basal heave) according to Terzaghi's method and Bjerrum and Eide's method, respectively.

Solution

The failure surface should be tangent to the second clay layer.
According to Terzaghi's method,

$$F_b = \frac{Q_u}{W - s_{u1} H_e} = \frac{5.7 s_{u2} D}{\gamma H_e D - s_{u1} H_e} = \frac{5.7 \times 45 \times 3}{17 \times 9 \times 3 - 45 \times 9} = 14.25$$

$\dfrac{D}{B} = \dfrac{3}{10} = 0.3$. From Figure 5.12a, we have $N_{c,s} = 7.15$

Moreover, $\dfrac{H_e}{B} = \dfrac{9}{10} = 0.9$. From Figure 5.12b, we have $f_d = 1.22$ $\dfrac{B}{L} = \dfrac{10}{30} = 0.3$. Following Eq. 5.14, we have $f_s = 1.06$. Thus,

$$F_b = \frac{s_u N_{c,s} f_d f_s}{\gamma H_e} = \frac{45 \times 7.15 \times 1.22 \times 1.06}{17 \times 9} = 2.72$$

5.8 BASE SHEAR FAILURE OF CANTILEVER WALLS

The cantilever wall is usually applicable to sand, gravel, or stiff clay. The stability of a cantilever wall counts on the soil reaction below the excavation bottom. The design is therefore confined to the fixed earth support method, and the free earth support method is inapplicable. Figure 5.20a illustrates a cantilever wall rotating about point O in a ultimate state. Figure 5.20b shows the possible earth pressure on the retaining wall. For the simplification of analysis, assume that the active and passive earth pressures above and below point O are fully mobilized, and therefore, the earth pressure distribution is discontinuous around point O as shown in Figure 5.20c.

As shown in Figure 5.20c, H_p and L are unknown. With the horizontal force equilibrium and the moment equilibrium, we can obtain the required penetration depth.

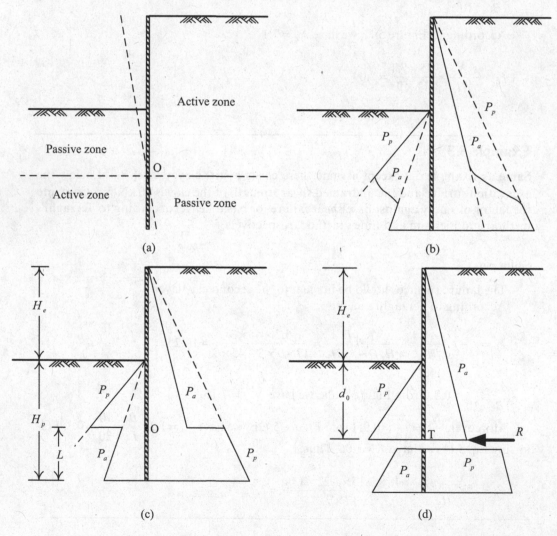

Figure 5.20 Analysis of a cantilever wall by gross pressure method: (a) deformation of the retaining wall, (b) real distribution of lateral earth pressure, (c) idealized distribution of lateral earth pressure, and (d) simplified analysis method.

Since both the horizontal force equilibrium and the moment equilibrium will generate quadratic and cubic equations, it is not easy to solve the equations directly. It is necessary to simplify the analysis method for practical use.

Figure 5.20d illustrates the simplified earth pressure distribution where the resultant of the pressures below the transition point T is replaced by a concentration force, R, acting on the transition point. It is necessary that R exists to keep the horizontal force equilibrium of the wall above point T. Based on the moment equilibrium against point T, we can find the value d_0. Because of the simplification of the analysis, d_0 should be slightly smaller than the actually required penetration depth and has to be increased properly (up to 20%). The increment has to be examined to ensure that the difference between the passive earth pressure and active earth pressure is greater than or equal to the concentrated force R. The detailed computing process can be found in Example 5.4. The simplified analysis method as shown in Figure 5.20d has been commonly adopted in engineering design (Padfield and Mair, 1984).

Excavation in a sand of gravel with groundwater on both sides of the retaining wall should be analyzed with the net pore water pressure distribution as discussed in Section 5.5.1, viewed as the driving force, following the method shown in Figure 5.20.

To compute the penetration depth with the gross earth pressure distribution, the load factor equilibrium method is usually adopted to estimate the factor of safety. Sometimes, the strength factor method or the dimension factor method is used. Figure 5.21 shows the net earth pressure distribution. According to the characteristics of the deformation of the cantilever wall, we can see that the earth pressures at point **b** (excavation surface) on the front and back of the retaining wall should achieve the passive earth and the active earth pressures, respectively. The net value of the earth pressure at the point is CD (Figure 5.21b). Point **c** is close to point **b**, and the earth

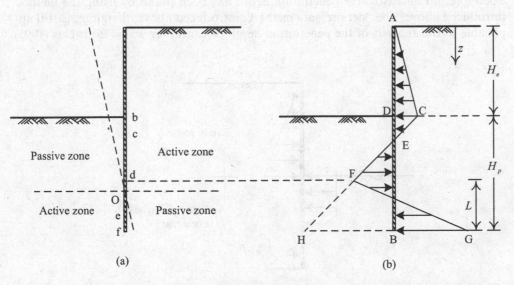

Figure 5.21 Analysis of a cantilever wall by net pressure method: (a) deformation of the wall and (b) net earth pressure distribution.

pressures on the front and back of the wall should also achieve the passive and active earth pressures separately. Thus, the net earth pressure at point **c** should be

$$\sigma_{h,c} = \left[\gamma H_e + \gamma(z - H_e)\right]K_a - \gamma(z - H_e)K_p = \gamma H_e K_a - \gamma(z - H_e)(K_p - K_a) \quad (5.16)$$

We can see from the above equation that the slope of line CF is $1 : \gamma(K_p - K_a)$ (the vertical: the horizontal). The net earth pressure with the slope $1 : \gamma(K_p - K_a)$ maintains till point **d**, where the passive and active earth pressures "begin" not to be fully mobilized. Assuming the lateral movement of the bottom of the wall is large enough, the soil in front and back of the wall will reach the active and passive states, respectively, and the net value of the earth pressure here is BG. The soil strength between points **f** and **d** in front and in back of the wall may not be fully mobilized, and to simplify the analysis, we assume that the net earth pressures between points **d** and **f** are of a linear relation, as line FG. Figure 5.21b shows an assumed net earth pressure distribution. According to the relation of the horizontal force equilibrium in Figure 5.21b, we know

Area ACE − Area EFHB + Area FHBG = 0

We can find L according to the above equation. Substitute L in the moment equilibrium equation with regard to point **f**, and we have the quadratic equation with H_p as a single unknown variable. To solve H_p, the trial-and-error method is recommended. When performing trial and error, we usually begin from $H_p = 0.75H_e$. For the detailed computation, please refer to Example 5.5.

The earth pressure distribution of a cantilever wall in sand is similar to that in Figure 5.21b. For one in clay, refer to Figure 5.22.

Though in Section 5.5.1 we have mentioned that the net pressure method is not applicable to the stability analysis of strutted walls, the net pressure method as above introduced, however, is not the real net pressure method as introduced in Section 5.5.1. Therefore, no unreasonable penetration depth has been found by using the method introduced above. The "net pressure method" introduced above, therefore, is still applicable to the analysis of the penetration depth of cantilever walls. Both Das (1995)

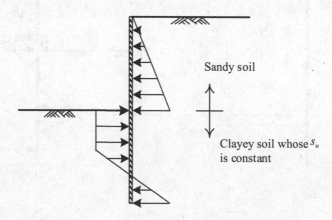

Figure 5.22 Net earth pressure distribution on a cantilever wall in clay.

and Bowles (1998) adopt the net pressure method and the dimension factor method to estimate the penetration depth of a cantilever wall.

When the computed safety factor does not satisfy the requirement, a measure similar to those in Section 5.6.3 can be adopted to ensure the safety of the excavation.

Example 5.4

Figure 5.23 shows a 4.0-m-deep excavation with a cantilever wall. The groundwater level is very deep. The unit weight of soil is $\gamma = 20$ kN/m^3. The effective strength parameters $c' = 0$ and $\phi' = 25°$. Adopt the simplified gross pressure method to compute the penetration depth instead.

Solution

Suppose that the friction angles between the retaining wall and soil on the active and passive sides are, respectively, $\delta = 2\phi'/3$ and $\delta = \phi'/2$. According to Coulomb's earth pressure theory and Caquot–Kerisel's earth pressure theory (consult Eq. 4.14 and Figure 4.10), we have $K_a = 0.37$ and $K_p = 3.55$. Thus, their horizontal components are as follows: $K_{a,h} = 0.37 \cos(2\phi'/3) = 0.36$, $K_{p,h} = 3.55 \cos(\phi'/2) = 3.47$. Figure 5.24 shows the gross pressure distribution on the wall where P_{fa} and P_{fp} are the resultants of the active and passive earth pressures above point T, respectively. Their values and arms are separate.

$$P_{fa} = \frac{1}{2} \cdot K_{a,h} \cdot \gamma \cdot (H_e + d_0)^2$$

$$P_{fp} = \frac{1}{2} \cdot K_{p,h} \cdot \gamma \cdot d_0^2$$

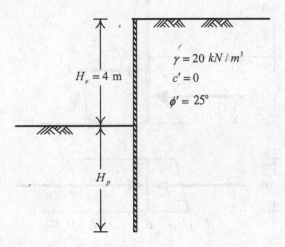

$H_e = 4$ m

$\gamma = 20 \ kN/m^3$

$c' = 0$

$\phi' = 25°$

H_p

Figure 5.23 Excavation of a cantilever wall in sand.

$$L_a = \frac{1}{3} \cdot (H_e + d_0)$$

$$L_p = \frac{1}{3} \cdot d_0$$

Assuming the factor of safety to be 1.5, take the moment with regard to point T. Then, we have

$$F_p = \frac{M_r}{M_d} = \frac{P_{fp} \cdot L_p}{P_{fa} \cdot L_a} = 1.5$$

Thus,

$$d_0 = 4.7 \text{ m}$$

Considering the inaccuracy due to simplification, the required penetration depth for design is about 1.2 times as large as d_0. Therefore,

$$H_p = 1.2 d_0 = 5.6 \text{ m}$$

From the horizontal force equilibrium above point T, we can compute R as

$$R = \frac{1}{2} \cdot K_{p,h} \cdot \gamma \cdot d_0{}^2 - \frac{1}{2} \cdot K_{a,h} \cdot \gamma \cdot (H_e + d_0)^2 = 494.0 \text{ kN}$$

According to Figure 5.24, compute the resultant below point T

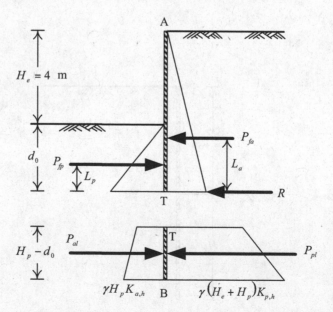

Figure 5.24 Simplified gross pressure analysis method.

$$K_{p,h}\gamma\left[\frac{(d_0+H_e)+(1.2d_0+H_e)}{2}\right](1.2-1)d_0 - K_{a,h}\gamma\frac{d_0}{2}(1+1.2)(1-1.2)d_0 = 563.2 \text{ kN}$$

To maintain the equilibrium of the retaining wall above point T, the resultant below point T should be larger or equal to R.

As shown above, the resultant 563.2 kN $> R = 494$ kN. The $H_p = 5.6$ m is sufficient.

Example 5.5

Same as Example 5.4, compute the required penetration depth following the net pressure method instead,

Solution

From Example 5.4, we have $K_{a,h} = 0.36$ and $K_{p,h} = 3.47$. Figure 5.25 shows the net pressure distribution. First, we have to find the point at which the net pressure is none. As shown in the figure, the earth pressure at the depth of z from the ground surface would be

$$p_a = \left[\gamma H_e + \gamma(z-H_e)\right]K_{a,h} = 7.2z$$

$$p_p = \gamma(z-H_e)K_{p,h} = 69.4z - 277.6$$

The point where the net pressure is none (point C) can be found as follows:

$$p = p_a - p_p = \left[\gamma H_e + \gamma(z-H_e)\right]K_{a,h} - \gamma(z-H_e)K_{p,h} = 0$$

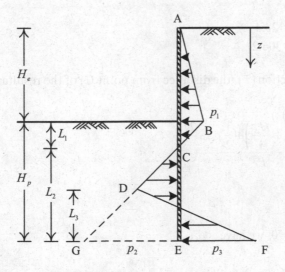

Figure 5.25 Net earth pressure distribution.

After reduction, we have

$$-62.2z + 277.6 = 0$$

$$z = 4.46 \text{ m}$$

$$L_1 = z - H_e = 0.46 \text{ m}$$

$$p_1 = \gamma H_e K_{a,h} = 28.8 \text{ kN/m}^2$$

As discussed in this section, the slope of line BD is $1 : \gamma(K_p - K_a)$ (the vertical: the horizontal). Thus,

$$p_2 = L_2(K_{p,h} - K_{a,h})\gamma = 62.2L_2$$

$$p_3 = \gamma H_e K_{p,h} + (K_{p,h} - K_{a,h})\gamma(L_1 + L_2) = 306.2 + 62.2L_2$$

The area of $\triangle$ ABC would be

$$P_f = \frac{1}{2}(H_e + L_1)p_1 = \frac{1}{2}\gamma H_e K_{a,h}(H_e + L_1) = 64.2 \text{ kN/m}^2$$

The earth pressure on the retaining wall should satisfy the horizontal force equilibrium. Therefore, we have
The area of $\triangle$ ABC-the area of $\triangle$ GEC+ the area of $\triangle$ DGF $= 0$

$$P_f - \frac{1}{2}p_2 L_2 + \frac{1}{2}L_3(p_2 + p_3) = 64.2 - 31.1L_2^2 + \frac{1}{2}L_3(124.4L_2 + 306.2) = 0$$

After reduction, we have

$$L_3 = \frac{62.2L_2^2 - 128.4}{124.4L_2 + 306.2}$$

The point of action $(\bar{z})$ (the distance from point C) of the resultant $(P_f$, the area of $\triangle$ ABC) would be

$$\bar{z} = \frac{\frac{1}{2}p_1 H_e\left(L_1 + \frac{H_e}{3}\right) + \frac{1}{2}p_1 L_1 \frac{2L_1}{3}}{P_f}$$

$$\bar{z} = 1.64 \text{ m}$$

Since $\sum M_E = 0$, we have

$$P_f(L_2 + \bar{z}) - \left(\frac{1}{2}p_2 L_2\right)\frac{L_2}{3} + \frac{1}{2}L_3(p_2 + p_3)\frac{L_3}{3} = 0$$

After reduction, we have

$$L_2 = 3.38\ \text{m}$$

$$H_p = L_2 + L_1 = 3.38 + 0.46 = 3.84\ \text{m}$$

Because the computation in this example does not follow the real net pressure distribution, it is not workable to adopt the factor of safety of the load factor method to compute the penetration depth as in the previous example. Instead, the factor of safety of the dimension factor method is used to compute the required penetration depth, which is

$$H_p = 1.5 \times 3.84 = 5.76\ \text{m}$$

As illustrated by the above two examples, the penetration depths computed with the simplified gross pressure method and the net pressure method are 5.6 and 5.76 m, respectively. The difference between the two is little. Therefore, the simplified gross pressure method is most commonly adopted.

5.9 UPHEAVAL FAILURE

If below the excavation surface there exists a permeable layer (such as sand or gravel soils) underlying an impermeable layer, the impermeable layer has a tendency to be lifted by the pore water pressure from the permeable layer. The safety, against upheaval, of the impermeable layer should be examined. As shown in Figure 5.26, the factor of safety against upheaval is

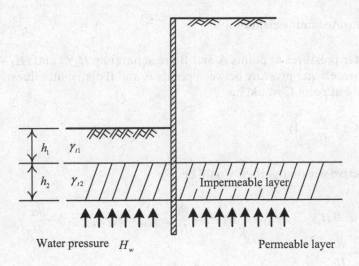

Figure 5.26 Analysis of upheaval failure.

$$F_{up} = \frac{\sum_i \gamma_{ti} \cdot h_i}{H_w \cdot \gamma_w} \tag{5.17}$$

where

F_{up} = factor of safety against upheaval

γ_{ti} = unit weight of soil in each layer above the bottom of the impermeable layer

h_i = thickness of each soil layer above the bottom of the impermeable layer

H_w = head of the pore water pressure in the permeable layer

γ_w = unit weight of the groundwater.

The factor of safety against upheaval F_{up} should be larger than or equal to 1.2.

To safeguard the safety of excavation construction, the possibilities (of the occurrence) of upheaval at each stage of excavation should be analyzed. If boring within the excavation zone is required (for example, in order to place piezometers or build a well), the possible paths of water flow should be examined and the possible upheaval induced by boring should be prevented to secure the excavation. Please see Section 5.11.

When the computed safety factor does not satisfy the requirement, it is necessary to reduce the water pressure below the impermeable layer by dewatering. Execution of soil improvement for the soil below the impermeable layer to increase the weight of the soils, i.e., numerator in Eq. 5.17, can be an alternative.

5.10   SAND BOILING

As shown in Figure 5.27, suppose there exists an upward water flow passing through a sandy layer. The total stress, σ, at the depth of z (point C) would be

$$\sigma = H_1\gamma_w + z\gamma_{sat} \tag{5.18}$$

where

γ_{sat} = saturated unit weight of soil.

Pore water pressures at points A and B are separately $H_1\gamma_w$ and $(H_1 + H_2 + h)\gamma_w$. Suppose the pore water pressure between points A and B distributes linearly. The pore water pressure at point C would be

$$u = \left(H_1 + z + \frac{h}{H_2}z \right)\gamma_w \tag{5.19}$$

The effective stress at point C would be

$$\sigma' = \sigma - u = \left(H_1\gamma_w + z\gamma_{sat} \right) - \left(H_1 + z + \frac{h}{H_2}z \right)\gamma_w = z(\gamma_{sat} - \gamma_w) - \frac{hz}{H_2}\gamma_w$$

$$= z\gamma' - \frac{hz}{H_2}\gamma_w \tag{5.20}$$

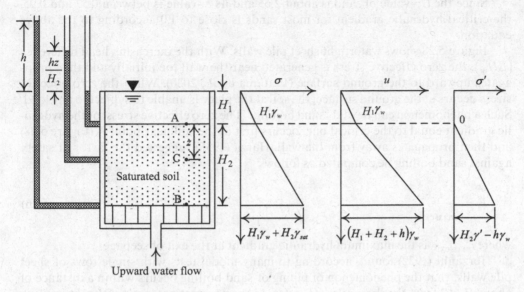

Figure 5.27 Mechanism of sand boiling.

As shown in the above equation, the upward water flow may cause the effective stress at point C to be 0, meaning that the effective stress is equal to 0 in the entire sandy soil. The sand is unable to bear any load, and this phenomenon is called sand boiling. Thus,

$$\sigma' = 0 = z\gamma' - \frac{hz}{H_2}\gamma_w \tag{5.21}$$

$$\frac{h}{H_2} = \frac{\gamma'}{\gamma_w} \tag{5.22}$$

The hydraulic gradient when the effective stress equals 0 is called the critical hydraulic gradient, i_{cr}, which can be expressed as follows:

$$i_{cr} = \frac{(h/H_2)z}{z} = \frac{h}{H_2} = \frac{\gamma'}{\gamma_w} \tag{5.23}$$

Besides, according to the phase relationship of soil, the submerged unit weight is

$$\gamma' = \left(\frac{G_s - 1}{1 + e}\right)\gamma_w \tag{5.24}$$

where G_s is the specific gravity of soil and e is the void ratio. The critical hydraulic gradient is then

$$i_{cr} = \frac{G_s - 1}{1 + e} \tag{5.25}$$

Since the G_s-value of sand is about 2.65 and its e-value is between 0.57 and 0.95, the critical hydraulic gradient for most sands is close to 1.0 according to the above equation.

Figure 5.28 shows watertight sheet pile walls. With the increasing head difference (ΔH_w), the zero effective stress is generated near the wall toe initially and then propagates upward to the ground surface (Pratama et al., 2020). When the zero effective stress occurs at the ground surface, the soil at the place is unable to subject to any load. Such a phenomenon can be said "sand boiling." The zero effective stress, or the hydraulic gradient equal to the critical one, occurs first in the soil near point **A** (Figure 5.28) and then propagates away from the wall. Harza (1935) suggested the factor of safety against sand boiling be computed as follows:

$$F_s = \frac{i_{cr}}{i_{max(exit)}} \tag{5.26}$$

where $i_{max(exit)}$ is the maximum hydraulic gradient at the exit of seepage.

Terzaghi (1922) found, according to many model tests with single rows of sheet pile walls, that the phenomenon of piping or sand boiling occurs within a distance of about $H_p/2$ from the sheet piles (H_p refers to the penetration depth of the sheet piles). Thus, to analyze the stability of single rows of sheet piles, the soil column $H_p \times H_p/2$ in front of the sheet pile is taken as an analytic object, as shown in Figure 5.28. The uplift force on the soil column would be

$$U = (\text{the volume of the soil column}) \times (i_{avg}\gamma_w) = \frac{1}{2}H_p^2 i_{avg}\gamma_w \tag{5.27}$$

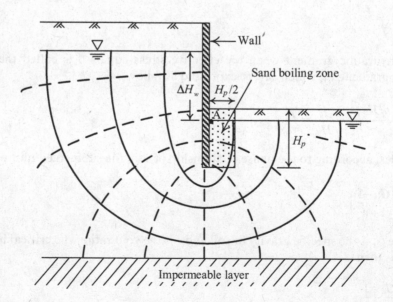

Figure 5.28 Seepage in soil below sheet piles.

where i_{avg} is the average hydraulic gradient of the soil column. The downward force of the soil column (i.e., the submerged weight) is

$$W' = \frac{1}{2} H_p^2 (\gamma_{sat} - \gamma_w) = \frac{1}{2} H_p^2 \gamma' \tag{5.28}$$

Therefore, the factor of safety is

$$F_s = \frac{W'}{U} = \frac{\gamma'}{i_{avg} \gamma_w} \tag{5.29}$$

According to Eq. 5.29, provided the computed factor of safety is too small, we can consider placing filters at the exits of seepage. Assuming the weight of the filters is Q, the factor of safety will be

$$F_s = \frac{W' + Q}{U} \tag{5.30}$$

By assuming one-dimensional flow, the $i_{max(exit)}$ in Eq. (5.26) can be approximated as a ratio of ΔH_w to the length of the shortest flow path $(2H_p + H_e - d_i - d_j)$. Harza's method can then be simplified as follows (referred to as the simplified Harza's method):

$$F_s = \frac{i_{cr}}{i_{max(exit)}} \approx \frac{i_{cr}(2H_p + H_e - d_i - d_j)}{\Delta H_w} \tag{5.31}$$

where d_i and d_j are the groundwater level to the excavation bottom and that to the ground surface level, respectively (refer to Figure 4.21).

Similarly, with the assumption of one-dimensional flow and $i_{avg} \approx (\Delta H_w / 2H_p)$, Terzaghi's method can then be simplified as follows (referred to as the simplified Terzaghi's method):

$$F_s = \frac{W'}{U} \approx \frac{2i_{cr}H_p}{\Delta H_w} \tag{5.32}$$

In addition to Harza's and Terzaghi's methods, their simplified versions are also adopted by some countries as their country codes due to their simplicity.

Table 5.1 Coefficients for Correction Factor f

	a_0	a_1	a_2	a_3	a_4	a_5	a_6	a_7	a_8	a_9	a_{10}
Simplified Harza	-0.5081	-0.1504	-0.1424	0.4950	-0.0863	-0.3688	-0.1693	2.4550	-0.1176	0.1869	-0.0543
Simplified Terzaghi	-0.0782	-0.3780	-0.0363	0.4994	-0.0944	-0.0700	-0.1714	1.7083	-0.1165	0.1884	-0.0578

Table 5.2 Required Factors of Safety for Various Methods

	Harza	Terzaghi	Simplified Harza	Simplified Terzaghi
Medium sand	2.04	1.76	1.79	1.80
Very dense sand	1.61	1.44	1.46	1.47

Both $i_{max(exit)}$ in Harza's method and i_{avg} in Terzaghi's method should be obtained with the flow net method or two-dimensional numerical analysis. However, the hydraulic gradients, $i_{max(exit)}$ and i_{avg}, in both simplified versions, i.e., Eqs. 5.31 and 5.32, are estimated without consideration of the excavation width and the location of impermeable soil but only the head difference. Therefore, the results from Eqs. 5.31 and 5.32 should be modified. Based on the reliability approaches by the author and his working team (Pratama et al., 2020), the correction factor (f) is

$$f = \exp\left(\begin{array}{l} a_0 + a_1 \ln\left(\dfrac{H_p}{H_e}\right) + a_2 \ln\left(\dfrac{H_p}{H_e}\right)^2 + a_3 \ln\left(\dfrac{B}{H_e}\right) + a_4 \ln\left(\dfrac{B}{H_e}\right)^2 \\[3mm] + a_5 \ln\left(\dfrac{\Delta H_w}{H_e}\right) + a_6 \ln\left(\dfrac{D}{\Delta H_w}\right) + a_7 \left(\dfrac{d_i}{H_p}\right) + a_8 \ln(i_{cr}) \\[3mm] + a_9 \ln\left(\dfrac{B}{H_e}\right)\ln\left(\dfrac{H_p}{H_e}\right) + a_{10} \ln\left(\dfrac{D}{\Delta H_w}\right)\ln\left(\dfrac{H_p}{H_e}\right) \end{array}\right) \quad (5.33)$$

where B is the excavation width, D is the depth of impermeable layer measured from the wall toe, and $a_0, a_1, ..., a_{10}$ are coefficients, which can be found from Table 5.1.

The multiplication of f and F_s from Eqs. 5.31 and 5.32 should be equal to or greater than the required factor of safety. Again, based on the reliability approaches, the required factors of safety for Harza's method and Terzaghi's method and their simplified versions are listed in Table 5.2.

When the factor of safety is insufficient, sand boiling likely occurs. Increasing the wall penetration depth, penetrating the wall into impermeable layer, and lowering the water level behind the wall are possible measures.

5.11   CASE STUDY OF SAND BOILING

Figure 5.29a shows the excavation profile and the geological profile of the Siemen Station of the Taipei Rapid Transit System. The excavation was 23.25 m wide, 560 m long, and 24.5 m deep. The retaining-strut system was a diaphragm wall 100 cm thick and 44.0 m deep with eight levels of struts. As shown in Figure 5.29a, 49.1 m down from the ground surface is a gravel layer known as the Chingmei formation. Above it are six alternated layers of sand and clay known as the Sungshan formation.

After the last stage of excavation, the foundation slabs partly cast and the rest laid with the steel reinforcements, the engineers found one of the piezometers placed on the diaphragm wall damaged and unable to take any measurement. According to

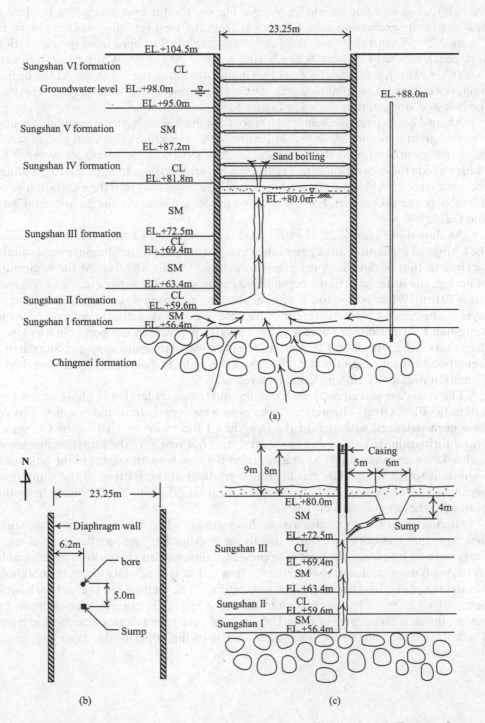

Figure 5.29 Excavation of the Siemen Station of the Taipei Rapid Transit System:
(a) excavation and geological profiles, (b) plan view, and (c) occurrence of
sand boiling (Chou and Ou, 1999).

the contract, a new one should be placed. The contractor thus was going to place a new one in the excavation zone 6.2 m away from the west retaining wall as shown in Figure 5.29b. To install it, a drill machine was employed to bore from the excavation surface (elevation EL.+80 m). When boring reached 20.5 m below the excavation surface (EL.+59.5 m), groundwater began springing out along the borehole. At the beginning, since there was not much water, the contractor put a 9.00-m-long steel case onto the hole and stopped it as shown in Figure 5.29c.

About 4 hours later, the sump, 5.00 m south of the borehole, began springing sandy water in great amounts, as shown in Figure 5.29c. The contractor then placed sandbags and grouted around the water source. The situation, however, got worse. The roads outside the excavation zone began cracking and settling. In addition to grouting, the contractor also blocked the roads and refilled coarse soils into the excavation zone to stabilize the excavation. For the detailed process, please see the document (Chou and Ou, 1999).

As shown in Figure 5.29, the silty sand layer (Sungshan I formation) right above the Chingmei formation was a permeable layer, whose pore water pressure head should be close to that of the Chingmei gravel formation, i.e., EL.+88.0 m. At the beginning of boring, the water level in the borehole was at most as high as the excavation surface (EL.+80.0 m). When boring to EL.+59.5 m, though the boring was working in the clayey layer (Sungshan II formation), either because the boring actually entered into the Sungshan I formation, or because the pore water pressure in the borehole was lower than that of the Sungshan I formation, the pore water pressure from the Sungshan I formation broke through the last 0.1-m-thick clayey layer and water in the Sungshan I formation began to gush up along the borehole.

The steel case placed onto the borehole, and the water level in the hole went up by 8.00 m (to El.+88.0 m). Theoretically, the pore water pressure in the borehole should have been balanced with that in the Sungshan I formation or that in the Chingmei gravel formation. However, due to the steel case not reaching the impermeable layer, with a 4.0-m-deep sump just 5.0 m away from the borehole, the water in the hole took a shortcut to the sump. The exit hydraulic gradient at the bottom of the sump thus became so large as to exceed the critical hydraulic gradient. Sand boiling from the bottom of the sump thus occurred.

Obviously, as the above discussions have shown, when boring within a construction zone that has been excavated, due to the overburden pressure decreasing, engineers have to be especially careful to prevent failures when there exists a permeable layer. Applying a steel case to obstruct the flow of the groundwater along the borehole has to take care of the occurrence of sand boiling. The methods to prevent such accidents are to lengthen the water path to lower the hydraulic gradient (for example, by having the case go deeper) or to extend the steel case to reach an impermeable layer. The latter method has to consider the thickness of the impermeable layer to prevent upheaval.

PROBLEMS

5.1 As shown in Figure P5.1, no groundwater exists at the site. Assume the friction angle between the retaining wall and soil $\delta = 0$. Use the factor of safety $F_b = 1.5$ against base shear failure to compute the required penetration depth following the dimension factor method.

5.2 Same as Problem 5.1. Assume the penetration depth as computed above. Compute the factor of safety against base shear failure following the earth pressure equilibrium method (load factor). Compare the result with the factor of safety as given in Problem 5.1.

5.3 Same as Problem 5.1. Use the factor of safety $F_b = 1.5$ against base shear failure to compute the required penetration depth following the earth pressure equilibrium method (strength factor).

5.4 As shown in Figure P5.4, an excavation with the groundwater level at the ground surface behind the wall and that at the excavation surface in front of the wall. If the friction angle between the retaining wall and soil $\delta = 0.5\phi'$, use the factor of safety $F_b = 1.2$ against base shear failure to compute the required penetration depth following the earth pressure equilibrium method (load factor) as well as the earth pressure equilibrium method (strength factor).

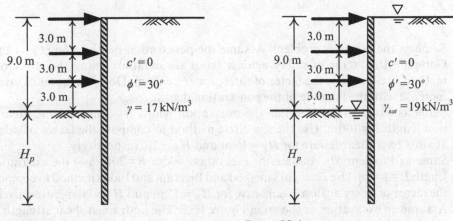

Figure P5.1   Figure P5.4

5.5 Figure P5.5 shows an excavation. Compute the required penetration depth against base shear failure for $F_b = 1.2$ using the slip circle method, the earth pressure equilibrium method (load factor), and the earth pressure equilibrium method (strength factor) with $c_w / s_u = 1.0$ between the retaining wall and clay.

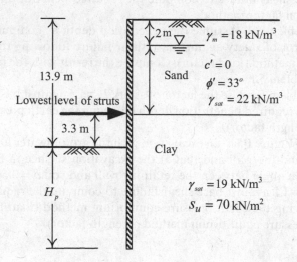

Figure P5.5

5.6 Same as the previous problem. Assume the penetration depth to be $H_p = 15$ m. Compute the factor of safety against basal shear failure using the slip circle method. Re-compute the factor of safety for $H_p = 20$ m. Does the factor of safety increase with the increase of the penetration depth?

5.7 Same as Problem 5.6. Assume the excavation width $B = 20$ m and the excavation length $L = 100$ m. Use the slip circle method to compute the factor of safety against base shear failure for $H_p = 15$ m and $H_p = 20$, respectively.

5.8 Same as Problem 5.6. Assume the excavation width $B = 20$ m and the excavation length $L = 100$ m. Use Terzaghi's method and Bjerrum and Eide's method to compute the factor of safety against basal heave for $H_p = 15$ m and $H_p = 20$ m, respectively.

5.9 Assume an excavation as shown in Figure P5.9. The undrained shear strength of clay has the normalized behavior. Compute the required penetration depth of the retaining wall against base shear with $F_b = 1.2$ using the slip circle method and the earth pressure equilibrium method (load factor and strength factor), considering $c_w / s_u = 1.0$.

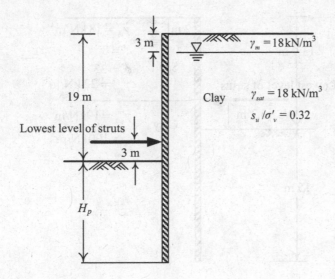

3 m

$\gamma_m = 18\,kN/m^3$

19 m

Clay $\gamma_{sat} = 18\,kN/m^3$

$s_u/\sigma'_v = 0.32$

Lowest level of struts

3 m

H_p

Figure P5.9

5.10 Same as Problem 5.9. Assume the penetration depth $H_p = 15$ m and $H_p = 20$ m, respectively. Compute the factors of safety against base shear using the slip circle method. Does the factor of safety increase with the increase of the penetration depth?

5.11 Same as Problem 5.9. Assume the excavation width $B = 20$ m and the excavation length $L = 100$ m. Compute the factor of safety against base shear failure for $H_p = 15$ m and $H_p = 20$ m, respectively, using Terzaghi's method.

5.12 Same as Problem 5.9. Assume the excavation width $B = 20$ m and the excavation length $L = 100$ m. Use Bjerrum and Eide's method to compute the factor of safety against base shear failure when $H_p = 15$ m and $H_p = 20$ m, respectively.

5.13 See Figure P5.13. Assume the excavation width $B = 50$ m and the excavation length $L = 100$ m. The penetration depth of the diaphragm wall $H_p = 15$ m. A gravel soil exists 39 m below the ground surface. Use Terzaghi's method and Bjerrum and Eide's method, respectively, to compute the factor of safety against base shear failure.

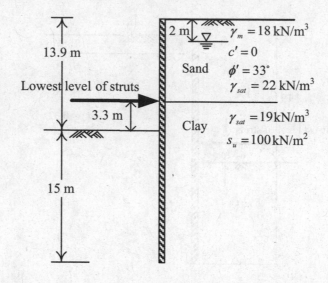

Gravel formation

Figure P5.13

5.14 See Figure P5.14. Between 45 and 60 m deep from the ground surface is a soft soil. Assume the excavation width $B = 50$ m and the excavation length $L = 100$ m. The penetration depth $H_p = 18$ m. Use the slip circle method to compute the factor of safety against base shear failure.

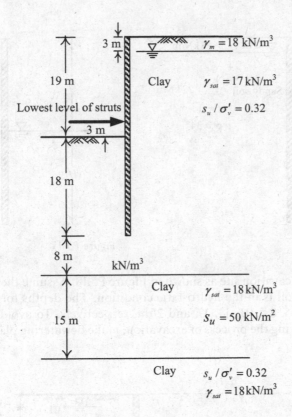

$\gamma_m = 18 \text{ kN/m}^3$

3 m

19 m

Clay $\gamma_{sat} = 17 \text{ kN/m}^3$

Lowest level of struts $s_u / \sigma_v' = 0.32$

3 m

18 m

8 m

kN/m^3

Clay $\gamma_{sat} = 18 \text{ kN/m}^3$

15 m $S_u = 50 \text{ kN/m}^2$

Clay $s_u / \sigma_v' = 0.32$

$\gamma_{sat} = 18 \text{kN/m}^3$

Figure P5.14

5.15 Same as the previous problem. Use Terzaghi's method and Bjerrum and Eide's method, respectively, to compute the factor of safety against base shear failure.

5.16 See Figure P5.16. The retaining wall is a diaphragm wall. The excavation width is 20 m, and the depth to the impermeable layer measured from the wall toe is 30 m. Compute the factor of safety against sand boiling using (1) Terzaghi's method, (2) Harza's method, (3) simplified Terzaghi's method, and (4) simplified Harza's method.

5.17 Figure P5.17 is a cantilever wall in saturated sandy soil. The groundwater levels inside and outside the excavation zone are at the excavation surface and the ground surface, respectively. The excavation depth $H_e = 3.5$ m. The saturated unit weight of sand $\gamma_{sat} = 22$ kN/m^3. The effective stress parameters $c' = 0$ and $\phi' = 35°$. The friction angles between the wall and soil inside and outside the excavation zone are the same, $\delta = 2\phi'/3$. Compute the required penetration depth with the factor of safety to be 1.5 using the simplified gross earth pressure method (to simplify analysis, ignore the seepage effect).

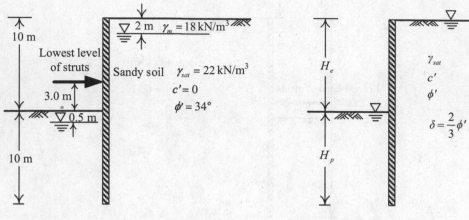

Figure P5.16

Figure P5.17

5.18 Here is an excavation site as shown in Figure P5.18. Assume the pore water pressure in the soil is in the hydrostatic condition. The depths for each excavation stage are 2, 5, 8.5, 11, 14, 17, and 20 m, respectively. To avoid sand boiling or upheaval during the process of excavation, make dewatering plans for each excavation stage.

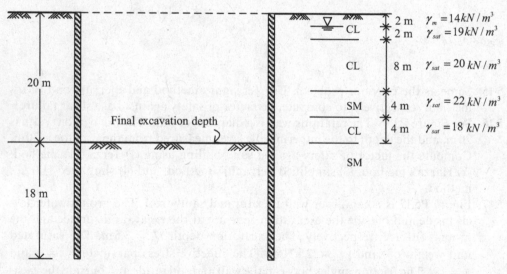

Figure P5.18

5.19 Same as Problem 5.17. Use the net earth pressure distribution to compute the required penetration depth for the retaining wall with the factor of safety to be 1.5 (to simplify analysis, ignore the seepage effect).

5.20 Figure P5.20 shows a cantilever wall. Assume the groundwater level is at the interface between sand and clay. $H_e = 3.3$ m, $\gamma = 17.6$ kN/m^3, $c' = 0$, $\phi' = 34°$ for sand. $\gamma_{sat} = 22$ kN/m^3, $s_u = 60$ kN/m^2 for clay. Assume there exists no friction between sand and the wall. Nor does there exist adhesion between clay and the wall. Compute the required penetration depth, with the factor of safety to be 1.5, using the simplified gross earth pressure method.

5.21 Same as the previous problem. Compute the required penetration depth with the factor of safety to be 1.5 using the net earth pressure distribution.

5.22 Figure P.5.22 shows a cantilever wall. Assume the groundwater level is at the ground surface. $H_e = 3.0$ m, $P = 600$ kN/m, $\gamma_{sat} = 19.2$ kN/m^3, $s_u = 70$ kN/m^2. Assume no adhesion exists between the wall and clay. Compute the required penetration depth with the factor of safety to be 1.5 using the simplified gross earth pressure method.

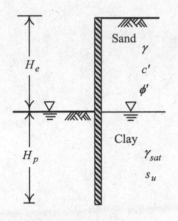

Figure P5.20

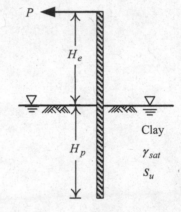

Figure P5.22

5.23 Same as the previous problem. Compute the required penetration depth with the factor of safety to be 1.5 using the net earth pressure distribution.

5.24 See Figure P5.17. $H_e = 9$ m, $H_p = 10$ m, and $B = 8.0$ m. The groundwater levels in front of and behind the wall are shown in the figure. The impermeable layer is very deep. The soil is loose sand where $c' = 0$ and $\phi' = 30°$. The saturated unit weight of sand $\gamma_{sat} = 19$ kN/m^3. Compute the factor of safety against sand boiling using (1) Terzaghi's method, (2) Harza's method, (3) simplified Terzaghi's method, and (4) simplified Harza's method.

5.25 Redo Problem 5.24 but the impermeable layer locates 18 m below the wall toe.

Chapter 6

Stress and deformation analysis

Simplified method

6.1 INTRODUCTION

Stress analysis is necessary for the design of structural components, and deformation analysis aims at predicting the wall deflections and soil movements caused by excavation to protect adjacent properties. The stress and deformation analysis methods for excavations include the simplified method, the beam on elastic foundation method, and the finite element method, which are introduced in Chapters 6–8, respectively. This chapter only introduces the simplified method.

Generally speaking, simplified methods employ the monitoring results of excavation case histories and then sort them into the stress and deformation characteristics of retaining walls and soils. The characteristics are useful not only to help understand the actual excavation behavior but to be used to derive simplified methods to predict the movement induced by excavation. The simplified method as inferred from field measurements represents the effect of every relevant element on deformation. Therefore, it can lead to effective predictions, without much complexity, for similar excavation projects, in terms of soil conditions, construction methods, and engineering designs. Besides, some people have also conducted systematic parametric studies using numerical methods and induced the deformation characteristics of excavation, which can be used for the prediction of wall deformation and (ground) surface settlement. The methods inferred in this way are also called the simplified method. The results of deformation analyses following the simplified method are generally good for common excavations.

The retaining-strut system of an excavation is, in nature, a highly static indeterminate structure and thereby is difficult to analyze by hand calculation unless the loading pattern, boundary conditions, and analysis method are simplified. The stress analysis methods that will be introduced in this chapter are induced from accumulated experience or observations by designers and are more applicable to common excavations. For special excavations (for example, large scale or great depth), the numerical method, introduced in Chapters 7 and 8, is recommended.

6.2 ANALYSIS OF SETTLEMENT INDUCED BY THE CONSTRUCTION OF DIAPHRAGM WALLS

As discussed in Chapter 3, the construction of a diaphragm wall first partitions the wall into several panels. The construction process of each panel includes guided

DOI: 10.1201/9780367853853-6

trench excavation, guided wall construction, trench excavation (for diaphragm wall), reinforcement placement, and concrete casting (please see Section 3.3.4).

The depth of a guided trench is generally about 2~3m, sometimes 5m. Considering that no significant settlement occurs in this stage, this chapter will not delve into the subject.

The stress condition of soil in the vicinity of trenches during diaphragm wall construction is rather complicated. Take the construction of a diaphragm wall panel, for example. To keep the trench wall stable and not collapsed, it is necessary to fill the trench with stabilizing fluid during excavation process of the trench panel. Under normal construction conditions, trench excavation will cause the stress states of the soil around the trench to change from the original K_0 to the balanced state of the fluid pressure of stabilizing fluid. However, the fluid pressure of stabilizing fluid is normally smaller than the earth and water pressures behind the trench. The lateral movement and surface settlement of the soil in the vicinity of the trench thus occur. During concrete casting, the lateral pressure of the concrete in the panel should be greater than the fluid pressure and also slightly greater than the earth and water pressures behind the trench. Therefore, the lateral movement and surface settlement rebound slightly.

The soil deformation behavior caused by trench excavation is not the same as that caused by main excavation because excavation geometric shapes and strut methods are different. The ratio of the depth of a trench panel to its width and that of the depth to length are both much larger than those in main excavations. What's more, there is the influence of stabilizing fluid, employed to counteract the lateral earth pressure and to ensure the stability of trench walls. Nevertheless, in spite of the differences in geometric shapes and construction techniques, the excavation of a trench panel is also a type of excavation. The shape of surface settlement is basically similar to that induced by main excavation.

As shown in Figure 6.1, the maximum settlement induced by the construction of diaphragm walls is in the range of 0.13% H_t and 0.15% H_t where H_t is the depth of the trench (Clough and O'Rourke, 1990; Ou and Yang, 2011). Settlement became less observable beyond the distance of 1.5–2 H_t from the diaphragm wall. The envelopes shown in Figure 6.1 can be used as a first estimation of the settlement induced by the construction of diaphragm walls under normal conditions. If the construction of a single diaphragm

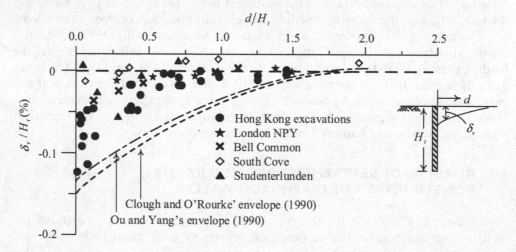

Figure 6.1 Envelopes of ground surface settlements induced by the construction of diaphragm wall (Clough and O'Rourke, 1990; Ou and Yang, 2011).

wall panel takes more than 1 day, for example, for a larger thickness and depth of a diaphragm wall, soil creep may become severer. The settlement beyond the envelopes in Figure 6.1 would be expected. It certainly needs more studies for such conditions.

6.3 CHARACTERISTICS OF WALL MOVEMENT INDUCED BY EXCAVATION

The magnitude of wall movement is related to the excavation-induced unbalanced forces, the stiffness of the retaining-strut system, the excavation stability, etc. The unbalanced forces are synthetic results of many factors such as the excavation width, the excavation depth, and the preload. The relations of these factors with the deformation of a retaining wall can be inferred logically. For example, the thicker the retaining wall, the narrower and the shallower the excavation, the stronger the strut stiffness, the larger the preload, and the greater the safety factor of stability, the smaller the wall deformation. Some factors have complicated relations with the deformation, and this section is going to explore them in detail.

6.3.1 Safety factors of stability

Based on the finite element parametric study results, Clough and O'Rourke (1990) developed a relationship between the maximum wall deflection (δ_{hm}) and system stiffness (S_w) for various factors of safety (F_b) against basal heave (base shear failure) as shown in Figure 6.2 where $S_w = EI/(\gamma_w h^4)$, EI= wall stiffness, h=average vertical spacing of struts, and F_b is Terzaghi's factor of safety against basal heave or base shear failure (Eq. 5.9).

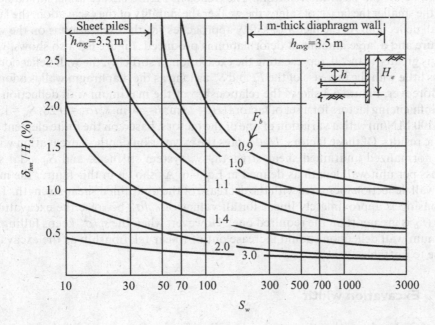

Figure 6.2 Relationship between the maximum wall deflection, system stiffness, and factors of safety against base shear failure or basal heave (*EI* denotes wall stiffness, γ_w denotes unit weight of water, $EI/(\gamma_w h^4)$ represents system stiffness) (Clough and O'Rourke, 1990).

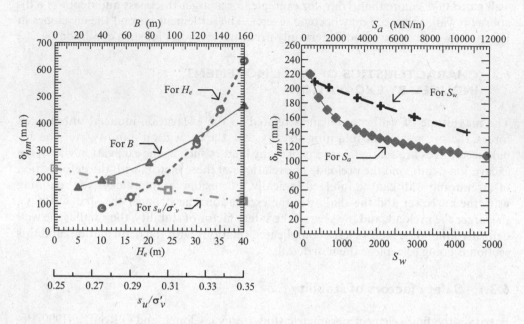

Figure 6.3 Variation of the maximum wall deflection with influencing factors.

As discussed in Section 5.5.4, Terzaghi's F_b does not take the influence of the penetration depth and stiffness of the retaining wall into account, and it only considers the soil strength below the excavation bottom. The trends in Figure 6.2 are understandable because the smaller the factor of safety, the weaker the stability of the excavation, the larger the wall deflection. As the factor of safety approaches 1.0, the excavation is on the verge of failure and a large amount of deformation is produced. Figure 6.2 also shows, when the F_b is greater than 2.0, representing the excavation in stiff clay, the wall deflection decreases little with the increase of the F_b and S_w as long as the diaphragm wall is adopted.

Moreover, Figure 6.3 shows the relationship of the maximum wall deflection and some influencing factors for a set of factors ($H_e = 22$ m, $B = 40$ m, $s_u/\sigma'_v = 0.28$, $S_w = 1,480$, $S_a = 1,100$ MN/m) with a variation of one of the factors, based on the finite element parametric results. Of those factors, H_e denotes the excavation depth, B excavation width, s_u/σ'_v normalized undrained shear strength, S_w system stiffness, and S_a axial strut stiffness per unit wall length as defined in Eq. 6.6. As shown in this figure, the maximum wall deflection increases with the decrease of the undrained shear strength. Their relationship is approximately linear for all values of s_u/σ'_v because the excavation is stable (F_b is greater than the required one). However, when the s_u/σ'_v keeps falling, the maximum wall deflection should increase out of linear relationship as the excavation is close to unstable.

6.3.2   Excavation width

Figure 6.3 also shows that the maximum wall deflection increases with the increase of the excavation width. Furthermore, Figure 6.4a shows the wall deflections for three different excavation widths computed using the finite element method. The result is

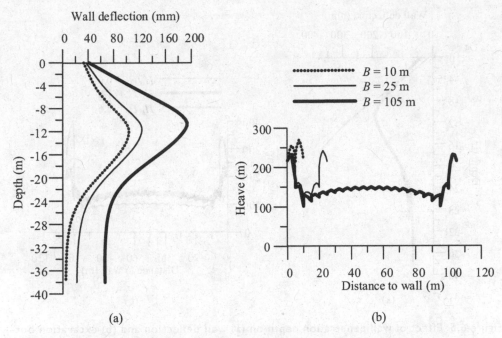

Figure 6.4 Effect of excavation width on (a) wall deflection and (b) excavation bottom movement.

also understandable. The larger the excavation width, the larger the unbalanced forces; the larger the unbalanced forces, the greater the wall deformation.

6.3.3   Excavation depth

As shown in Figure 6.3, the maximum deflection increases with the excavation depth. The relationship is nonlinear because for the same undrained shear strength, when excavation goes deeper, more soil below the excavation bottom is subject to plastic response, leading to the increase of the wall deflection.

6.3.4   Wall penetration depth

Figure 6.5a shows the relationship between computed wall deformations engendered by a 10.5-m-deep excavation and penetration depths for $s_u/\sigma'_v = 0.22$ using the finite element method. As shown in the figure, the wall deformation for $H_p/H_e = 0.6$ is very large, with its the maximum value occurring at the bottom of the wall, implying that the excavation is close to base shear (or basal heave) failure. As H_p/H_e increases to be 1.8, the wall deformation reduces significantly. The factor of safety against base shear failure satisfies the requirement at this state. When H_p/H_e increases to be 2.6, the wall deformation remains unchanged.

We can see from the above discussion that as long as the wall is in a stable state, the growth of the penetration depth does not affect the deformation of the wall.

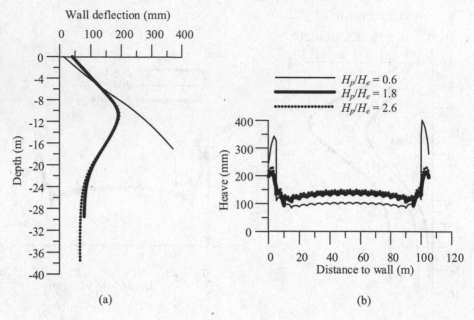

Figure 6.5 Effect of wall penetration depth on (a) wall deflection and (b) excavation bottom movement.

6.3.5 Wall stiffness

Figure 6.2 also shows that, when the F_b is greater than 1.4 with the system stiffness exceeding a certain value, the reduction of the maximum deformation of the retaining wall is very small even with the increase of system stiffness by increasing wall thickness or reducing the vertical strut spacing. Figures 6.6a and b further shows the relationship of wall deflection and wall thickness for $H_p/H_e = 0.6$ and 2.6, respectively ($H_e = 10.5\,\mathrm{m}$, $s_u/\sigma_v' = 0.22$). As shown in Figure 6.6a, when the penetration depth is insufficient, for example, $H_p/H_e = 0.6$, corresponding to $F_b = 0.8$, the excavation is on the verge of failure and the bottom of the wall cannot be restrained so that increase of the wall thickness is unable to reduce the wall deflection. On the other hand, when the penetration depth is sufficient, for example, $H_p/H_e = 2.6$ ($F_b = 1.66$), the bottom of the wall is restrained so that increase of the wall thickness can reduce the wall deflection, but just to a certain extent.

6.3.6 Strut stiffness

Figure 6.7a shows the relationship between computed wall deformations induced by a 10.5-m-deep excavation and penetration depths for $s_u/\sigma_v' = 0.3$ using the finite element method. Str = 100% denotes the "normal" strut axial stiffness, and Str = 250% and 25% represent a 2.5 and 0.25 times the normal strut axial stiffness, respectively. It can be seen from this figure, when the strut stiffness is increased to Str = 250%, the wall deflection decreases slightly, whereas when it is decreased to Str = 0.25, the wall deflection exhibits a great deal of increase.

As shown in Figure 6.8a, with the start of the first stage of excavation, wall movement will be produced and form a cantilever shape. The second stage of excavation

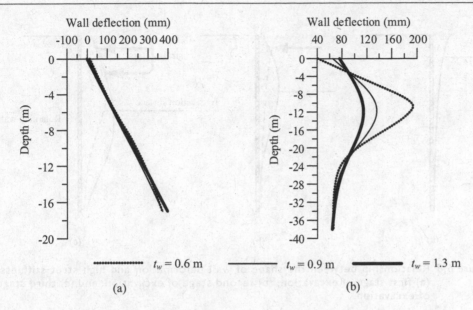

Figure 6.6 Effect of wall thickness on (a) $H_p / H_e = 0.6$ and (b) $H_p / H_e = 2.6$.

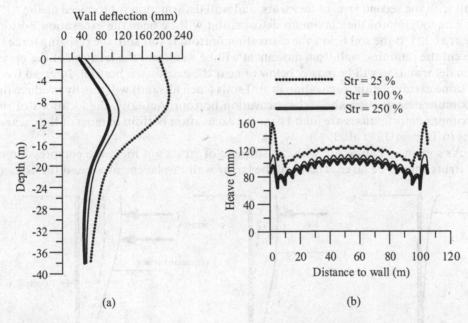

Figure 6.7 Effect of strut stiffness on (a) wall deflection and (b) excavation bottom movement.

starts after the installation of the first level of struts. If the stiffness of the struts is high enough, the compression of the struts will be rather small, so that the retaining wall will rotate about the contact point between the struts and the wall, and wall deformation is thus generated. The maximum wall deformation will occur near the excavation bottom as shown in Figure 6.8b. With the completion of the second level of struts, the third stage of excavation starts. Suppose the stiffness of the second level of

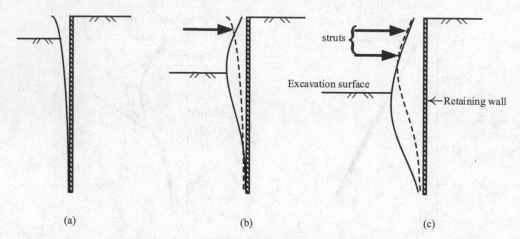

Figure 6.8 Relationship between the shape of wall deformation and high strut stiffness: (a) first stage of excavation, (b) second stage of excavation, and (c) third stage of excavation.

struts is also strong enough. The retaining wall will continue rotating about the contact point with the second level of the struts, and wall deformation is produced again.

The location of the maximum deformation will be near the excavation bottom (Figure 6.8c). If the soil below the excavation bottom is soft soil, the resisting force to prevent the retaining wall from movement will be weak and the location of the maximum deformation will be mostly below or near the excavation bottom. Inferred from the same extrapolation, excavation in stiff soils (such as sand) will mostly produce the maximum deformation above the excavation bottom. Actually, the locations of the maximum deformations are found near the excavation bottom in most of the excavations in Taipei (Ou et al., 1993).

As shown in Figure 6.9, with the stiffness of struts not high, the compression of the struts should be large. There will be larger wall displacement around the contact

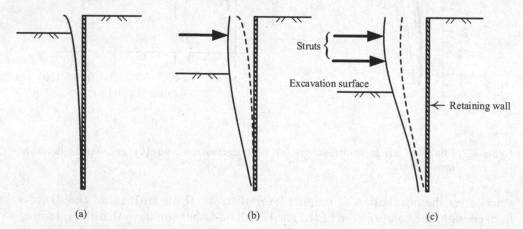

Figure 6.9 Relationship between the shape of wall deformation and low strut stiffness: (a) first stage of excavation, (b) second stage of excavation, and (c) third stage of excavation.

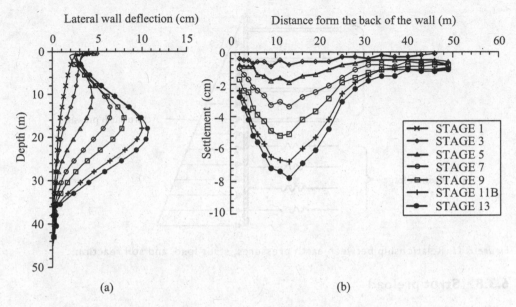

Figure 6.10 Lateral wall deflections and ground surface settlements in the TNEC excavation: (a) wall deflections and (b) surface settlements (Ou et al., 1998).

points during the second and the third stages of excavation. The final deformation pattern of the retaining wall will be close to that of the cantilever type, and the maximum deformation will be produced at the top of the retaining wall.

Figure 6.10 shows the lateral deformation of the retaining wall at each excavation stage in the TNEC excavation (Ou et al., 1998). Since the top-down construction method was employed in this case, the axial stiffness of floor slabs was quite high and the deformation behavior was similar to that shown in Figure 6.8, in which the maximum deformation occurs near the excavation bottom.

6.3.7   Strut spacing

The problem of strut spacing can be distinguished into that of horizontal spacing and vertical spacing. Narrowing the horizontal spacing can increase the stiffness of the struts per unit width. The effect will be the same as discussed in Section 6.3.6, and the problem will not be further explored here.

Shortening the vertical spacing of struts can effectively decrease the deformation of a retaining wall because the stiffness of the strut system is raised. As the system stiffness raised, the deformation of a retaining wall certainly declines (Figure 6.2). Put another way, since the deformation of a retaining wall is the accumulated deformation throughout all the excavation stages, with the unsupported length generated in each stage reduced due to the shortening of the vertical spacing, the deformation of a retaining wall will decline as a result. The "unsupported length" refers to the distance between the lowest level of strut and the excavation bottom.

For the detailed mechanism of the unsupported length affecting excavation deformation, please see Section 11.4.1.

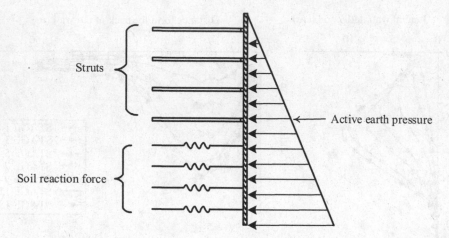

Figure 6.11 Relationship between earth pressures, strut load, and soil reaction.

6.3.8   Strut preload

When applying the braced excavation method (or the anchored excavation method), preload is often exerted onto struts. Suppose the struts are placed at shallower levels. Thus, under normal conditions (with the preload not too small), the preload will be capable of pushing the retaining wall backward. If the struts are placed at deeper levels, with the earth pressure growing with the depth, the preload of struts will not be able to push the wall backward easily (Ou et al., 1998).

Actually, no matter whether the preload is capable of causing a retaining wall to move, preload is always helpful to reduce the displacement of a retaining wall or the surface settlement. The reason for this can be explained by Figure 6.11. As explicated in Section 5.4, the design of a braced excavation is based on the concept of the free earth support method. Therefore, the retaining wall will inevitably move toward the excavation zone once excavation is started and the phenomenon will make the earth pressure on the retaining wall approach the active pressure. Figure 6.11 shows the way the struts and soil resist, collectively, against the active earth pressure acting on the retaining wall. According to the concept of the force equilibrium of the wall, when the struts bear more lateral earth pressure because of preload, the soil below the excavation bottom will then bear less pressure, leading to less wall deformation and surface settlement.

6.4   CHARACTERISTICS OF GROUND MOVEMENT INDUCED BY EXCAVATION

Observing the shapes of wall deflection (Figures 6.8 and 6.9), we can see that soil in back of the retaining wall moves forward and down with the retaining wall deforming under normal conditions and surface settlement will thus be produced. Thus, the factors causing wall deformation will also influence surface settlement. This section will not introduce those influence factors repeatedly (i.e., the stiffness of struts, the preload, the factor of safety of stability, the excavation depth, and the excavation width). This section will focus on the characteristics of surface settlement.

6.4.1   Shapes and types of surface settlement

The author found that the shapes or types of the surface settlement engendered by excavation can be categorized into the spandrel type and the concave type, as shown in Figure 6.12 (Hsieh and Ou, 1998). The main factors responsible for these two types of surface settlements are the magnitude and shape of deformation of a retaining wall.

If the first stage of excavation has generated more deflection of the retaining wall than the later excavation stages or excavation has continued to produce the cantilever type of deflection in the later stages, the spandrel type of settlement will be more likely to occur and the maximum surface settlement will be found near the retaining wall. If the wall has a deep inward movement as shown in Figure 6.8, the maximum surface settlement will be found located at a distance in back of the wall, i.e., a concave type of settlement will be produced.

Under normal construction conditions, excavation in soft clay will produce more deflection of the retaining wall and tend to bring about deep inward movement, which likely leads to the concave type of settlement. Excavation in sandy grounds or stiff clay, on the other hand, will produce less deformation of the retaining wall, and the spandrel type of settlement may be produced (Clough and O'Rourke, 1990).

As discussed above, the shape of surface settlement is supposed to correlate with the area of the cantilever component and the area of the deep inward component of the lateral wall movement. To predict the type of surface settlement based on the shape of lateral wall movement, the author defined A_c as the area of the cantilever component and A_s as the area of the total wall movement subtracting the area of the cantilever component (Hsieh and Ou, 1998), as shown in Figure 6.13. A_c is determined as follows:

$$A_c = \max(A_{c1}, A_{c2}) \qquad (6.1)$$

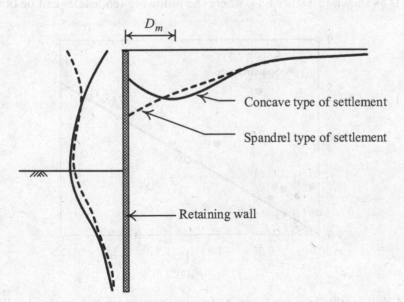

Figure 6.12 Types of surface settlement.

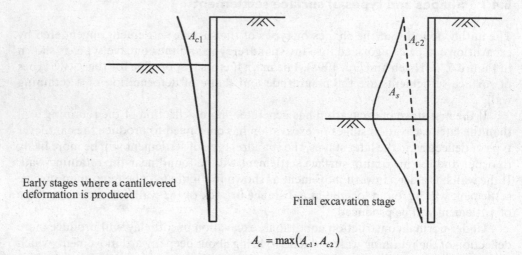

$$A_c = \max(A_{c1}, A_{c2})$$

Figure 6.13 Definition of the area of the deep inward part and the cantilever part of wall deformation.

where A_{c1} = area of the cantilever deformation of the wall at the beginning of excavation
A_{c2} = area of the cantilever component of the lateral wall deformation at the final stage of excavation

To study the relationship between the shape of the lateral wall displacement and that of surface settlement, the author took the monitoring results of 16 surface settlement cases of the concave type and seven of the spandrel type to compute their A_c and A_s and established the relationships between the type of surface settlement and the shape of the lateral wall movement. The relationship between A_c and A_s of the 16 case histories is as shown in Figure 6.14, where the following tendencies can be observed:

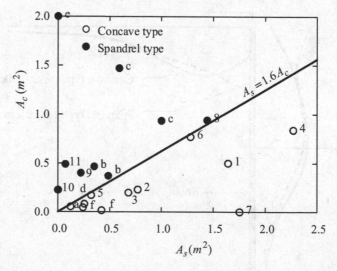

Figure 6.14 Relationship between the type of surface settlement and shapes of wall deflection (English alphabets labels refer to excavation cases from other countries, while Arabic numbers labels to cases from Taiwan) (Hsieh and Ou, 1998).

When A_s is greater than 1.6 A_c, the shape of surface settlement tends to be the concave type. Otherwise, the spandrel type.

Therefore, to predict the type of surface settlement, we can refer to the relationship between A_c and A_s. When $A_s < 1.6A_c$, we can predict that the surface settlement is of the spandrel type. When $A_s \geq 1.6A_c$, we can predict that the surface settlement is of the concave type.

6.4.2   Influence zones of settlement

Peck (1969) proposed that the influence zone of settlement should be 2 or 3 times the excavation depth. Clough and O'Rourke (1990) proposed that excavation in sandy soils may induce an influence zone of settlement about twice the excavation depth. As for stiff to very stiff clay, three times the excavation depth. Soft or medium soft clay, twice the excavation depth. In addition to the above, there are various other suggested values of the influence zone (Nicholson, 1987, for example). However, most of them are lacking in formal definitions of influence zone and the excavation depth is the only parameter in estimating the influence zone. Nevertheless, according to the author's numerical analyses and studies of the characteristics of surface settlement from excavation case histories, the influence zone of settlement does not relate exclusively to the excavation depth. It also relates to the excavation width, the location of the hard soil, etc.

As discussed earlier, the types of settlement induced by excavation include the spandrel type and the concave type. The author (Hsieh and Ou, 1998) has proposed the conception of the primary influence zone (PIZ) and the secondary influence zone (SIZ) on the basis of the principles of mechanics and regression analysis of excavation case histories. We claim that the influence zone may extend very far and the distribution of the settlement curve includes the PIZ and the SIZ no matter whether the settlement belongs to the spandrel type or the concave type. The curve is steeper in the PIZ where buildings receive more influence. In the SIZ, the slope of the curve is gentler and the influence on buildings is less. The range of the SIZ is approximately equal to that of the PIZ. Settlement might still happen beyond the SIZ, but its magnitude is so gentle that is beyond perception or evenly distributed. Under normal conditions, its influence on buildings is ignorable.

Take the surface settlement of the TNEC excavation, as shown in Figure 6.10b, as an example to illustrate the characteristics of influence zones.

1. As soon as excavation was started, settlement began occurring and its influence zone was large. The excavation depth of the third stage was 4.9 m, while the influence zone, for settlement, reached as far as 50 m away from the retaining wall. At this stage, no distinction between the PIZ and the SIZ was discerned.
2. At the fifth stage of excavation (excavation depth = 8.6 m), the PIZ, which was about as far as 32 m away from the retaining wall, began to be distinguishable from the settlement trough.
3. After the fifth stage, though the excavation depth kept growing, the range of the PIZ was not enlarged accordingly.

According to the loading behavior of soil, strain in soil will increase greatly when soil is going to yield or fail. The movement or strain within the PIZ is rather large, and we can reasonably assume that the PIZ is the potential failure zone. Excavation in soft

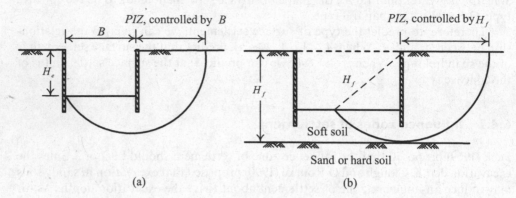

Figure 6.15 Primary influence zone produced by the failure surface of potential basal heave or base shear.

clay can induce base shear failure or basal heave. As shown in Figure 6.15, the potential basal heave failure surface can be the one with the wall top as axis and the excavation width as a radius (B in Figure 6.15). If the formation of the potential basal heave failure surface is not restricted by sandy or hard soil, the range of the potential basal heave failure in back of the wall will be close to the excavation width. If the sandy or the hard soil is located shallower, the potential basal heave failure surface will then be tangential to the sandy or hard soil and the potential basal heave failure range in back of the wall will be close to the distance to the soft clay bottom from the ground surface. Thus, the potential failure zone based on the basal heave failure mode can be determined as follows (Ou and Hsieh, 2011):

$$PIZ_1 = \min\left(H_f, B\right) \tag{6.2}$$

where
   H_f = depth of the soft clay bottom
   B = excavation width

Since the design of the strutted wall is based on the conception of the free earth support method, not only the bottom of the retaining wall has movement but also the soil below the wall bottom may exhibit movement. If we suppose that the movement of the soil is not restricted (i.e., hard soils are very far underneath), the formation of the active failure in back of the retaining wall will not be obstructed and the active failure zone will be about twice as wide as the excavation depth ($2H_e$). When the hard soil is shallow enough to restrict the movement of the soil, the range of the active failure zone will be about the same as the depth of the hard soil. Under such conditions, the potential failure zone based on the active failure zone can be determined as follows (Figure 6.16):

$$PIZ_2 = \min\left(2H_e, H_g\right) \tag{6.3}$$

where H_g = depth of the hard soil

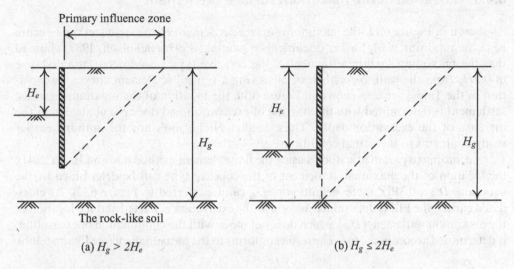

(a) $H_g > 2H_e$ (b) $H_g \leq 2H_e$

Figure 6.16 Primary influence zone produced by active failure surface.

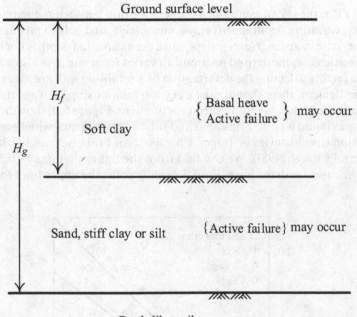

Figure 6.17 The difference in H_f and H_g.

Figure 6.17 further explains the difference in H_f and H_g. Both PIZ$_1$ and PIZ$_2$ are potential failure zones. The PIZ of settlement is the maximum of the potential failure zones. Thus, the PIZ is the larger of PIZ$_1$ or PIZ$_2$:

$$PIZ = \max(PIZ_1, PIZ_2) \tag{6.4}$$

6.4.3 Location of the maximum surface settlement

As shown in Figure 6.12, the maximum surface settlement of the spandrel type occurs near the retaining wall. Earlier documents (Ou et al., 1993; Nicholson, 1987) claimed that the maximum surface settlement of the concave type would occur at a distance of $0.5H_e$ from the wall. Nevertheless, observing a typical settlement curve of excavation in the Taipei area, as shown in Figure 6.10, the location of the maximum surface settlement is determined with the starting of excavation and does not change with the increase of the excavation depth. Thus, neither Nicholson's nor the author's earlier study conforms to the actual conditions.

According to parametric studies using the finite element method (Ou and Hsieh, 2011), the location of the maximum settlement of the concave type can be determined by the equation: $D_m = 0.3PIZ$ (note: definition of D_m can be referred to Figure 6.12). As elucidated earlier, the PIZ is determined as soon as the excavation is started and the location of the maximum settlement (D_m), which does not move with the continuation of excavation, is determined accordingly, too. The result conforms to the measurement (see Figure 6.10b).

6.4.4 Magnitude of the maximum surface settlement

Clough and O'Rourke (1990) established the relationship between the maximum settlement and the excavation depth in stiff clays, sandy soils, and soft to medium soft clays, based on many case histories. Nevertheless, since the excavation depth is not the only factor affecting settlement, the derived relationship varies from one case history to another.

Since the factors affecting the deformation of a retaining wall are also those affecting surface settlement, there should exist a certain relationship between the maximum wall deformation and the maximum surface settlement. Figure 6.18 shows relationships between the maximum wall deformations and surface settlements, which were obtained from excavation case histories in Taipei, Chicago, San Francisco, and Oslo (Ou et al., 1993; Mana and Clough, 1981). We can find from the figure that $\delta_{vm} = (0.5 \sim 0.75)\delta_{hm}$ for most of the cases, with the lower limit for sandy soils, the upper limit for clays, and

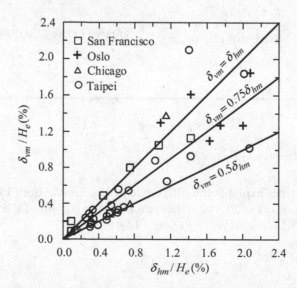

Figure 6.18 Maximum surface settlement and wall deflection (Ou et al., 1993).

somewhere between the two for alternating soils of sand and clay. For very soft soils, δ_{vm} might run beyond $1.0\,\delta_{hm}$.

Thus, to estimate the maximum surface settlement, we can follow the beam on elastic foundation method or empirical formulas to estimate the maximum value of wall deflection and then estimate the maximum surface settlement with the help of Figure 6.18.

6.4.5 Relationship between surface settlements and soil movements

The magnitude and distribution of settlements introduced so far refer to those of the ground surface. Actually, most foundations are embedded at a certain depth below the ground surface. How to estimate the settlement of the foundations will be something interesting that remains to be resolved. Due to the absence of reliable study results, some engineers assume that the maximum settlement occurs on the ground surface and the settlement decreases with growing depth.

Figure 6.19 shows the measurement values of the displacement vectors of soil below the ground surface and in back of the retaining wall at the last stage in the TNEC

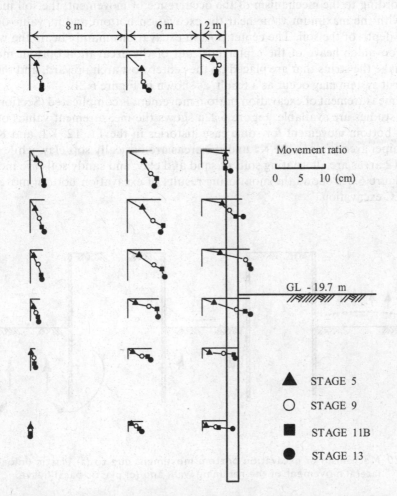

Movement ratio

0   5   10 (cm)

GL - 19.7 m

▲ STAGE 5

○ STAGE 9

■ STAGE 11B

● STAGE 13

Figure 6.19 Displacement vectors below the ground surface in the TNEC excavation (Ou et al., 2000).

excavation (Ou et al., 2000). We can see from the figure that the soil settles downward and moves toward the excavation zone. The soil near the excavation bottom moves toward the excavation zone (horizontally) more than moving down (vertically). The vertical displacements of the soil above the excavation bottom are basically close. On the other hand, the vertical displacements of the soil below the excavation bottom decrease with the increase of the depth from the excavation bottom. Data provided in Figure 6.19 can be used to estimate the settlement of embedded foundations.

6.5   CHARACTERISTICS OF EXCAVATION BOTTOM MOVEMENT INDUCED BY EXCAVATION

The excavation bottom movement may come from elastic unloading, the lateral displacement of the embedded part of the retaining wall, or the plastic deformation of the soil below the excavation bottom (Figure 6.20). For wide excavations, the vertical movement of the soil near the retaining wall is larger than that at the center because the soil is pushed directly by the embedded wall, as shown in Figures 6.4b, 6.5b, and 6.7b.

According to the mechanism of the occurrence of movement, the soil movement occurs with the maximum value near the excavation bottom and its value decreases with the depth of the soil. The center post in excavations mainly bears the weight of struts. Too much heave of the center post out of the excavation bottom movement would make the struts that are placed on the center post arch upward, and the failure of the strut system may occur as a result, as shown in Figure 6.21.

The measurement of excavation bottom movement is complicated (Section 12.4.4), and few studies are available. Figure 6.22a shows the measurement values of the excavation bottom movement for some case histories in the T1, T2, K1, and K2 areas of the Taipei area (soils in the K1 and K2 areas are basically soft clay, while those in T1 and T2 areas are alternating soils of sand and clay, and sandy soils are more common). Figure 6.22b shows the monitoring results of excavation bottom movement in the TNEC excavation.

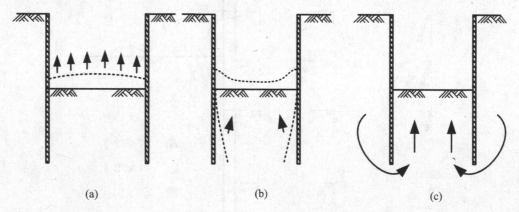

(a)   (b)   (c)

Figure 6.20 Mechanism of excavation bottom movement due to (a) elastic unloading, (b) lateral movement of the retaining wall, and (c) plastic basal heave.

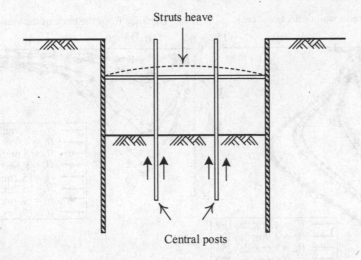

Struts heave

Central posts

Figure 6.21 Influence of the heave of the central post on the strut system.

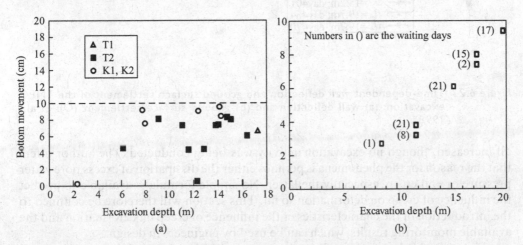

Figure 6.22 Relations between excavation bottom heaves and excavation depths: (a) case histories in Taipei (Woo and Moh, 1990) and (b) TNEC excavation (Ou et al., 1998).

6.6 TIME-DEPENDENT MOVEMENT

The top-down construction method requires long periods to erect molds, cast floor slabs, and wait for them to gain enough strength before proceeding to the next stage of excavation. Sometimes, for the convenience of construction, the next stage of excavation is not to be started until the construction of the superstructures has come to a scheduled stage. Take the TNEC excavation, for example; each stage took a waiting period of 30–60 days (the waiting time refers to the period between the completion of an excavation stage and the start of the next). As the field measurements showed (Figures 6.22b and 6.23a, b), during the waiting periods, the lateral displacement of the retaining wall, the surface settlement, and the movement of the excavation bottom

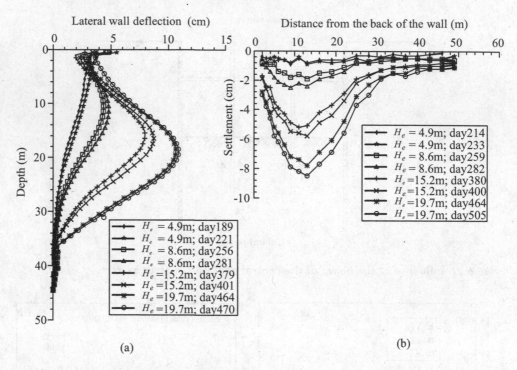

Figure 6.23 Time-dependent wall deflection and ground surface settlement of the TNEC excavation: (a) wall deflection and (b) ground surface settlement (Ou et al., 1998).

all increased, though no excavation activity was being conducted. The author infers that the reason for the phenomena is perhaps either the dissipation of excess pore water pressure or soil creep, especially the latter. No simple formula is available to predict the influence of creep on deformation so far. This section will therefore be confined to the introduction of the characteristics of the influence of creep on deformation and the available monitoring results, which can be used by engineers in design.

Figure 6.23a shows a significant increase in displacement of the retaining wall during waiting periods. If we take the excavation depth of 8.6 m, for example, the period between the 256th day and the 281st day of construction, the maximum deflection increases from the original 4.40 to 4.81 cm. Figure 6.24 shows the relationship between the maximum wall deflection rates $(\Delta\delta / \Delta t)_{max}$ and the excavation depths. The maximum wall deflection rate is defined as the maximum wall deflection increment divided by the waiting time. We can see from the figure that the deeper the excavation the higher the deflection rate. Besides, inclinometer casings were also installed in the soil outside the excavation zone and the monitoring results showed that the characteristics of the lateral deflection rate of the soil were similar to those of the retaining wall. The farther the soil is from the retaining wall, the lower the lateral deflection rate is. The reason may be that the stress level grows smaller with the increase in distance from the retaining wall (the stress level is the ratio of the shear stress in soil to the shear strength; at failure, the stress level equals 1.0). The lower the stress level is, the less obvious the creep behavior is. Ou et al. (1998) found from case studies of the TNEC

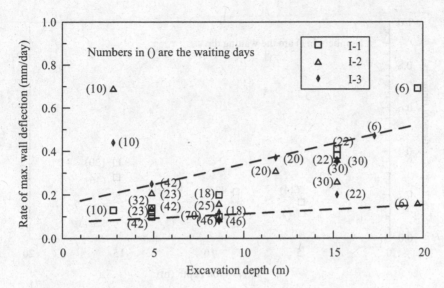

Figure 6.24 Relationship between the rates of maximum wall deflection and excavation depths in the TNEC excavation (Ou et al., 1998).

excavation that the deflection of the wall generated during the waiting periods added up to 30%–35% of the total deflection, a fairly high percentage.

Mana and Clough (1981) found that when the factor of safety against basal heave is small, creep often happens, based on case studies of excavations in Chicago and San Francisco. Their study also pointed out that the deflection rate of walls was about 0.3–30 mm/day in these cases, which is far larger than the rate at the TNEC excavation. The reason for the significant difference should have a lot to do with the stress level of the soil. As far as the objects of Mana and Clough's study are concerned, the steel sheet pile was employed as the retaining system. Though the excavation depths were relatively small, between 9.1 and 13.5 m, the lateral deflections of the retaining walls were all so large that the soil in the vicinity of the excavation zone could be seen to be close to the yield. Thus, the stress levels were much larger than those in the TNEC excavation, and thereby, higher deflection rates were produced.

Figure 6.23b shows the settlements during the waiting periods in the various excavation stages. We can see that the settlement grew with the increase of waiting time for a certain excavation depth. Their tendencies were similar to the deformation behaviors of the retaining wall. Figure 6.25 shows the relationship between the rate of surface settlement ($\Delta s/\Delta t$) of the soil 13 m from the retaining wall and the excavation depth. The settlement rate refers to the change of settlement during the waiting time divided by the waiting time. As shown in the figure, the settlement rate ($\Delta s/\Delta t$) grows larger with the increase of excavation depth. Ou et al.'s study also found that the settlement produced during the waiting periods added up to 43% of the total settlement, which is a significant proportion.

Figure 6.22b also shows that the magnitude of excavation bottom movement increases with waiting time. If we judge from the causes of soil movement, we can tell that the rate of creep-induced movement increases with the increase of excavation depth and decreases with the increase of the safety factor against basal heave failure.

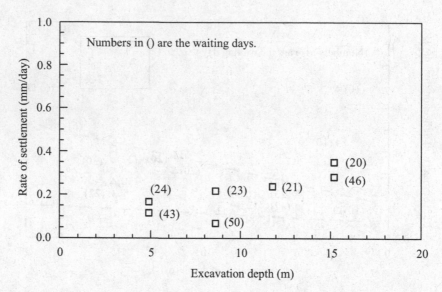

Figure 6.25 Relationship between the rates of ground surface settlement and excavation depths in the TNEC excavation (Ou et al., 1998).

6.7 ANALYSIS OF WALL MOVEMENTS INDUCED BY EXCAVATION

As discussed in Section 6.3, in addition to soil factors, factors influencing the deformation of a retaining wall include the excavation width, the excavation depth, the safety factor of stability, the penetration depth of a retaining wall, the strut stiffness, and the strut preload. Thus, Clough and O'Rourke's design chart (1990), as shown in Figure 6.2, can be used for the estimation of the maximum wall deformation for excavations in soft to medium soft clay.

Ou et al. (1993) studied the deformation of excavations in Taipei and found that the deformation of a retaining wall in soft clay is generally greater than that in sand. According to their study, the maximum deformation (δ_{hm}) can be estimated by the following equation:

$$\delta_{hm} = (0.2\% - 0.5\%)H_e \tag{6.5}$$

where H_e = excavation depth.

The upper limit in Eq. 6.5 is recommended for soft clay, while the lower limit should be used for sand. An averaged value is suggested for alternating layers of sand and clay. It is notable that both Clough and O'Rourke's design chart (Figure 6.2) and Eq. 6.5 are only suitable for the excavation width around 40 m.

Based on a series of finite element parametric studies, the following equation was derived to predict the maximum wall deflection in clay (Ou and Chen, 2015):

$$\delta_{hm} = \exp(X_m) \tag{6.6}$$

$$X_m = 0.0678H_e + 0.0076B - 5.2803\exp(s_u / \sigma_v') - 0.000099S_w$$
$$- 0.2316\ln(S_a) + 5.2828 \tag{6.6a}$$

where
   δ_{hm} = the maximum wall deflection under the plane strain condition (unit: m)
   B = excavation width (unit: m)
   s_u / σ_v' = normalized undrained shear strength
   S_a = axial strut stiffness per unit wall length (unit: MN/m), $S_a = A_{st} E_{st} / s$ where A_{st}
      is the average cross-sectional area for all levels of struts, E_{st} is Young's modulus of struts, and s is the horizontal spacing of struts.

6.8   ANALYSIS OF SURFACE SETTLEMENTS INDUCED BY EXCAVATION

This section will introduce empirical formulas to predict surface settlement and the characteristics of soil movement. Though many empirical formulas have been proposed, only four of the most well-known ones among them are to be discussed.

6.8.1   Peck's method

Peck (1969) was the first to propose a method to predict excavation-induced surface settlement, based on field observations. He mainly employed the monitoring results of case histories in Chicago and Oslo and established the relation curves between the surface settlement (δ_v) and the distance from the wall (d) for different types of soil, as shown in Figure 6.26. The method classifies soil into three types according to the characteristics of soil:

Type I: sand and soft to stiff clay, average workmanship
Type II: very soft to soft clay

   1.   Limited depth of clay below the excavation bottom
   2.   Significant depth of clay below the excavation bottom but $N_b < N_{cb}$.

Type III: very soft to soft clay to a significant depth below the excavation bottom and $N_b \geq N_{cb}$.

where N_b, the stability number of soil, is defined as $\gamma H_e / s_u$, where γ is the unit weight of the soil, H_e is the excavation depth, and s_u is the undrained shear strength of soil, and N_{cb} is the critical stability number against basal heave.

   Since Peck's method took the monitoring results of case histories before 1969, most of which employed steel sheet piles or soldier piles with laggings as the retaining wall, quite different from the more advanced design and construction methods (for example, the diaphragm wall that offers higher stiffness) employed in excavation projects in recent years, the relation curves proposed by Peck are not necessarily applicable to all excavations.

   Basically, the curves derived from Peck's method are envelopes.

6.8.2   Bowles' method

Bowles (1988) proposed a procedure to estimate excavation-induced surface settlements, which can be described as follows (Figure 6.27):

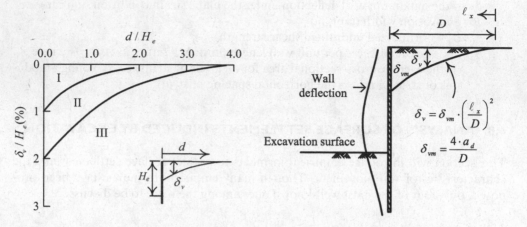

Figure 6.26 Peck's method (1969) for estimating surface settlement.

Figure 6.27 Bowles' method (1986) for estimating
surface settlement.

1. Compute the lateral displacement of the wall using the finite element method or
    the beam on elastic foundation method.
2. Compute the area of the lateral wall deflection (a_d)
3. Estimate the influence range of surface settlement (D) following Caspe's method
    (1966):

$$D = (H_e + H_d)\tan\left(45° - \frac{\phi}{2}\right) \tag{6.7}$$

where H_e = the excavation depth; $H_d = B$ if $\phi = 0$ and $H_d = 0.5B\tan(45° + \phi/2)$ if
$\phi \geq 0$, where B = the excavation width and ϕ = the parameter of soil strength.

4. Suppose the maximum surface settlement is located at the intersection of the wall
    and ground surface. Estimate the maximum surface settlement (δ_{vm}):

$$\delta_{vm} = \frac{4a_d}{D} \tag{6.8}$$

5. Suppose the surface settlement exhibits parabolic distribution. The settlement (δ_v)
    at ℓ_x can be computed as follows:

$$\delta_v = \delta_{vm}\left(\frac{\ell_x}{D}\right)^2 \tag{6.9}$$

where ℓ_x = distance from a point at the distance of D from the wall and δ_v = the
settlement at the distance of ℓ_x.

Theoretically, excavating in undrained saturated soft soils, the area of lateral wall
displacement should be about equal to that of surface settlement (Milligan, 1983).

Thus, δ_{vm} should equal $3a_d/D$ instead of $4a_d/D$. The value of δ_{vm} following Bowles' method is 1.33 times larger than that derived from the theoretical derivation. Bowles did not explain the reasons for using $4a_d/D$ rather than $3a_d/D$.

The method proposed by Bowles is obviously only applicable for the spandrel type of surface settlement.

6.8.3 Clough and O'Rourke's method

Clough and O'Rourke (1990) proposed various types of envelopes of excavation-induced surface settlements for different soils on the basis of case studies. According to their studies, excavation in sand or stiff clay will tend to produce triangular surface settlement. The maximum settlement will be found near the retaining wall. The envelopes of surface settlement are as shown in Figure 6.28a and b, whose influence ranges are separately $2H_e$ and $3H_e$ where H_e is the final excavation depth. Excavation in soft to medium clay will produce a trapezoidal envelope of surface settlement, as shown in Figure 6.28c. The maximum surface settlement occurs in the range of $0 \le d/H_e \le 0.75$, while $0.75 \le d/H_e \le 2.0$ is the transition zone where settlement decreases from the largest to almost none. Basically, the curves in Figure 6.28 are also envelopes.

6.8.4 Ou and Hsieh's method

Ou and Hsieh (2011) developed a method to predict the surface settlement on the basis of studies of the type of surface settlement, the influence zone, the location of the

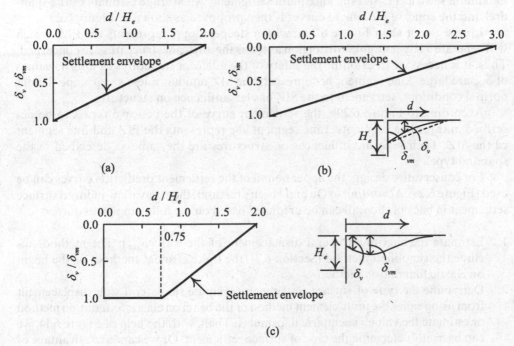

Figure 6.28 Clough and O'Rourke's method (1990) for estimating surface settlement: (a) sand, (b) stiff to very stiff clay, and (c) soft to medium soft clay.

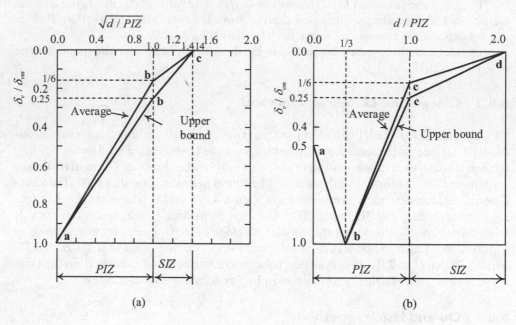

Figure 6.29 Ou and Hsieh's method (2011) for estimating surface settlement.

maximum settlement, and the maximum settlement. An average estimate of the spandrel and the concave settlement curves is then proposed as shown in Figure 6.29.

Line segment **ab** in Figure 6.29a, with a steeper slope, represents the PIZ, which will generate a larger angular distortion as far as the adjacent structures are concerned. Thus, it is necessary to examine the safety of the adjacent structures as long as values of δ_{vm} are large. Line segment **bc** represents the SIZ and has a less steep slope. Under normal conditions, settlement in the SIZ has less influence on structures.

According to Figure 6.29b, the settlement curve of the concave type can be described in three line segments. Line segment **abc** represents the PIZ and line segment **cd** the SIZ. Their separate influences on structures are the same as described in the spandrel type.

For conservative design, the upper bound of the settlement prediction curves can be used (Figure 6.29). According to Ou and Hsieh's method, the excavation-induced surface settlement in back of the wall can be predicted based on the following procedure:

1. Estimate the maximum lateral displacement of the wall (δ_{hm}): The methods include the simplified method (Section 6.7), the finite element method, or the beam on elastic foundation method.
2. Determine the type of surface settlement: Compute the lateral wall displacement from using either the finite element method or the beam on elastic foundation method or estimate the value via empirical formulas. Then, with the help of Figure 6.14, we can basically determine the type of surface settlement. Or we can take advantage of the monitoring results of settlement at the initial excavation stage because the type of settlement profile emerging at the initial stages often lasts until the final stage.

3. Estimate the value of δ_{vm} on the basis of the relations between the maximum settlement (δ_{vm}) and the maximum lateral displacement (δ_{hm}) as shown in Figure 6.18.
4. According to the type of surface settlement determined at the second step, compute various settlements occurring in different positions in back of the wall.

6.8.5   Comparison of the various analysis methods

This section will compare the predicted surface settlement profile and angular distortions using Peck's method, Bowles' method, Clough and O'Rourke's method, and Ou and Hsieh's method with both an average estimate and the upper bound, respectively, with the measurement data of three excavation case histories. For the convenience of comparison, both Ou and Hsieh's and Clough and O'Rourke's methods adopt the measurement value for δ_{vm}. For Bowles' method, its own formula to estimate the δ_{vm}-value is followed. The settlement profile in Peck's method will be computed adopting the boundary curve between Zone I and Zone II (the 1%-curve).

The first case history is the TNEC excavation. The excavation was 40 m wide. The retaining wall was a 35-m-deep and 90-cm-thick diaphragm wall. The final excavation depth was 19.7 m. The basement was constructed using the top-down construction method, which included 16 construction stages (note: seven excavation stages). For the soil properties, the construction process, and the monitoring results of the TNEC excavation, please see the index in Appendix B.

Concerning the potential base shear or basal heave, the depth of the bottom of the soft clay $\left(H_f\right)$ is 33 m, $B = 40$ m, $PIZ_1 = 33$ m. Considering the active failure condition, $2H_e = 39.4$ m, and treat the cobble-gravel soil as a hard soil, $H_g = 46$ m, $PIZ_2 = 39.4$ m. Following Ou and Hsieh's method, we have $PIZ = 39.4$ m. The location of the maximum settlement $D_m = 0.3PIZ = 11.8$ m. Figure 6.30a shows that the settlement profile computed from Ou and Hsieh's method with an average estimate satisfactorily conforms to the field measurement. The settlement profile derived from Clough and O'Rourke's method can also lead to a satisfactory settlement envelope as far as the PIZ is concerned, though the SIZ is obviously ignored.

The second case history is located in the K1 zone of the Taipei Basin. The excavation was 60 m long and 35 m wide. The diaphragm wall was 31 m deep and 80 cm thick. The final excavation depth was 18.45 m. The braced excavation method was adopted with seven excavation stages. The locations of the temporary steel struts and the soil profile are as shown in Figure 6.30b, according to which we have the following data: $B = 35$ m, $H_f = 31$ m, $PIZ_1 = 31$ m. $2H_e = 36.9$ m, $H_g = 31$ m, $PIZ_2 = 31$ m. Thus, we can conclude $PIZ = 31$ m. $D_m = 0.3PIZ = 9.3$ m.

Figure 6.30b also shows the comparison between the measured settlement profile and that computed, respectively, from Clough and O'Rourke's method and Ou and Hsieh's method with an average estimate. As shown in the figure, the settlement profile derived from Ou and Hsieh's method with an average estimate is close to the field measurement. The result computed using Clough and O'Rourke's method does not satisfactorily correspond to the field measurement: Only the settlement in the PIZ is enveloped.

The third excavation case is also located in the K1 zone of the Taipei Basin. The excavation, basically a trapezoid 133–136.5 m long and 63.8–88.15 m wide, was about

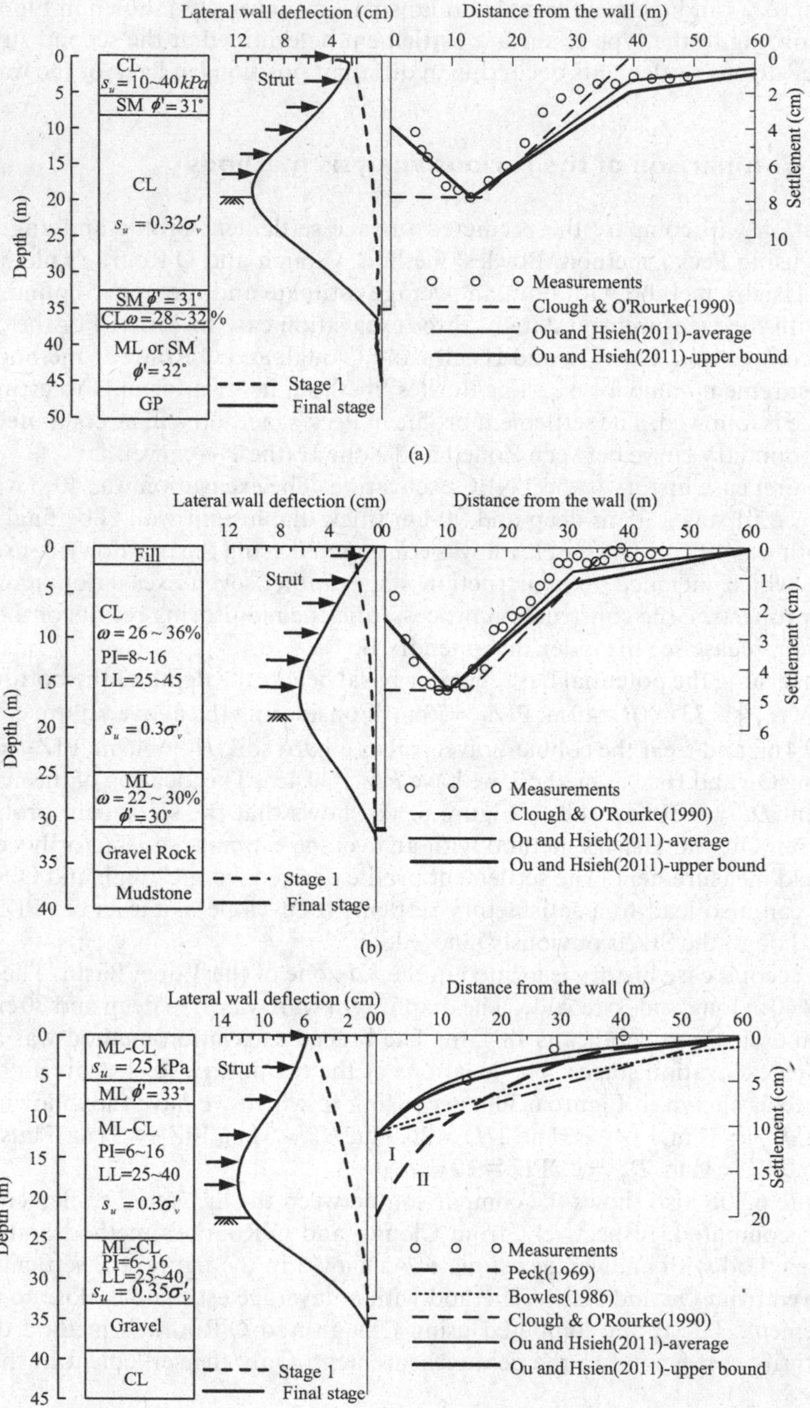

Figure 6.30 Comparisons of predicted and observed surface settlements: (a) Case I: the excavation of TNEC, (b) Case II: the excavation of a building, and (c) Case III: the excavation of a building (modified after Ou and Hsieh (2011)).

10,000 square meters. The diaphragm wall was 32 m deep and 70 cm thick with the final excavation depth of 20.0 m. The top-down construction method was adopted. The locations of the lateral supports (i.e., concrete floor slabs) and the soil profile are as shown in Figure 6.30c.

We can obtain the following data from Figure 6.30c: $B = 64$ m, $H_f = 32$ m, $PIZ_1 = 32$ m. $2H_e = 40$ m, $H_g = 32$ m, $PIZ_2 = 32$ m. Thus, $PIZ = 32$ m. Figure 6.30c also shows the comparison between the settlement profile computed separately from the four prediction methods and field measurement. As shown in the figure, Peck's method, represented by the 1%-*curve*, obviously overestimates the settlements. Bowles' method is good in estimating the maximum settlement δ_{vm} but overestimates the settlements in back of the wall and the influence range significantly. Clough and O'Rourke can hardly predict the actual conditions of settlement. The settlement profile derived from Ou and Hsieh's method with an average estimate satisfactorily conforms to the field monitoring results.

As shown in Figure 6.30, Ou and Hsieh's method with the upper bound is conservative in predicting the settlements.

Angular distortion is the main factor causing damage of structures when ground settles. Thus, this section will compute the angular distortion on the basis of the monitoring results of the above three case histories and compare the results with the computed values of Ou and Hsieh's method with an average estimate and other empirical formulas to examine their applicability.

As shown in Figure 6.31, the definition of the angular distortion (β_{12}) between the footings F_1 and F_2 induced by settlement is given as follows:

$$\beta_{12} = \frac{\delta_{12}}{L_{12}} \tag{6.10}$$

where δ_{12} is the differential settlement of the two footings and L_{12} is the distance between them.

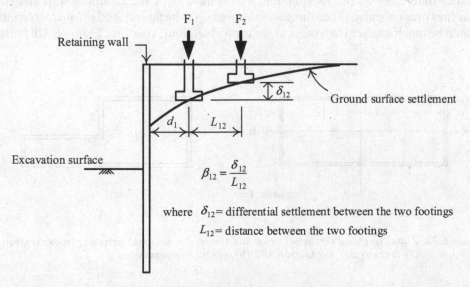

$$\beta_{12} = \frac{\delta_{12}}{L_{12}}$$

where δ_{12} = differential settlement between the two footings

L_{12} = distance between the two footings

Figure 6.31 Angular distortions of footings near an excavation.

Table 6.1 Comparisons of the Predicted and Observed Angular Distortions at the Final Excavation Stage of the Case

Case	Observation & prediction method	d_1/H_e			
		0.0	0.5	1.0	1.5
Case I	Observation	1/200	1/5,000	1/300	1/750
	Clough and O'Rourke	0	1/7,350	1/320	1/320
	Ou and Hsieh	1/300	1/4,100	1/400	1/400
Case II	Observation	1/150	1/870	1/660	1/1,400
	Clough and O'Rourke	0	1/6,800	1/530	1/530
	Ou and Hsieh	1/390	1/540	1/520	1/750
Case III	Observation	1/180	1/370	1/450	1/820
	Clough and O'Rourke	1/540	1/540	1/540	1/540
	Bowles	1/430	1/485	1/555	1/660
	Ou and Hsieh	1/125	1/390	1/530	1/1,100

Table 6.1 shows the values of angular distortion in relation to the locations of F_1 (represented by d_1/H_e) for the three cases, given $L_{12} = 5$ m. As shown in the table, Ou and Hsieh's method with an average estimate obtains good results in predicting the value of angular distortion and is better than other methods.

6.9　THREE-DIMENSIONAL EXCAVATION BEHAVIOR

Figure 6.32a shows a rectangular excavation where the deformations of the diaphragm wall near the central section of the longer side (the A-A section) are about the same. Thus, the behavior can be seen as two-dimensional plane strain behavior (for the definition of plane strain, see Appendix C). The deformation behavior at the central section of the shorter side (the B-B section), affected by concrete arching effect, is three-dimensional, and the displacements of both the diaphragm wall and soil are smaller than those at the A-A section. The behaviors of the retaining wall and soil near the corner are also three-dimensional (area C in the figure), and the displacements should be much smaller than those at the central section. Area D in Figure 6.32b is also

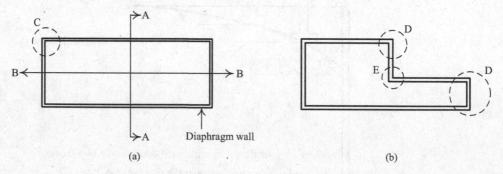

(a) (b)

Figure 6.32 Zones of plane strain behavior and three-dimensional behavior in excavations: (a) rectangular excavation and (b) irregular excavation.

under the influence of the arching effect, and the displacements of the wall and soil are smaller than those in the central section. For area E, the conditions are opposite: The displacement of the wall and soil should be larger than that in area D.

The characteristics of wall deformation, surface settlement, excavation bottom movement, and the related empirical formulas explicated earlier in this chapter all refer to the two-dimensional plane strain behavior and are not necessarily true in the vicinity of corners. With lateral stiffness, diaphragm walls have three-dimensional behaviors. Other retaining methods, such as the soldier pile, the steel sheet pile, and the column pile, do not have three-dimensional behaviors since they do not possess lateral stiffness. Three-dimensional analysis is required to obtain correct results for excavations possessing three-dimensional behavior. The beam on elastic foundation method that will be introduced in Chapter 7 is also a two-dimensional plane strain method. The only method capable of analyzing three-dimensional problems is the finite element method.

Ou et al. (1996) explored the three-dimensional behaviors of an excavation using a three-dimensional finite element analysis program and obtained the plane strain ratio (PSR) of the wall displacement in terms of the width–length ratio of the excavation and the distance from the corner, as shown in Figure 6.33a. The definition of

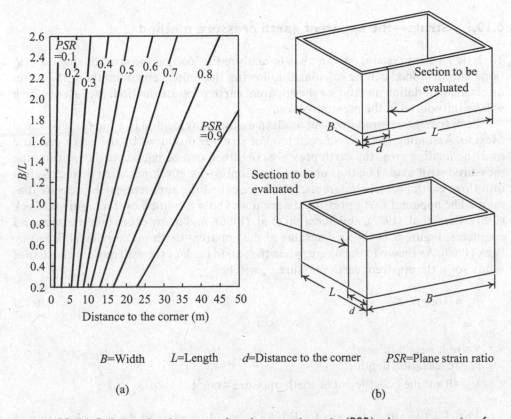

B=Width L=Length d=Distance to the corner PSR=Plane strain ratio

(a)

(b)

Figure 6.33 (a) Relationship between the plane strain ratio (PSR), the aspect ratio of an excavation geometry, and the distance from the corner and (b) symbol explanation (Ou et al., 1996).

the width–length ratio (B/L) is given as shown in Figure 6.33b while that of the plane strain ratio is given by

$$\text{PSR} = \frac{\delta_{hm,d}}{\delta_{hm,ps}} \qquad (6.11)$$

where

PSR = plane strain ratio

$\delta_{hm,d}$ = maximum wall deflection at the distance of d from the corner

$\delta_{hm,ps}$ = maximum wall deflection under the plane strain condition

In analysis, the maximum displacement of the wall in the plane strain condition $\delta_{hm,ps}$ has to be determined first, using the methods explicated in Section 6.7 and Chapters 7 or 8, and then, one should consult Figure 6.33 for PSR for the section, given the width–length ratio of the excavation and the distance from the corner. The maximum deflection, $\delta_{hm,d}$, on the section can then be computed by $\delta_{hm,d} = \text{PSR} \times \delta_{hm,ps}$.

6.10 STRESS ANALYSIS

6.10.1 Struts—the apparent earth pressure method

To design a strut system, one first has to analyze the load on the strut during excavation. The strut load can be calculated following the finite element method, the beam on elastic foundation method, or the apparent earth pressure method, the last of which will be introduced in the present section.

Peck (1969) measured the strut loads in excavations located in Chicago, Oslo, and Mexico. Assuming the load on each level of struts is produced by the earth pressure over the loading area, the earth pressures obtained can be back-calculated using the measured strut load. The thus obtained earth pressure envelopes were then classified into three earth pressure diagrams, which are called the apparent earth pressure diagram. The apparent earth pressure diagram was later modified by Terzaghi and Peck (1967), Peck et al. (1977), and Terzaghi et al. (1996), and many other investigators and engineers. Figure 6.34 shows diagrams of the apparent earth pressure established by Peck (1969). As shown in the figure, when the soil in back of the wall mainly consists of sandy soils, the apparent earth pressure p_a will be:

$$p_a = 0.65\gamma H_e K_a \qquad (6.12)$$

where

γ = unit weight of sand

H_e = excavation depth

K_a = Rankine's coefficient of earth pressure = $\tan^2\left(45^\circ - \phi/2\right)$.

The effective stress analysis for Eq. 6.12 should be adopted for the calculation of the apparent earth pressure, i.e., the friction angle of soil refers to the drained friction angle or the effective friction angle (ϕ'); the unit weight of soil above the groundwater

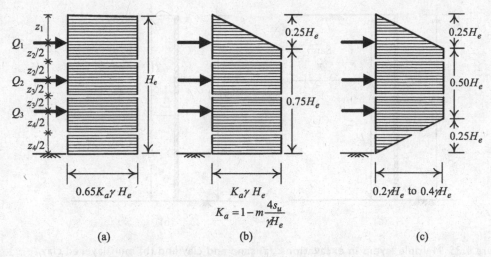

Figure 6.34 Peck's apparent earth pressure diagram: (a) sand, (b) soft to medium soft clay $(\gamma H_e / s_u > 4)$, and (c) stiff clay $(\gamma H_e / s_u \leq 4)$.

level refers to the moisture unit weight of soil (γ_m); the earth pressure and the water pressure below the groundwater level should be calculated separately and the unit weight of soil be taken as the submerged unit weight (γ').

If the soil in back of the wall is soft to medium soft clay (i.e., $\gamma H_e / s_u > 4$), the apparent earth pressure, p_a, would be the larger of

$$p_a = \gamma H_e \left(1 - m\frac{4s_u}{\gamma H_e}\right) \text{ or } p_a = 0.3\gamma H_e \qquad (6.13)$$

where s_u = undrained shear strength of soil within the range of the excavation depth,

m = an empirical coefficient, which is related to the stability number $N_b = \gamma H_e / s_{u,b}$, where $s_{u,b}$ = undrained shear strength of soil between the excavation bottom and the influence depth of excavation. The larger is N_b, the more possible is the occurrence of large displacement at the bottom of the retaining wall. When $N_b > 5.14$, $m = 0.4$ (Peck et al., 1974); when $N_b \leq 5.14$, $m = 1.0$ (Terzaghi et al., 1996).

Eq. 6.13 should be used based on the total stress method, that is, assumes $\phi = 0$ without the pore water pressure considered.

If the soil in back of the wall is stiff clay $(\gamma H_e / s_u \leq 4)$, the apparent earth pressure p_a would be:

$$p_a = 0.2\gamma H_e - 0.4\gamma H_e \left(\text{the average is } 0.3\,\gamma H_e\right) \qquad (6.14)$$

Similarly, Eq. 6.14 should be used based on the total stress method.

For alternating layers of sand and clay, two ways of calculating the apparent earth pressure are available. The first is to judge which, sand or clay, is the dominant soil within the range of the excavation depth and then consult Figure 6.34 to calculate the

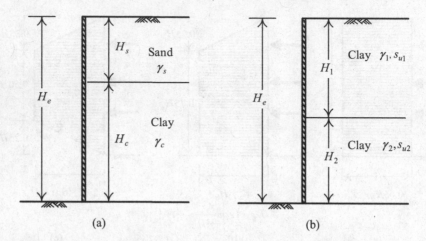

(a)   (b)

Figure 6.35 Multiple layers in excavations: (a) sand and clay and (b) multilayered clay.

apparent earth pressure. The other is to apply the concept of the equivalent cohesion to calculating the apparent earth pressure (Peck, 1943). Take an example for alternating layers of a sand layer above a clay layer (Figure 6.35a), and the equivalent undrained shear strength $\left(s_{u,eq}\right)$ and unit weight $\left(\gamma_{eq}\right)$ can be estimated as follows:

$$
\begin{aligned}
s_{u,eq} &= \frac{1}{H_e}\left[(K_s\gamma_s H_s\tan\delta)H_s/2 + H_c s_u\right] \\
&= \frac{1}{2H_e}\left[\gamma_s K_s H_s^2\tan\delta + 2H_c s_u\right]
\end{aligned}
\tag{6.15}
$$

$$
\gamma_{eq} = \frac{1}{H_e}\left[\gamma_s H_s + \gamma_c H_c\right]
\tag{6.16}
$$

where
  γ_s and γ_c = unit weight of sand and clay, respectively
  H_s and H_c = depth of the sand and clay, respectively
  K_s = coefficient of lateral earth pressure. For simplicity, K_s can be assumed to be equal to Rankine's K_a.
  δ = friction angle between sand and wall
  s_u = undrained shear strength of clay

When a clay is above a sand layer, the $s_{u,eq}$ and γ_{eq} can be derived in a way similar to the above procedure. Similarly, for layered clay layers, the concept of equivalent values can also be used to calculate the strut load, as shown in Figure 6.35b. The equivalent undrained shear strength $\left(s_{u,eq}\right)$ and the unit weight $\left(\gamma_{eq}\right)$ of clay would be

$$
s_{u,eq} = \frac{1}{H_e}\left(s_{u1}H_1 + s_{u2}H_2\right)
\tag{6.17}
$$

$$\gamma_{eq} = \frac{1}{H_e}\left[\gamma_1 H_1 + \gamma_2 H_2\right] \tag{6.18}$$

where

s_{u1}, s_{u2} = undrained shear strength of the first and second clay layers, respectively.
γ_1, γ_2 = unit weight of the first and second clay layers, respectively.
H_1 and H_2 = depth of the first and second clay layers, respectively.

Providing the equivalent undrained strength and the unit weight as derived above, consult Peck's apparent earth pressure diagram and choose a proper one. Then, calculate the load (e.g., Q_1, Q_2, and Q_3 in Figure 6.34) on each level of struts as the resultant of the earth pressure within the range covering half the vertical span between the upper and current struts and half the vertical span between the current and lower struts.

Since the apparent earth pressure diagram was established 20 or 30 years ago, both excavation scale and depth have become larger, and some people doubt the applicability, considering that excavation depth and retaining systems having remarkably advanced. According to many documents and empirical experience, nevertheless, the apparent earth pressure method is still useful for excavations less than 10 m.

6.10.2 Cantilever walls—the simplified gross pressure method

As explained in Sections 5.4 and 5.8, the design of a cantilever wall is based on the fixed earth support method, i.e., the embedded part of the wall is assumed to be fixed at a certain depth. Thus, in the ultimate state, the active earth pressure above the excavation bottom can develop fully, whereas the passive and active earth pressures near the fixed point cannot, as shown in Figure 5.20 in Chapter 5. On the other hand, under working load, though the active earth pressure above the excavation bottom may still fully develop, the passive and the active earth pressures near the fixed point are not fully developed. It is therefore difficult to estimate the distribution of earth pressure on the retaining wall. To be conservative and simplify the procedure of computing the stress of cantilever walls, we may assume that the earth pressure is in the ultimate state.

With the earth pressure on the wall known, we can then adopt the simplified gross pressure method for the analysis of the stress of a cantilever wall (Padfield, and Mair, 1984), as shown in Figure 6.36a (also see Section 5.8). The earth pressure distribution should not include the safety factors otherwise the distribution of earth pressures will be distorted. Though the depth of the wall used for analysis, as shown in Figure 6.36a, is not as deep as designed (Figure 6.36b), the maximum bending moment of the wall thus derived is closest to the real value, as shown in Figure 6.36b. If sheet piles or soldier piles are to be adopted, we can design according to the maximum bending moment without knowing the distribution of bending moments. If the retaining wall is RC wall (such as diaphragm walls or column piles), the bending moment derived from the simplified method has to be modified. The way to perform modification is as illustrated in Example 6.3.

The stress analyses of cantilever walls can also follow the net pressure method as introduced in Section 5.8. To analyze the stress/bending moment of a cantilever wall using the net pressure method is similar to that using the gross pressure method and is not to be discussed here.

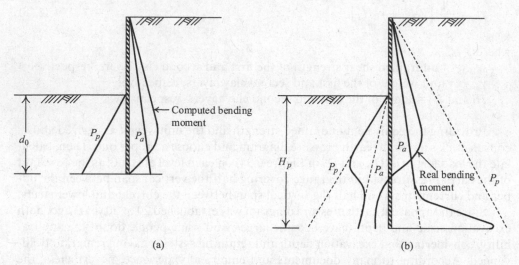

Figure 6.36 Computation of moment of cantilever walls: (a) moment diagram computed from simplified analysis and (b) real moment diagram.

Note that the earth pressure theories adopted for the analysis/or design should follow Caquot–Kerisel's earth pressure theory and those introduced in Section 4.6. Also, as discussed in Section 6.10.1, we can see that Peck's apparent earth pressure (Figure 6.34) is the earth pressure derived from the strut load rather than the true earth pressure. Thus, the apparent earth pressure can only be used to calculate the strut load. Though some engineers adopt the apparent earth pressure for the calculation of the stress/bending moment of the retaining wall, it is not recommended.

6.10.3   Strutted walls—the assumed support method

The design of a strutted wall is usually based on the concept of the free earth support method. Thus, in the ultimate state, both the active and passive earth pressures can fully develop. If multi-level strutted walls are regarded as continuous beams, though the earth pressures on the wall are given, the stress/bending moment of the strutted wall may not be able to be computed by way of simple structural mechanics or hand calculation. Moreover, under working load, the active earth pressure on the back of the wall may be fully developed, whereas the passive earth pressure rarely is. As a result, if we adopt the method of hand calculation to solve the problem, we have to first simplify the boundary conditions and loading pattern.

The assumed support method simplifies the interaction between the soil and wall as the lateral earth pressure on the wall, and the soil resisting against the wall movement as centered on an assumed support point. The retaining wall is simulated as a simply supported beam or continuous beam to solve the problem by simple structural mechanics. With the merits of simple computation of the highly indeterminate beam, in the past, the assumed support method was often adopted to obtain the stress/bending moment of the wall for deep excavation problems. Because many high-level analysis methods have been coded into computer programs, the assumed support method has lost its standing in engineering design. It can, nonetheless, still be helpful in some small-scale excavation projects in case that computer programs are not available. The procedure can be described in three parts:

1.   Distribution of lateral earth pressure

   As explained in Chapter 5, the design of a strutted wall is based on the concept of the free earth support method and the earth pressure on the back of the wall can be assumed as the active earth pressure. The distribution of the earth pressures should be as close as possible to the real earth pressure distribution, so it follows that no safety factors, which will distort the real distribution, are to be included. It is noted that the earth pressure theories adopted for the analysis/or design should follow Caquot–Kerisel's theory and those introduced in Section 4.6. Also, as stated in Section 6.10.2, Peck's apparent earth pressure diagrams (Figure 6.34) should not be used to calculate the stress of the wall and they only can be used for strut design.

2.   Location of the assumed support

   Retaining walls will move toward the excavation zone due to the lateral earth pressure behind the wall, and the soil in front of the wall will produce passive resistance accordingly. If the passive earth pressure is viewed as a concentrated reaction force, the action point of this concentrated reaction force is designated as the assumed support. The determination of the assumed support can follow one of the methods as below:

(1)   At each stage of excavation, with the moments produced by the active and passive earth pressures with respect to the lowest level of struts, respectively, in equilibrium, the action point of the passive earth resultant would be the assumed support. As shown in Figure 6.37, the location of the assumed support (ℓ) is determined as follows:

$$\ell = \frac{P_a \ell_a}{P_p} - s \tag{6.19}$$

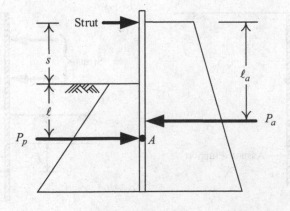

A is the location of the assumed support

Figure 6.37 Determination of the location of the assumed support by way of the moment equilibrium of earth pressures below the lowest level of struts.

(2) According to the stiffness of soil, the location of the assumed support can be determined as illustrated below:

	Sandy soils	Clayey soils	The locations of the assumed supports
Dense soils	$N > 50$	$N > 15$	$\ell = 0 \sim 0.5\,\mathrm{m}$
Medium dense soils	$10 \leq N \leq 50$	$4 \leq N \leq 15$	$\ell = 1.0 \sim 2.0\,\mathrm{m}$
Soft soils	$N < 10$	$N < 4$	$\ell = 3.0 \sim 4.0\,\mathrm{m}$

where N = standard penetration number

ℓ = depth of the assumed support from the excavation bottom. If a dense soil layer is right below a soft layer, as shown in Figure 6.38, ℓ is then the distance from the dense layer to the assumed support.

3. Computation procedure

With the distribution of earth pressures and location of the assumed support both determined, the retaining wall can be seen as a simply supported beam or a continuous beam, and the stress, bending moment, and deformation of the retaining wall can be obtained using simple structural mechanics. If the retaining wall is simulated as a continuous beam, the computation of stresses of the wall is complicated. Using a complicated method to solve the problem is meaningless if it cannot increase the analysis accuracy. Instead, the simply supported beam model is more often adopted.

To analyze the stress/bending moment of the wall with the simply supported beam model, the earth pressure can be assumed to be added onto the retaining wall once for all. The method is designated as the one-stage loading method, as shown in Figure 6.39.

The analysis based on the simply supported beam model is also able to simulate the procedure of excavation by loading the lateral earth pressure on a simply supported beam stage by stage and then compute the stress/bending moment and deformation of the wall at each stage, which are then added up to get the total

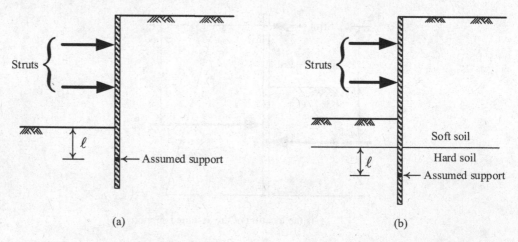

Struts

Assumed support

(a)

Struts

Soft soil

Hard soil

Assumed support

(b)

Figure 6.38 Locations of the assumed support: (a) homogeneous soil and (b) soft clay above a stiff layer.

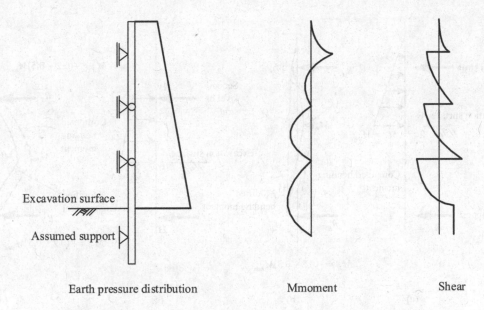

Earth pressure distribution   Mmoment   Shear

Figure 6.39 One-stage loading simply supported beam method (the simply supported beam model).

stress/bending moment and deformation, after excavation is finished. The method is designated as the phased loading method. The following are the principles of computation for the wall using the assumed support method with phased loading. There are also other structural analysis models to compute the stress/bending moment of the wall. Whichever one chooses, the method should avoid complexity. The procedure of using the assumed support method is as follows:

1. Compute the bending moment for the second stage of excavation using the simply supported beam model as illustrated in Figure 6.40a and b.
2. The computing of the bending moment after the second stage: Take the distribution of lateral earth pressure below the lowest level of struts to the excavation bottom for each excavation stage. The computing procedure is shown in Figure 6.40c and d.
3. The bending moment in the position where the assumed support is located can be assumed 20%–50% of the maximum bending moment at the stage. The greater is the soil strength at the position where the assumed support is located, the greater is the bending moment. The bending moment at the fulcrum of the strut can be similarly assumed to be 20%–50% of the maximum bending moment at the stage. With the bending moments at the assumed support and at the fulcrum of the strut both determined, modify the calculated bending moments by way of the method as shown in Figure 6.40. The modified maximum bending moment would be $M_E' = M_E - M_1$.
4. As shown in Figure 6.41, to compute the bending moment of the wall at the stage of strut demolition and floor slab construction, the strut above the one to be demolished and the floor slab or the base slab can be viewed as two fulcrums, the span between which is considered as a simply supported beam. The axial load of the strut to be demolished can be viewed as a concentrated force acting on the simply supported

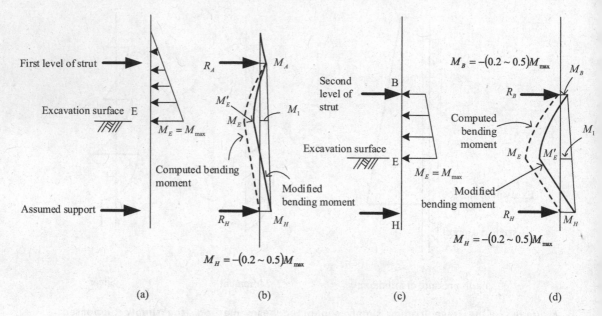

First level of strut

R_A M_A

Second level of strut   B

$M_B = -(0.2 \sim 0.5)M_{max}$ M_B

R_B

Excavation surface E

M'_E M_1

M_E

$M_E = M_{max}$

Computed bending moment

$M_E = M_{max}$

Computed bending moment

M_E M'_E M_1

Excavation surface E

Computed bending moment

Assumed support

Modified bending moment

Modified bending moment

R_H M_H

R_H M_H

R_H M_H H

$M_H = -(0.2 \sim 0.5)M_{max}$

$M_H = -(0.2 \sim 0.5)M_{max}$

(a)   (b)   (c)   (d)

Figure 6.40 Phased loading assumed support method: (a) earth pressure distribution at the second excavation stage, (b) computation of moment at the second excavation stage, (c) earth pressure distribution at the third excavation stage, and (d) computation of moment at the third excavation stage.

beam, whose acting direction is the same with that of the lateral earth pressure. The concentrated force is the sum of the strut load at the final stage of excavation and increment of strut load generated by demolition of the struts at the previous stage. The strut load at the final stage can be computed with the apparent earth pressure diagram. For example (refer to Figure 6.41), the concentrated force due to demolition of the third level of the struts (strut C) after the raft foundation construction is the strut load as computed with the apparent earth pressure diagram. The concentrated force due to demolition of the second level of struts (strut B) is the sum of the strut load as computed with the apparent earth pressure diagram and the axial increment produced by demolition of the third level of struts (strut C).

5. The bending moment or shear of the simply supported beam can be easily computed using structural mechanics. However, to match the actual boundary conditions, the bending moment at the upper fulcrum (where the upper struts are) can be assumed to be 30% of the maximum bending moment whereas that at the lower fulcrum (where the floor slabs are supported) can be assumed to be 50% of the maximum bending moment. With the bending moments at the fulcrums determined, modify the calculated values. The modified maximum bending moment would be $M'_c = M_c - M_1$ or $M'_B = M_B - M_2$ (refer to Figure 6.41).

Figure 6.42 is a calculating procedure for the assumed support method. Please note that only the bending moment and the shear derived from the method can be used for the design of a retaining wall. Though also feasible for the computing of wall deformation, with poor accuracy, the method may not work well for the protection of the adjacent property in deep excavations.

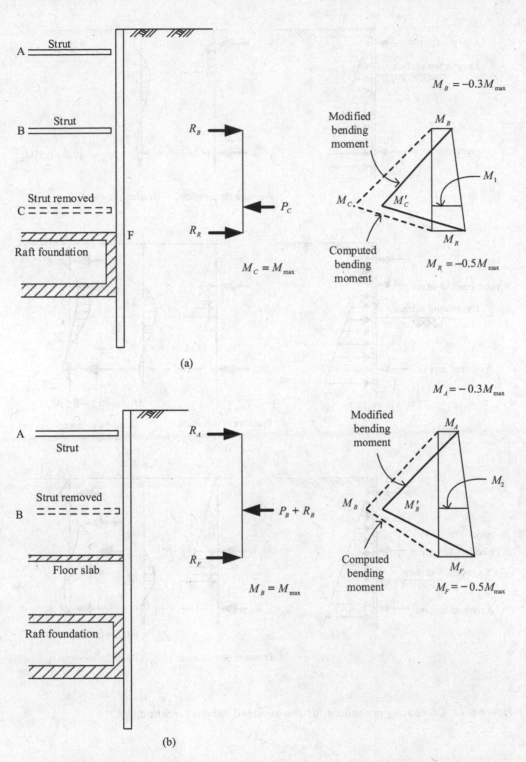

$$M_B = -0.3M_{max}$$

$$M_C = M_{max}$$

$$M_R = -0.5M_{max}$$

(a)

$$M_A = -0.3M_{max}$$

$$M_B = M_{max}$$

$$M_F = -0.5M_{max}$$

(b)

Figure 6.41 Phased loading assumed support at the stage of strut demolition (R_B, R_R, R_A, R_F are reaction forces due to demolition of the struts; P_C, P_B are strut loads at the final stage of excavation and can be computed using the apparent earth pressure diagram).

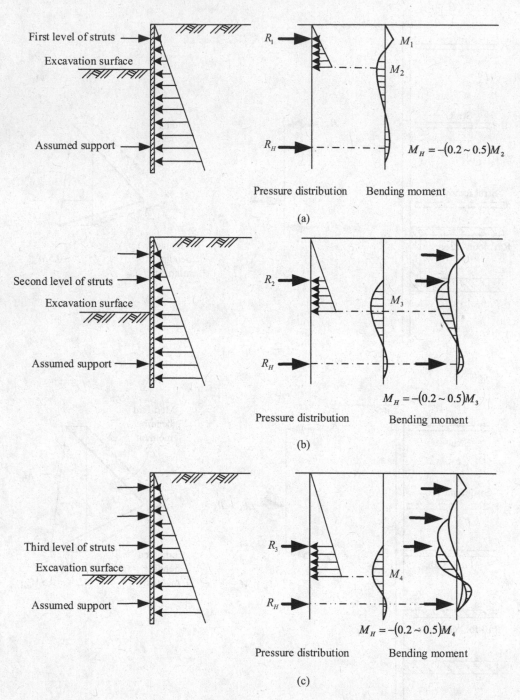

Figure 6.42 Computing procedure for the assumed support method.

Example 6.1

Figure 6.43 is the excavation profile of a building (Building P) in Taipei. The length of the excavation was 40 m, the width was 34 m, and the excavation depth was 12.3 m. The retaining wall was a 60-cm-thick and 23-m-deep diaphragm wall. The excavation was carried out in five stages. There were four levels of struts: With the completion of each excavation, one level of struts was then built. The type of the first level of struts was H250×250×9×14, the second, H300×300×10×15, and the third and the fourth, H350×350×12×19. The average distance between struts was 4.4 m. Figure 6.44 shows the wall deformations at the excavation stages measured at the central section of the longer side (the 40-m-long side). Estimate the type of excavation-induced surface settlement.

Solution

According to the figure of the lateral deformation of the wall, we have $A_{c1} = 3.7$ cm-m, $A_{c2} = 6.0$ cm-m. Thus, $A_c = \max(A_{c1}, A_{c2}) = 6.0$ cm-m, $A_s = 88$ cm-m; since $A_s \geq 1.6A_c$, the type of surface settlement can be predicted to be the concave type.

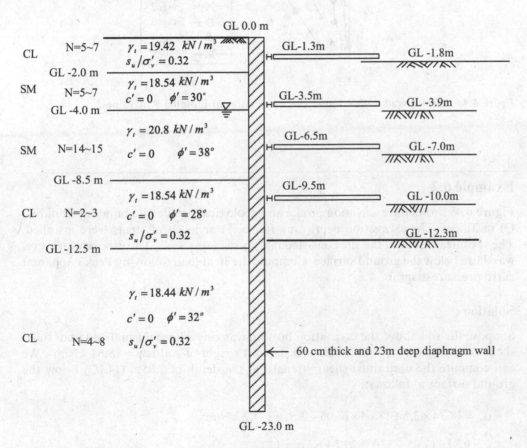

Figure 6.43 Soil and excavation profile of Building P in Taipei.

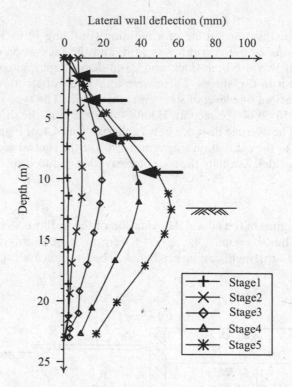

Figure 6.44 Lateral wall deflection in the excavation of Building P in Taipei.

Example 6.2

Figure 6.45 shows the excavation profile and geological profile of a building (Building Q) in Taipei. The excavation depth was 14.1 m. Four levels of struts were installed. The subsurface soils at the site consisted mainly of clayey soils. The groundwater level was 3.0 m below the ground surface. Compute the strut load following Peck's apparent earth pressure diagram.

Solution

Suppose the soil above the excavation bottom was clay whose normalized undrained shear strength is $s_u/\sigma'_v = 0.32$. The average unit weight of soil is $\gamma = 18.64$ kN/m³. We can compute the undrained shear strength at the depth of 7.05 m (14.1/2) below the ground surface as follows:

$$\sigma_v = 18.74 \times 3.5 + 18.64 \times (7.05 - 3.5) = 131.8 \text{ kN/m}^2$$

$$u = 9.81 \times (7.05 - 3.0) = 39.7 \text{ kN/m}^2$$

$$\sigma'_v = \sigma_v - u = 131.8 - 39.7 = 92.1 \text{ kN/m}^2$$

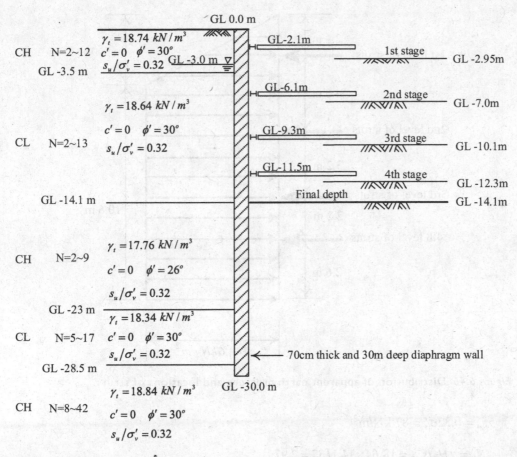

Figure 6.45 Soil and excavation profile of Building Q in Taipei.

$$s_u = 0.32\sigma'_v = 29.5 \text{ kN/m}^2$$

$$\gamma H_e / s_u = 18.64 \times 14.1 / 29.5 = 8.9 > 4.0$$

The soil at the site can then be categorized as soft to medium soft clay according to Peck's apparent earth pressure diagrams. The influence depth of excavation can be defined to be as deep as the excavation width. Since the excavation width $B = 28.8$ m, the influence depth is about from the excavation bottom down to $14.1 + 28.8 = 42.9$ m below it. Then, the stress in the soil at a depth of 28.5 m ($14.1 + 28.8/2$) is

$$\sigma_v = 131.8 + 18.64 \times 7.05 + 17.76 \times (23 - 14.1) + 18.34 \times (28.5 - 23) = 522.2 \text{ kN/m}^2$$

$$u = 9.81 \times (28.5 - 3.0) = 250.2 \text{ kN/m}^2$$

$$\sigma'_v = \sigma_v - u = 522.2 - 250.2 = 272 \text{ kN/m}^2$$

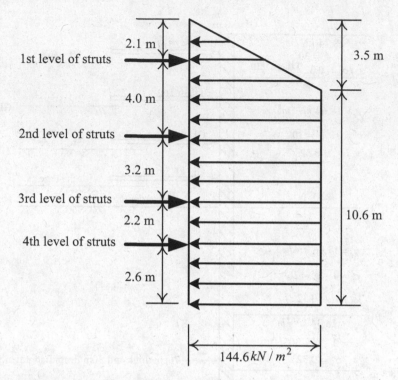

1st level of struts

2.1 m

4.0 m

2nd level of struts

3.2 m

3rd level of struts

2.2 m

4th level of struts

2.6 m

3.5 m

10.6 m

$144.6\,kN\,/\,m^2$

Figure 6.46 Distribution of apparent earth pressure and locations of struts.

$s_u = 0.32\sigma'_v = 87 \text{ kN/m}^2$

$N_b = \gamma H_e / s_{u,b} = 18.64 \times 14.1 / 87 = 2.97$

According to Eq. 6.13, we can reasonably assume that $m = 1.0$, given the value of N_b.

$p_a = \gamma H_e - 4s_u = 18.64 \times 14.1 - 4 \times 29.5 = 144.6 \text{ kN/m}^2$

or

$p_a = 0.3\gamma H_e = 78.8 \text{ kN/m}^2$

The apparent earth pressure is $p_a = 144.6 \text{ kN/m}^2$

The distribution of earth pressure is as shown in Figure 6.46. Referring to Figure 6.34, the load on each level of struts would be

The load on the first level of struts $= 144.6 \times 3.5/2 + 144.6 \times (2 + 2.1 - 3.5) = 339.8$ kN/m

The load on the second level of struts $= 144.6 \times (4.0/2 + 3.2/2) = 520.6 \text{ kN/m}$

The load on the third level of struts $= 144.6 \times (3.2/2 + 2.2/2) = 390.4 \text{ kN/m}$

The load on the fourth level of struts $= 144.6 \times (2.2/2 + 2.6/2) = 347 \text{ kN/m}$

Example 6.3

Assume a cantilever wall in a sandy soil, as shown in Figure 6.47a. Compute bending moment and shear using the simplified gross pressure method.

Solution

Assume the friction angles between the retaining wall and soil in the active side and passive side are, respectively, $\delta = 2\phi'/3$ and $\delta = \phi'/2$. According to Coulomb's and Caquot–Kerisel's earth pressure theories (consult Eq. 4.14 and Figure 4.10), we can

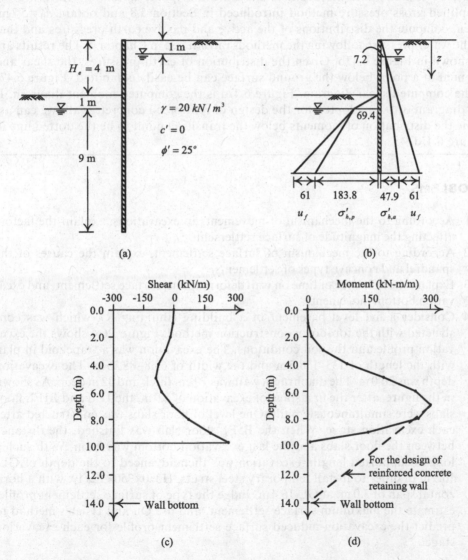

(a)

$H_e = 4$ m

1 m

$\gamma = 20 \ kN/m^3$

$c' = 0$

$\phi' = 25°$

9 m

7.2

69.4

61   183.8   47.9   61

u_f $\sigma'_{h,p}$ $\sigma'_{h,a}$ u_f

(b)

Shear (kN/m)

−300 −150   0   150   300

Depth (m)

0.0
2.0
4.0
6.0
8.0
10.0
12.0
14.0 ← Wall bottom

(c)

Moment (kN-m/m)

0   150   300

Depth (m)

0.0
2.0
4.0
6.0
8.0
10.0
12.0 ← For the design of reinforced concrete retaining wall
14.0 ← Wall bottom

(d)

Figure 6.47 Cantilever wall excavation: (a) excavation profile, (b) distribution of earth pressure for the simplified gross pressure method, (c) shear diagram, and (d) moment diagram.

obtain the coefficients of the active and passive earth pressures, which are 0.37 and 3.55, respectively. Thus, the coefficient of the horizontal active earth pressure would be $K_{a,h} = 0.37 \cos(2\phi'/3) = 0.36$ and that of the horizontal passive earth pressure would be $K_{p,h} = 3.55 \cos(\phi'/2) = 3.47$. Since there exists a difference between the water levels inside and outside the retaining wall, the after seepage water pressure and effective stress should be taken into account. The seepage-induced water pressure can be estimated using the method introduced in Section 4.6.6.

According to the elucidations in Section 6.10.2, the distribution of earth pressure for the stress analysis of a retaining wall has to exclude the factor of safety. Thus, we can compute the depth of the transition point d_0 for $F_b = 1.0$ using the simplified gross pressure method introduced in Section 5.8 and obtain $d_0 = 5.7\,\mathrm{m}$. Then, compute the distributions of the active and passive earth pressures and that of the water pressure following the methods introduced in Chapter 4. The results are as shown in Figure 6.47b. Given the distribution of earth pressures, the shear and moment at a point below the ground surface can be easily computed. Figure 6.47c is the computed shear diagram. Figure 6.47d is the computed moment diagram. If the diagram is to be adopted for the design of reinforced concrete wall, we can assume the distribution of moments below the transition point to be the dotted line in Figure 6.47d.

PROBLEMS

6.1 According to the mechanism of movements in excavations, explain the factors affecting the magnitude of surface settlement.

6.2 According to the mechanism of surface settlement, explain the causes of the spandrel and concave types of settlements.

6.3 Explain the influence of time on wall deformations, surface settlement, and excavation bottom movement.

6.4 Consider a five-level basement of a building (Building R), which was constructed with the top-down construction method. Figure P6.4 shows the excavation profile and the soil conditions. The excavation was a trapezoid in plan with the length of 133–136.5 m and the width of 63.8–88.15 m. The excavation depth was 20.0 m. The diaphragm wall was 70 cm thick and 32 m deep. As shown in the figure, after the first stage of excavation of 5.0 m, the 1FL and B1FL floor slabs were simultaneously built. One level of floor slabs was constructed after each excavation stage. When the B5FL floor slab was finished, the distance between the floor slabs and the last excavation bottom was 6.1 m. With such a long unsupported length, excavation was then advanced to the depth of GL-16.9 m in order to install temporary steel struts, H350×350×12×19, with a horizontal span of 3.0 m, at GL-16.4 m. Judge the type of surface settlement profile, estimate the maximum surface settlement, and use Ou and Hsieh's method to predict the excavation-induced surface settlement profile for each excavation stage.

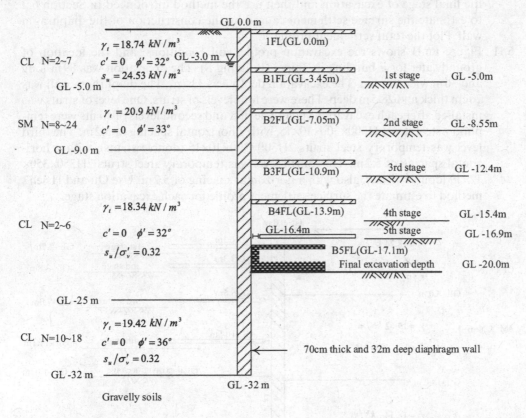

GL 0.0 m

1FL(GL 0.0m)

CL N=2~7 $\gamma_t = 18.74\ kN/m^3$
$c' = 0\quad \phi' = 32°$ GL -3.0 m ▽
$s_u = 24.53\ kN/m^2$

B1FL(GL-3.45m)   1st stage   GL -5.0m

GL -5.0 m

SM N=8~24 $\gamma_t = 20.9\ kN/m^3$
$c' = 0\quad \phi' = 33°$

B2FL(GL-7.05m)   2nd stage   GL -8.55m

GL -9.0 m

B3FL(GL-10.9m)   3rd stage   GL -12.4m

CL N=2~6 $\gamma_t = 18.34\ kN/m^3$

B4FL(GL-13.9m)   4th stage   GL -15.4m

$c' = 0\quad \phi' = 32°$

GL-16.4m   5th stage   GL -16.9m

$s_u/\sigma'_v = 0.32$

B5FL(GL-17.1m)
Final excavation depth   GL -20.0m

GL -25 m

CL N=10~18 $\gamma_t = 19.42\ kN/m^3$
$c' = 0\quad \phi' = 36°$
$s_u/\sigma'_v = 0.32$

70cm thick and 32m deep diaphragm wall

GL -32 m

GL -32 m

Gravelly soils

Figure P6.4

6.5 Same as Problem 6.4. Use Clough and O'Rourke's method to estimate the excavation-induced surface settlement profile at the final stage.

6.6 Same as Problem 6.4. Use Bowles' method to estimate the excavation-induced surface settlement profile at the final stage.

6.7 Compare the results of Problems 6.4–6.6 with the surface settlement tendencies in the excavation of TNEC (see Figure 6.10b) and explain which method to estimate the surface settlement is more rational. Why?

6.8 Same as Problem 6.4. Assume in back of the retaining wall exists an individual footing 10 m from the wall. The footing is 3.0 m deep below the ground surface. According to the results of Problem 6.4, estimate the settlement of the footing when the excavation reached the final depth.

6.9 Assume in back of the retaining wall, as shown in Figure P6.4, exist two individual footings that are separated by 5.0 m. One is 10.0 m from the wall, while the other is 15.0 m from the wall. Use Ou and Hsieh's method, Clough and O'Rourke's method, and Bowles' method to estimate the angular distortion of the foundation at the final stage of the excavation.

6.10 Same as shown in Figure P6.4. Use Ou and Hsieh's method, Clough and O'Rourke's method, and Bowles' method to compute the surface settlement at the final stage of excavation and then use the method introduced in Section 6.2 to estimate the surface settlement caused by the construction of the diaphragm wall. Plot the total settlement profile.

6.11 Figure P6.11 shows the excavation profile, soil conditions, and the location of groundwater for a building in Taipei (Building S). The excavation was 89 m long and 36 m wide in plan. The excavation depth was 13.7 m. The diaphragm wall was 70 cm thick and 28.5 m deep. There were four levels of struts. One level of struts was installed after each excavation stage. The first and second levels of struts were temporary steel struts H300×300×10×15, with a horizontal spacing of 5.2 m. The third level was temporary steel struts 2H300×300×10×15 (double struts), with a horizontal spacing of 5.2 m. The fourth level was temporary steel struts 2H350×350×12×19 (double struts), also with a horizontal spacing of 5.2 m. Use Ou and Hsieh's method to estimate the surface settlement profile for each excavation stage.

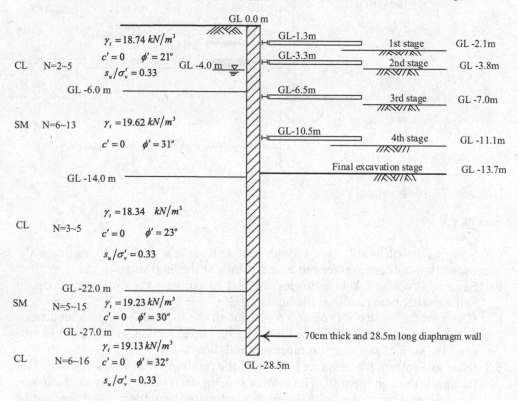

Figure P6.11

6.12 Same as the previous problem. Estimate the surface settlement profile for each excavation stage using Clough and O'Rourke's method.

6.13 Same as Problem 6.11. Use Bowles' method to estimate the surface settlement profile for each excavation stage.

6.14 Compare the results of Problems 6.11–6.13 with the tendency of the surface settlement in the TNEC excavation (see Figure 6.10b) and evaluate which method is the most rational. Why?

6.15 Same as Problem 6.11. Assume in back of the retaining wall exists an individual footing 6.0 m from the wall. The footing is 3.0 m deep below the ground surface.

According to the result of Problem 6.11, estimate the settlement of the footing after the completion of the last excavation stage.

6.16 Assume two individual footings, 4.0 m from each other, are located 10.0 and 14.0 m in back of the retaining wall as shown in Figure P6.11. Use the calculated results with Ou and Hsieh's method, Clough and O'Rourke's method, and Bowles' method separately to estimate the angular distortions after the completion of the last excavation stage.

6.17 As shown in Figure P6.11, use Ou and Hsieh's method, Clough and O'Rourke's method, and Bowles' method separately to compute the final surface settlement and use the method introduced in Section 6.2 to estimate the surface settlement caused by the construction of the diaphragm wall. Then, plot the total surface settlement profile.

6.18 The top-down construction method is characterized by using concrete floor slabs as lateral supports during excavation. The distance between floor slabs and the excavation bottom should be at least 1.5 m. After the completion of each excavation stage, excavation is usually suspended for a period before proceeding to the next stage. On the other hand, the bottom-up method is characterized by using H steel as lateral struts. The distance between struts and the excavation bottom is usually 0.5 m. When using the bottom-up excavation method, it is not necessary to wait before proceeding to the next stage. The axial stiffness of a floor slab is far larger than that of a H steel. Under the same excavation conditions, which will generate larger surface settlement when excavating in soft clay? Why?

6.19 Figure P6.19 shows the plan of an excavation and its adjacent building. Assume the excavation profile is as shown in Figure P6.4 and the soil is sandy soil. The groundwater level is 2.0 m deep below the ground surface. The effective stress parameters for the sand: $c' = 0$ and $\phi' = 32°$. Use the simplified method to estimate the maximum surface settlement at the section where the adjacent building is located.

6.20 Figure P6.20 shows the plan of an excavation and its adjacent building. Assume the excavation profile is as shown in Figure P6.11 and the soil is clayey soil. The groundwater level is 2.0 m below the ground surface. The normalized undrained shear strength $s_u/\sigma'_v = 0.3$. Use the simplified method to estimate the maximum surface settlement at the section where the adjacent building is located.

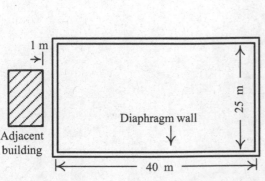

Figure P6.19

Figure P6.20

6.21 Assume an excavation as shown in Figure P6.11. The soil is clayey soil. The groundwater level is 2.0 m below the ground surface. The moisture unit weight and the saturated unit weight of clay are $\gamma_m = 16$ kN/m^3 and $\gamma_{\text{sat}} = 18$ kN/m^3, respectively. The normalized undrained shear strength $s_u / \sigma'_v = 0.3$. Use Peck's apparent earth pressure method to compute the strut load.

6.22 Same as the previous problem except that the soil is sandy soil. The groundwater level is 2.0 m below the ground surface. The moisture unit weight and the saturated unit weight of sand are $\gamma_m = 15$ kN/m^3 and $\gamma_{\text{sat}} = 19$ kN/m^3, respectively. The effective stress parameters: $c' = 0$ and $\phi' = 32°$. Use Peck's apparent earth pressure method to compute the strut load.

6.23 Assume an excavation profile as shown in Figure P6.11. The soil condition is also identical with that as shown in the figure, alternating layers of sand and clay. Use Peck's apparent earth pressure method to compute the strut load.

6.24 Same as Problem 6.21. Use the assumed support method with one-stage loading to compute the bending moment of the retaining wall.

6.25 Same as Problem 6.21. Use the assumed support method with phased loading to compute the bending moment of the retaining wall.

6.26 Same as Problem 6.22. Use the assumed support method with one-stage loading to compute the bending moment of the retaining wall.

6.27 Same as Problem 6.22. Use the assumed support method with phased loading to compute the bending moment of the retaining wall.

6.28 See Problem 5.17 and Figure P5.17. Assume $H_p = 7.0$ m. Use the net earth pressure to compute the bending moment and shear of the retaining wall.

6.29 Same as the previous problem. Use the simplified gross earth pressure method to compute the bending moment and shear of the retaining wall.

6.30 See Problem 5.20 and Figure P5.20. Use the simplified gross earth pressure method to compute the bending moment and shear of the retaining wall.

6.31 Same as the previous problem. Use the net earth pressure method to compute the bending moment and shear of the retaining wall.

Chapter 7

Stress and deformation analysis

Beam on elastic foundation method

7.1 INTRODUCTION

Under normal excavation conditions, the stress and deformation of walls and ground surface settlement all relate to the unbalanced forces acting on the walls, the stiffness of the wall-strut system, the stability of the excavations, construction sequence, and so on. The unbalanced forces are in turn related to the soil conditions, groundwater levels, water pressures, the excavation width, the excavation depth, the excavation area, etc. To analyze excavation-induced deformation using the numerical method, the analysis method has to simulate all these factors rationally. In addition to the finite element method, the beam on elastic foundation method is also able to model all those factors. Many practicing engineers prefer to use the beam on the elastic foundation method to compute the stress and deformation of the wall and the strut load because of its simplicity, succinct input parameters, shorter computation, and data processing time.

As elucidated in Chapter 6, a complete stress analysis for structural design of excavations includes detailed stress analyses of many structural components such as the end brace, corner brace, wale (see Figure 3.3), and the central post. These contents will be introduced in Chapter 10.

7.2 BASIC PRINCIPLES

In foundation engineering, the soil–structure interaction problem is often simulated as a beam on an infinite number of elastic springs to simplify analysis, which is often referred to as Winkler's model (Winkler, 1867).

As shown in Figure 7.1, the basic assumption of Winkler's model is, given that the foundation is a structure with stiffness and soils are of elastic foundation, their interaction can be simulated as a beam on an infinite number of elastic springs. The spring constant is the ratio of pressure (p) to displacement (δ), which can be expressed as follows:

$$k_s = \frac{p}{\delta} \tag{7.1}$$

where k_s is called the coefficient of subgrade reaction, the modulus of subgrade reaction, or the soil spring constant, the unit of which is (force) × (dimension)$^{-3}$. The strength of

DOI: 10.1201/9780367853853-7

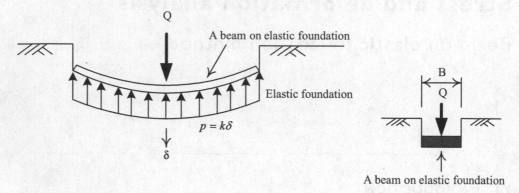

Figure 7.1 Beam on elastic foundation: (a) an infinitely long beam and (b) profile of Section AA.

Winkler's model is that it greatly simplifies analysis, for it assumes that the springs are individually acting without interaction.

The beam on elastic foundation method assumes the retaining wall as a beam on an elastic foundation. The interaction between the retaining wall and adjacent soils is simulated by a series of soil springs (Figure 7.2a), while the earth pressures acting on both sides of the wall before excavation are taken to be the at-rest earth pressure (K_0 -condition) (Figure 7.2b). When excavation is conducted, unloading induced by excavation will generate unbalanced forces, which is the difference between the at-rest earth pressure in front of and behind the wall, assuming that the wall remains unmoved (Figure 7.2c). The unbalanced force then acts on the structural system, including the retaining wall, struts, and soil spring, as shown in Figure 7.2a.

The deformation mechanism of soil springs subject to pressure is as shown in Figure 7.3. Before yield or failure of the soil, soil springs are simulated as elastic material with constant k_h (note: the subscript "h" denotes horizontal). When the retaining wall moves forward due to unbalanced force, the wall movement will cause the soil springs to be compressed. A sufficient large compression will cause the soil springs to be in the failure or plastic state. The pressure on the soil springs is called the passive earth pressure (p_p). Simultaneously, the pressure on the soil springs behind the wall will decrease from the at-rest earth pressure. Similarly, a sufficient large deformation will make the soil springs be in the failure or plastic state where the pressure on the soil springs is called active earth pressure (p_a).

As shown in Figure 7.2d, acted on by the unbalanced forces, the wall is displaced and will change the distribution of earth pressures. The earth pressure on the back of the wall is decreased to $p_0 - k_h\delta$ (δ is the lateral displacement of the wall) with the increase of displacement. The minimum lateral earth pressure is the active earth pressure. The earth pressure on the front of the wall is increased to $p_0 + k_h\delta$ due to the forward displacement of the wall. When soil springs develop up to the passive condition, the soil reaction on the passive side ceases increasing and stays at the passive earth pressure. This state is called the plastic state. When the reaction forces of soil springs are smaller than the passive earth pressure at a point, this state is called the elastic state.

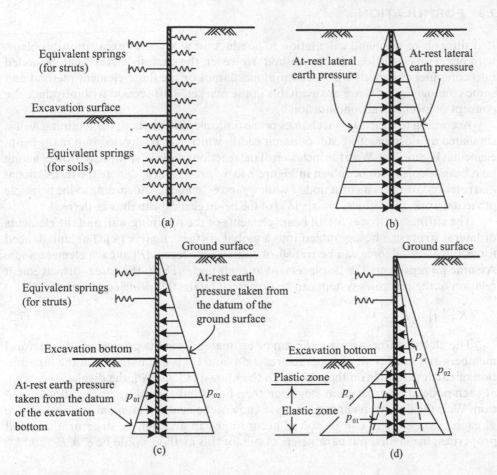

Figure 7.2 Beam on elastic foundation method: (a) springs placed at both sides of a beam, (b) before excavation, (c) after excavation while no wall movement, and (d) earth pressures after wall movement.

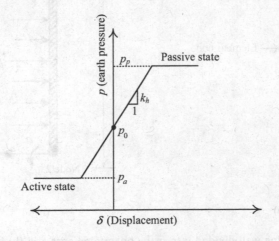

Figure 7.3 Relationship of pressure on the spring and displacement.

7.3 FORMULATION

It is difficult to use hand calculation to conduct an analysis of excavation problems with beam on elastic foundation method. However, the analysis method can be coded into computer programs with structural mechanics or the finite element method and some commercial software are available in the market. This section will introduce the concept of mechanical computation.

According to structural mechanics or the finite element method, the retaining wall is simulated as a continuous beam of a unit width, which will be divided into many beam elements (Figure 7.4a). With the inclusion of the reaction from soil springs, the forces acting on a beam element can be shown in Figure 7.4b, where q_1 and θ_1 denote the translational and rotational displacement at node 1 while q_2 and θ_2 at node 2. According to the principle of virtual work, the stiffness matrix $[K_E]$ of the beam element can thus be derived.

The stiffness matrices for all beam elements of the retaining wall and the elements of lateral struts can be assembled into a global stiffness matrix $[K]$. The unbalanced force and external force can be transformed to be the forces $[P]$ at each element node. Assume $[q]$ represents the displacement at each node. Then, the force–displacement relation of the retaining system can be expressed by the following equation:

$$[K] \cdot [q] = [P] \tag{7.2}$$

The global stiffness matrix $[K]$ can be estimated from the properties of structural members and coefficient of subgrade reaction, and $[P]$ is estimated from the distribution of earth pressures on the wall. With the known $[K]$ and $[P]$, the displacement, $[q]$, at each node at an excavation stage can then be solved according to the above equation. With the displacement at each node known, the bending moment and shear force of each beam element can be solved accordingly. In addition to structural material properties, the main input parameters of soil for this method would be c, ϕ, K_0, and k_h.

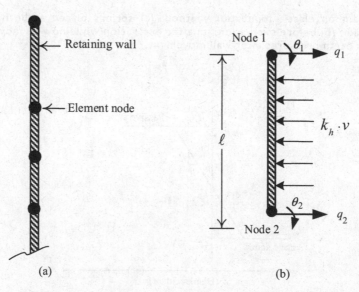

(a)   (b)

Figure 7.4 (a) Retaining wall divided into many beam elements and (b) beam element with the reaction from soil spring.

The distribution of earth pressures on a retaining wall used in the analysis should be as close to the actual distribution of earth pressures on the wall as possible. For granular soils such as sand or gravel, Coulomb's and Caquot and Kerisel's earth pressure theories (Eq. 4.14 and Figure 4.10) should be adopted for active and passive earth pressures, respectively. For cohesive soil such as clay, the earth pressures on the active and passive sides of the wall can be adopted using Eqs. 4.16–4.19 where the adhesion (c_w) between the retaining wall and soils is taken into account. For more details on the discussion of the earth pressure theories, refer to Sections 4.5.3 and 4.6.

Since the magnitude of the friction angle (δ) or the adhesion c_w between the retaining wall and soil is highly to the value of the earth pressure, the selection of the δ and c_w should be as reasonable or as close to the real condition as possible. For more discussion on the δ and c_w, refer to Sections 4.6.2 and 4.6.1, respectively.

7.4   SIMULATION OF CONSTRUCTION SEQUENCE

Soil is with nonlinear and plastic behavior. Any construction activity would cause the stress changes in soil. Therefore, numerical analysis should be conducted with the simulation of each construction activity. Basically, one activity, one phase of analysis. For a typical excavation problem, numerical analysis or beam on elastic foundation analysis should be conducted with the simulation of each excavation step as well as simulation of strut demolish and floor slab construction.

Take a construction of two basements in sandy soil, for example. As shown in Figure 7.5, the foundation slab and concrete floor slabs are assumed to be 0.5 and 0.3 m, respectively. The groundwater level is at the ground surface. The excavation depth is 8.1 m. For conducting the analysis, one should first select a retaining-strut system based on experiences. Then, three levels of struts against the retaining wall are designed. The struts are placed initially as shown in Figure 7.5. A 0.5-m-thick and 14.6-m-deep diaphragm wall is selected, which will also be used as an outer wall of the basement after the completion of basement construction.

As shown in Figure 7.6, a total of 16 steps should be simulated, which include dewatering, excavation, strut installation, preloading, strut demolish, and floor slab

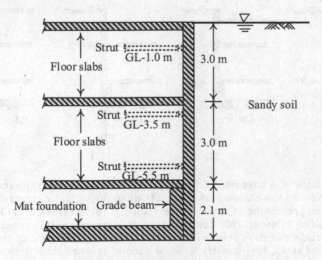

Figure 7.5 Profile of a two-story basement with mat foundation.

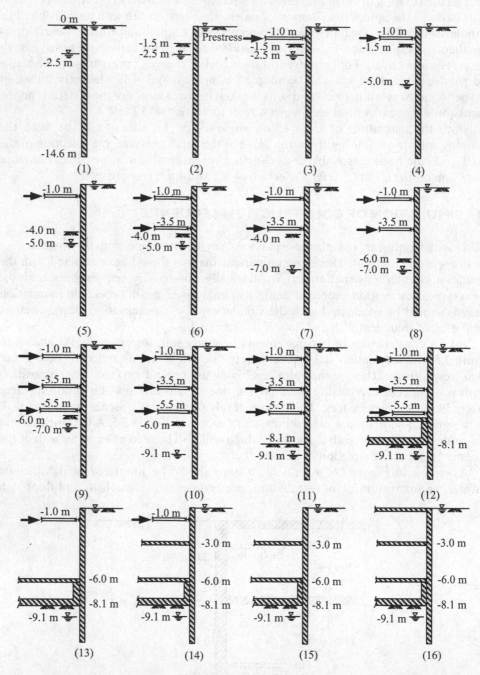

Figure 7.6 Procedure of a basement construction: (1) dewatering, (2) excavation, (3) installation and preloading of strut, (4) dewatering, (5) excavation, (6) installation and preloading of strut, (7) dewatering, (8) excavation, (9) installation and preloading of strut, (10) dewatering, (11) excavation, (12) construction of mat foundation and third level of floor slab, (13) demolishment of second and third levels of strut, (14) construction of second level of floor slab, (15) demolishment of first level of strut, and (16) construction of first level of floor slab.

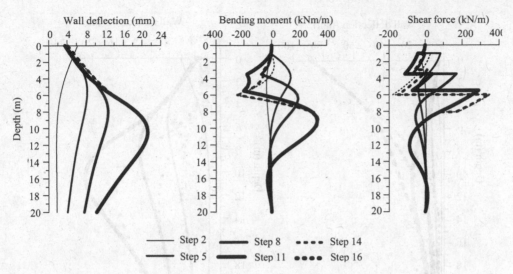

Figure 7.7 Deformation, bending moment, and shear force diagrams of the retaining wall computed at main stages.

construction. Each step should be treated as an individual phase of analysis. Figure 7.7 shows the computed deformation, bending moment, and shear force diagrams of the retaining wall at some main stages using the beam on elastic foundation method. As shown in this figure, the wall deformation, bending moment, and shear force mostly come from the excavation process. Strut demolish steps produce an insignificant change in the deformation and stresses of the retaining wall.

Although one activity is suggested to be treated as one analysis phase, sometimes for simplification, several steps of excavation or construction are often merged into one analysis phase. Taking Figure 7.6 as an example, dewatering and excavation may be combined into one analysis phase, steps 1 and 2 one analysis phase, steps 4 and 5 another analysis phase, and so on. The analysis results show that the wall deformations for the combined steps are the same as those with individual steps (Figure 7.6). Combination of dewatering and excavation steps into one analysis phase is feasible.

Some merge the strut installation, strut preloading, dewatering, and excavation into one analysis phase. For example, steps 1 and 2 are merged into one analysis phase, steps 3–5 into one phase, and so on. Figure 7.8a shows that the computed deflections of the retaining wall for those combined steps are very different from those without combined. Combination of strut installation, strut preloading, dewatering, and excavation into one analysis phase is obviously unreasonable. Some merge dewatering, excavation, strut installation, and preloading into one analysis phase. For example, steps 1–3 are combined into one analysis phase. As shown in Figure 7.8b, the wall deflections for those combined are smaller than those without combined. Such a combination is also unreasonable.

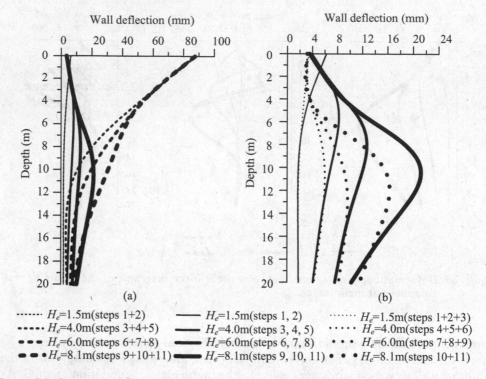

Figure 7.8 Comparison of wall deflections with the simulation of individual activity and combined activities.

7.5 ESTIMATION OF COEFFICIENT OF SUBGRADE REACTION

Vesic (1961) derived the coefficient of subgrade reaction for an infinitely long beam with the width, B, acted on by a concentrated load (see Figure 7.1) as

$$k_s = 0.65 \cdot \sqrt[12]{\frac{E_s B^4}{E_b I}} \frac{E_s}{B(1-\mu_s^2)} \tag{7.3}$$

where

E_s = secant Young's modulus of soils
E_b = Young's modulus of the beam
μ_s = Poisson's ratio of the soil
B = width of the beam
I = moment inertia of the beam

As shown in Figure 7.1b, the width of beam also can be regarded as the loading area. According to Eq. 7.3, we can find that the coefficient of subgrade reaction is not basic soil properties because it is related to not only the properties of soil but also the loading area (B) and the properties of the beam such as Young's modulus (E_b) and moment inertia (I).

Since excavations are usually carried out in stages, the penetration depth, that is, the loaded area, decreases with the progress of excavation. Therefore, the coefficient of horizontal subgrade reaction (k_h) for the soils in front of the retaining wall would increase with the progress of excavation. On the other hand, the effective stress of the soil below the excavation bottom in front of wall should decrease with excavation, which certainly reduces the k_h. Currently, there are no rigorous formulas available to consider all of the influencing factors in excavation problems, such as penetration depth, and basic soil and structural properties. However, some investigators have proposed empirical equations to estimate the k_h, correlated with the strength parameters of soils and other basic soil parameters. The author performed a series of back analyses of many excavation histories and proposed the following empirical formula to estimate k_h by

$$\text{For clay } k_h = \left(40 \text{ to } 50\right) s_u \tag{7.4}$$

$$\text{For sand } k_h = \left(700 \text{ to } 1,000\right) N \tag{7.5}$$

where

k_h = coefficient of subgrade reaction. The unit is kN/m^3
s_u = undrained shear strength of soils. The unit is kN/m^2
N = SPT value

The beam on elastic foundation method does not take the excavation width into account. Therefore, the computed wall deformations are all the same for different excavation widths. However, as shown in Figure 6.4, the wall deformation increases with the excavation width. In general, the wall deformation is inversely proportional to the excavation width. In other words, the coefficient of subgrade reaction decreases with the excavation width, which can be expressed as

$$k_h \propto \frac{1}{B} \tag{7.6}$$

where B is the excavation width.

Eqs. 7.4 and 7.5 are applicable to the excavation width of about 40 m. If the excavation width is different from 40 m, the k_h value should be adjusted according to Eq. 7.6.

In addition to Eqs. 7.4 and 7.5, one can obtain the coefficient of subgrade reaction of an excavation by back analyzing case histories with well-documented field monitoring data and with similar geological formation, construction sequence, excavation depth, and excavation width. Back analysis can also be applied to staged construction. The k_h can be back analyzed through the earlier stages of excavation and use the obtained k_h to predict the deformation at the final or critical stage of excavation. For further information regarding back analysis, refer to Section 7.8.

7.6   ESTIMATION OF COEFFICIENT OF THE AT-REST EARTH PRESSURE

The coefficient of earth pressure at rest, K_0, is required to input to obtain an initial earth pressure. For the drained material, the K_0 value can be estimated directly from Jaky's equation (Eq. 4.3).

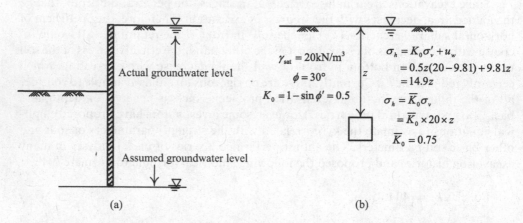

Figure 7.9 Estimation of initial stresses in clay: (a) assumption of groundwater level and (b) modification of the coefficient of the at-rest earth pressure.

For the undrained material such as clay, the total stress analysis should be used. The groundwater level must be assumed to not exist by placing it at the bottom of the wall (Figure 7.9a), but the saturated unit weight, γ_{sat}, for the soil below the groundwater level should be adopted, as stated in Section 2.9.3.2 The total stress parameters such as $c = s_u$, $\phi = 0$, and $K_a = K_p = 1$ should be used. The coefficient of at-rest earth pressure should be modified as the ratio of the total lateral earth pressure to the total overburden pressure, which is different from that of the finite element method (Section 8.6.1). As shown in Figure 7.9b, if we assume the saturated unit weight of clay to be 20kN/m^3 and $K_0 = 0.5$, the modified coefficient of at-rest earth pressure is then $\overline{K}_0 = \sigma_h / \sigma_v = 0.75$.

7.7 ESTIMATION OF STRUCTURAL PARAMETERS

As discussed in Section 3.3.1, H steels and rail piles are commonly used soldier piles. For the dimensions and related properties of H steels and rail piles, please refer to books on steel structures or steel structure design manuals. The nominal Young's modulus for solider piles is 2.04×10^5 MPa. Theoretically, the stiffness (EI) does not need reduction in analysis. Considering the repeated use of solider piles, which decreases their stiffness as a result, therefore, the nominal Young's modulus is usually reduced by 20%.

The nominal Young's modulus for sheet piles is also 2.04×10^5 MPa. Some people consider that sheet piles are not rigidly joined together and advise the nominal moment of inertia per unit width be reduced by 40%. The author, however, does not think it necessary to take the question of joining into consideration since it is an analysis on the basis of plane strain, that is, only the vertical stiffness is to be considered. Therefore, the nominal stiffness is recommended for use. Considering the repeated use of sheet piles, however, the stiffness can be assumed to be 80% of the nominal value in analysis. When analyzing the three-dimensional behaviors of sheet piles, on the other hand, the joining should be considered and a suitable reduction factor for the horizontal stiffness should be taken into account.

Young's modulus for diaphragm walls is basically determined according to the compressive strength of concrete. According to the ACI Code, Young's modulus for concrete can be estimated using the equation: $E = 4,700\sqrt{f_c'}$ MPa where f_c' is the 28th-day compressive strength of concrete. Considering the possibility of bending moment-induced cracking in concrete and the reduction of the sectional modulus accordingly, the stiffness (EI) is usually reduced by 20%–40% in analysis.

Figure 7.10 is the wall bending moments in the main observation section of the TNEC excavation (Ou et al., 1998). The solid line refers to the bending moments obtained from strain gauges of main steel reinforcements in the diaphragm wall. Because the cracking of the diaphragm wall will influence the measurement results of the strain gauges, the computed bending moments contain the effect of cracking of the diaphragm wall. The dotted line in the figure represents the bending moment computed from the deformation curve of the diaphragm wall. The computation is as follows: First, compute the radius of the curvature by differentiating twice the multinomial function simulating the deformation curve. Assuming that EI is not reduced, the moment at a certain depth of the diaphragm wall can be obtained using the equation: $M = EI / r$, where r is the radius of curvature.

The bending moment computed from the wall deformation curve excludes the effect of the diaphragm wall cracking. The ratios of the bending moments from the strain gauges to those from wall deformation curves are then the reduction factors (R), as shown in Figure 7.11. From the figure, we can see that the reduction factors at different depths are different. Basically, the reduction factors at the top and bottom of the wall are close to 1.0 and those near the excavation bottom are as low as 0.5. In analysis, we can then assign different reduction factors for different depths of the diaphragm wall.

As elucidated in Section 3.3.3, column piles can be distinguished into packed-in-place (PIP) piles, reinforced concrete piles, and mixed piles. Reinforced concrete piles can further be classified into reverse circulation drill piles and all-casing piles. The value of f_c' for PIP piles is about 17 MPa and that for reinforced concrete piles is about 28 MPa. In analysis, it can be reduced by 50%–70%. Which value of the reduction factor is to be taken, however, depends on the construction quality of the column pile. Besides, if considering the different degrees of cracking of concrete at different depths of the column pile, we can assign different reduction factors at different depths of the column pile, which is similar to the approach for diaphragm walls.

The value of f_c' for mixed piles is about 0.5 MPa. With rather low strength, the stiffness of mixed piles can be ignored and we can, instead, only consider the stiffness of the H steel (or W section) within mixed piles.

Struts or floor slabs only bear axial forces. Thus, they can be simulated as springs, whose stiffness can be estimated as follows:

$$k = \frac{AE}{L} \tag{7.7}$$

where A = cross-sectional area of a strut or floor slab, E = Young's modulus, and
    L = length of the strut or floor slab, usually half of the excavation width

In the field, struts are installed by splicing H steels together, which are thus not easily lined up straight to achieve nominal axial stiffness. What's more, the bending

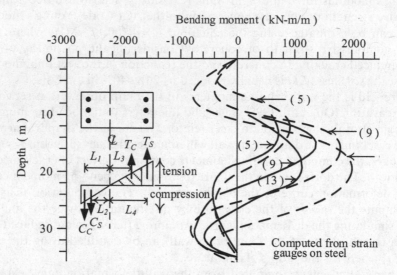

Figure 7.10 Variations of bending moments of the diaphragm wall in the TNEC excavation.

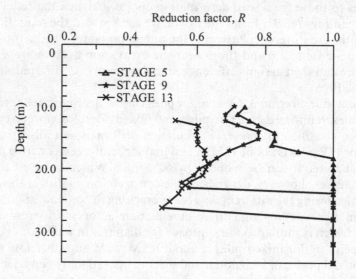

Figure 7.11 Reduction factors of bending moments for the diaphragm wall in the TNEC excavation.

phenomena caused by heave of the center posts will reduce their axial stiffness. In analysis, the axial stiffness of a strut has to be reduced.

The axial stiffness of floor slabs of the top-down construction method has also to be reduced. The reasons are two. One involves whether the compressive strength of concrete in the field agrees with the design strength. The other is cracking due to concrete shrinkage.

Under the same conditions, the axial stiffness of floor slabs is less affected by construction quality, compared to struts. The axial stiffness of floor slabs is thus less reduced. According to experience, the stiffness of a floor slab is about 80% of the nominal value whereas that of a steel strut is 50%–70% of the nominal value.

7.8   DIRECT ANALYSIS AND BACK ANALYSIS

Input the soil parameters derived from soil tests into the computer program, and the thus derived results can represent the behavior of the excavation. The method is called the direct analysis. In theory, if the soil model and soil parameters can thoroughly simulate the soil behavior, the direct analysis can result in reasonable and accurate results.

However, the stress–strain behavior of soils is in nature anisotropic and is influenced by stress paths. The existing testing methods may not fully simulate the behavior of in situ soils. The available soil models may not appropriately simulate soil behaviors, either. Therefore, the soil parameters derived from soil tests sometimes have to be adjusted. If the adjustment procedure has a certain regularity, reproducibility, and consistency and is also applicable to any soil and construction conditions, the method is also called the direct analysis method.

On the other hand, we can take measurement data as an object, modify the parameters of the soil model to make the analysis results match the measurements, and then use the same soil parameters for the prediction of excavations with similar geological conditions, construction sequence, and procedures, which usually result in satisfactory results. The analysis method is called the back analysis. Back analysis can also be applied to staged construction. That is, one uses the observations at the initial stages as an object to back analyze the soil parameters, which are then used for the prediction of behavior at the final or critical construction stage. The method can also obtain fairly satisfying results. As early as 1969, Peck utilized the method to analyze geotechnical problems and designated it as the observational method.

As a matter of fact, any soil models usually require some parameters. Besides, soils below the ground surface are usually multi-layered. As a result, if we assume that all of the parameters are unknown, to carry out a back analysis will be time-consuming and expensive. A reasonable way is to focus on those parameters that cannot be derived from tests or cannot be reasonably estimated. Those parameters that are relatively reliable should be obtained from soil tests. For example, with the beam on elastic foundation method, the s_u value of saturated clay can be obtained from soil tests. The only parameter that has to be back analyzed is k_h. The procedure of back analysis can then be simplified substantially.

7.9   COMPUTATION OF GROUND SURFACE SETTLEMENT

The beam on elastic foundation method can only yield the deformation of a retaining wall but is not capable of computing the ground settlement. As discussed in Chapter 6 about the characteristics of ground settlement, under normal conditions, the factors influencing wall deformation also influence ground settlement. These factors include strut stiffness, strut preload, safety factor for stability, excavation depth, and excavation width. The beam on elastic foundation method takes these factors all into consideration and thus is capable of analyzing wall deformation. Since deformation of a retaining wall relates closely to ground settlement, we can estimate the ground settlement on the basis of the computed wall deformation (see Section 6.8).

Chapter 6 introduces Peck's method, Bowles' method, Clough and O'Rourke's method, and Ou and Hsieh's method to estimate ground settlement. The last three methods have to estimate the lateral deformation or the maximum lateral deformation of the retaining wall before analysis. The ground surface settlement caused by excavation can be classified into the spandrel and concave types. Peck's method does not separate the two types. Bowles' method can only predict the spandrel type of ground surface settlement. Though separating the spandrel and concave types, Clough and O'Rourke's method does not provide a method to distinguish the two. Ou and Hsieh's method, on the other hand, offers a quantified method to distinguish between the spandrel and concave types.

In prediction, Bowles' method can only be used with the lateral wall deformation given. Clough and O'Rourke's method needs the maximum lateral wall deformation. Ou and Hsieh's method needs the lateral wall deformation. These data can all be obtained using the beam on elastic foundation method. According to the case studies (see Section 6.8.4), Ou and Hsieh's method can reasonably predict the excavation-induced ground surface settlement (see Figure 6.30). In this section, we will elucidate the procedure of prediction using Ou and Hsieh's method as follows:

1. Predict the lateral wall deformation using the beam on elastic foundation method.
2. The lateral wall deformation known, compute the areas of the cantilevered part (A_c) and the deep inward movement (A_s). Determine the type of ground surface settlement with the help of Figure 6.14.
3. Estimate the maximum ground surface settlement (δ_{vm}) according to its relation with the maximum lateral wall deformation (δ_{hm}), as shown in Figure 6.18.
4. According to the settlement type determined in step 2, compute the ground surface settlements in different positions behind the retaining wall using Figure 6.29.

The above analysis process is all elucidated in detail in Chapter 6.

7.10   LIMITATIONS OF BEAM ON ELASTIC FOUNDATION METHOD

Basically, the beam on elastic foundation method is an analysis method based on the plane strain condition (whose definition is given in Appendix C). As discussed in Section 6.9, however, in excavations with short sides, influenced by the corner effect, the deformations of the retaining wall or the ground surface settlements will be smaller than those in the plane strain condition. In analysis, we can first analyze the wall deformation under the plane strain condition using the beam on elastic foundation method and then modify the analysis result using the concept of plane strain ratio (PSR) and Figure 6.33. Or, we can also assume that the corners of the retaining wall are restrained and adopt the techniques of structural analysis to analyze the difference between the deformation of the "object section" and that of the section under the plane strain condition. Then, modify the results of the beam on elastic foundation method.

Figure 7.12a illustrates soil improvement over the whole excavation area within a certain depth. From the excavation profile as shown in Figure 7.12b, we can compute the

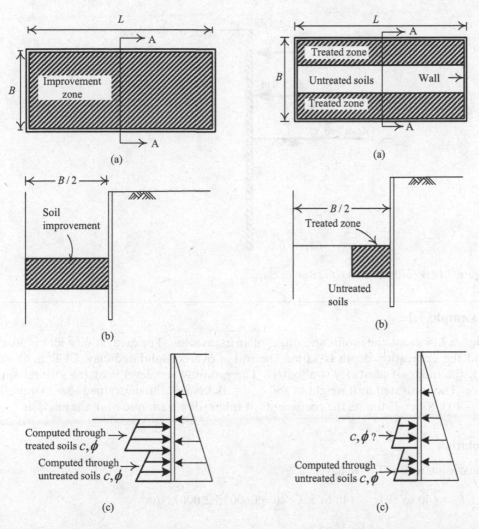

Figure 7.12 Soil improvement implemented to the whole site: (a) plan, (b) profile, and (c) distribution of earth pressure.

Figure 7.13 Soil improvement implemented to a part of the excavation zone: (a) plan, (b) profile, and (c) distribution of earth pressure.

distribution of earth pressures as shown in Figure 7.12c, adopting the parameters of the treated soils within the range of the depth where the soil is treated. Figure 7.13a shows the condition where only a part of the excavation site is treated. From the excavation profile as shown in Figure 7.13b, though the deformation of the retaining wall and ground surface settlement in the central section are under the plane strain condition, the strength parameters needed for the computation of earth pressures and coefficient of subgrade reaction may not be computed directly from the parameters of the treated soils (Figure 7.13c). They are not necessarily obtained through weighting with regard to comparative areas, either. The finite element method certainly can be used to analyze the excavation behavior.

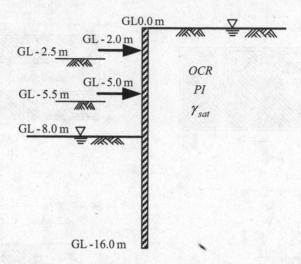

GL0.0 m

GL - 2.0 m

GL - 2.5 m

GL - 5.0 m

GL - 5.5 m

GL - 8.0 m

GL - 16.0 m

OCR

PI

γ_{sat}

Figure 7.14 Profile of an excavation in clay.

Example 7.1

Figure 7.14 shows the profile and stages of an excavation. The excavation width is 30 m, and the excavation depth is 8.0 m. The soil is overconsolidated clay. OCR is about 1.5; the index of plasticity is about 17. The groundwater level is at the ground surface. The saturated unit weight of soil $\gamma_{sat} = 18$ kN/m³; the undrained shear strength $s_u = 40$ kN/m². Estimate the coefficients of subgrade reaction used in the analysis.

Solution

According to Eq. 7.4, we have

$$k_h = (40 \text{ to } 50)s_u = (40 \text{ to } 50) \times 40 = 1,600 \text{ to } 2,000 \text{ kN/m}^3$$

Example 7.2

Figure 7.15 shows the profile and stages of an excavation. The soil is sandy, and the N values from the standard penetration test are as shown in the figure. The unit weight of soil $\gamma = 15$ kN/m³, and the effective strength parameters $c' = 0$ and $\phi' = 32°$. Estimate the coefficient of subgrade reaction used in the analysis.

Solution

According to Eq. 7.5, the coefficient of subgrade reaction can be estimated asAt a depth of 4 m

$$k_h = (700 \text{ to } 1,000)N = (700 \text{ to } 1,000) \times 8 = (5,600 \text{ to } 8,000) \text{ kN/m}^3$$

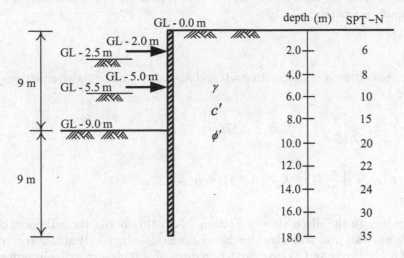

Figure 7.15 Profile of an excavation in sand.

At a depth of 10 m

$$k_h = (700 \text{ to } 1{,}000)N = (700 \text{ to } 1{,}000) \times 8 = (14{,}000 \text{ to } 20{,}000) \text{kN/m}^3$$

At a depth of 14 m

$$k_h = (700 \text{ to } 1{,}000)N = (700 \text{ to } 1{,}000) \times 8 = (16{,}800 \text{ to } 24{,}000) \text{kN/m}^3$$

At a depth of 18 m

$$k_h = (700 \text{ to } 1{,}000)N = (700 \text{ to } 1{,}000) \times 8 = (24{,}500 \text{ to } 35{,}000) \text{kN/m}^3$$

Example 7.3

Suppose an excavation with a diaphragm wall in soft clay. A computer program based on the beam on elastic foundation method is to be used for analysis. From the user's manual, the active and passive earth pressures used in the computer program are as follows:

$$\sigma_a = K_a \sigma_v - \frac{c}{\tan\varphi}\left[\frac{\cos\delta - \sin\varphi\cos\gamma}{1 + \sin\varphi} \cdot e^{-(\gamma-\delta)\tan\varphi}\cos\delta - 1\right]$$

$$\sigma_p = K_p \sigma_v + \frac{c}{\tan\varphi}\left[\frac{\cos\delta + \sin\varphi\cos\gamma}{1 - \sin\varphi} \cdot e^{(\gamma+\delta)\tan\varphi}\cos\delta - 1\right]$$

where K_a and K_p represent the coefficients of the active and passive earth pressures, respectively. σ_v = the overburden pressure; c = cohesion; δ = the friction angle between the wall and soils; $\sin\gamma = \sin\delta / \sin\phi$, where γ is between $0°$ and $90°$. If we use Caquot and Kerisel's earth pressure theory, as introduced in Section 4.5, how should we modify the input parameters?

Solution

Since the excavation is in clay, with $\phi = 0$ and $K_a = K_p = 1$, the above equations can be simplified as

$$\sigma_a = K_a\sigma_v - c\left[\frac{\pi}{2}+1\right] = K_a\sigma_v - 2.57c = \sigma_v - 2.57c$$

$$\sigma_p = K_p\sigma_v + c\left[\frac{\pi}{2}+1\right] = K_p\sigma_v + 2.57c = \sigma_v + 2.57c$$

According to the discussion in Section 5.5 in this book, the adhesion between a diaphragm wall and soft clay can be assumed $c_w = 2c/3$. Besides, according to Section 4.6, the active and passive earth pressures of soft clay on a diaphragm wall are

$$\sigma_a = K_a\sigma_v - 2c\sqrt{\left(1+\frac{c_w}{c}\right)K_a} = \sigma_v - 2.58c$$

$$\sigma_p = K_p\sigma_v + 2c\sqrt{\left(1+\frac{c_w}{c}\right)K_p} = \sigma_v + 2.58c$$

As discussed above, when feeding the computer program with parameters $\phi = 0$ and $K_a = K_p = 1$, the earth pressures computed by the computer program will be close to the commonly computed earth pressure with the assumption $c_w = 2c/3$.

PROBLEMS

7.1 The profile and stages of an excavation are shown in Figure 7.14. Assume that the soil is overconsolidated clay whose OCR is about 1.7. The groundwater level is at the ground surface. The plasticity index is about 30. The unit weight of the soil $\gamma_{sat} = 16.7$ kN/m^3. The undrained shear strength $s_u = 34.3$ kN/m^2. Estimate the coefficient of subgrade reaction for the soil in front of the wall.

7.2 Same as Problem 7.1. Suppose the stiffness of the two levels of struts is the same, i.e., 35,000 kN/m^2 (the reduced value). The stiffness of the retaining wall (reduced value) is 160,000 kN/m^3. Analyze the excavation-induced deformation of the retaining wall using the coefficients of subgrade reaction computed in Problem 7.1 and a computer program based on the beam on elastic foundation method, and compare the result with those obtained from the empirical equation in Chapter 6 to judge the rationality of the result.

7.3 Assume that the profile and stages of an excavation are as shown in Figure 7.14. The excavation depth is 8.0 m. The soil is sandy, and the groundwater level is at the ground surface. The saturated unit weight of the soil $\gamma_{sat} = 21.6$ kN/m^3. The effective strength parameters $c' = 0$ and $\phi' = 32°$. The SPT-N value at a depth of 12 m below the ground surface before excavation $N = 15$. Estimate the coefficient of subgrade reaction for the soil in front of the wall (to simulate the actual conditions, the seepage should be considered in analysis).

7.4 Same as Problem 7.3. Assume that the stiffness of the first and second levels of struts is 20,000 and 30,000 kN/m² (reduced values). The stiffness of the retaining wall (EI) is 150,000 kN/m³ (reduced value). Analyze the deformation of the retaining wall using the coefficient of subgrade reaction obtained in Problem 7.3 and a computer program based on the beam on elastic foundation method, and compare the result with the empirical equations introduced in Chapter 6 to judge the rationality of the analysis result.

7.5 Same as Example 6.1. Figure 6.43 shows the profile and stages of the excavation of Building P in Taipei. The staged deformations of the retaining wall are as shown in Figure 6.44. Estimate the coefficient of subgrade reaction, and analyze the excavation-induced deformation using a computer program based on the beam on elastic foundation method. Then, compare the result with Figure 6.44.

7.6 Same as Problem 6.11. Figure P6.11 shows the geological and excavation profiles for the Building S excavation. Figure P7.6 shows the measured lateral deformations of the retaining wall in the central section of the longer side (89 m side) of the site. Estimate the coefficient of subgrade reaction, and analyze the excavation-induced deformation using a computer program based on the beam on elastic foundation method. Compare the result with empirical value.

7.7 Same as Problem 6.4. Figure P6.4 shows the excavation profiles of the top-down construction of Building R in Taipei along with geological profile. Figure P7.7 shows the measured lateral displacements of the retaining wall in the central section of the longer side of the site. Analyze the lateral displacements of the retaining wall using a beam on elastic foundation computer program with the suggested method introduced in this chapter, and compare the results with Figure P7.7.

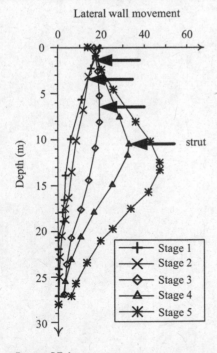

Figure P7.6

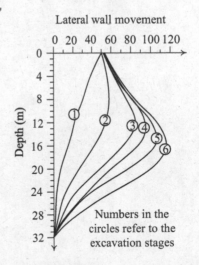

Figure P7.7

Chapter 8

Stress and deformation analysis

Finite element method

8.1 INTRODUCTION

As explained in Chapter 7, under normal excavation conditions, excavation-induced stress and deformation are engendered by unbalanced forces acting on the wall due to the removal of soils within the excavation zone. The magnitude of the unbalanced forces is influenced by many factors: the conditions of soil layers, the level and pressures of groundwater, the excavation depth, the excavation width, and so on. The finite element method is capable of simulating these factors, and therefore, the results derived from the method would be more accurate than those derived from simplified methods (Chapter 6) or the beam on elastic foundation method (Chapter 7). This chapter starts with the introduction of the basis of the finite element method (FEM) for elastic material. Since soil is with nonlinear and plastic behavior and rate of dissipation of excess pore water pressure causes the soils with drained or undrained behavior, all those would affect the analysis results significantly. Formulation of those behavior in the finite element analysis procedure is extremely complicated, and this chapter tends to explain them in a simple and understandable way. The persons equipped with those basic knowledge will have good sense in performing finite element analysis of geotechnical or excavation problems. Moreover, the commonly used soil models such as Mohr-Coulomb model, Duncan and Chang model, hardening soil model are introduced in an understandable way. Evaluation of the parameters used for analysis for those models under drained or undrained condition are also provided. Following those procedures, the persons performing an analysis are expected to obtain a reasonable analysis results.

Besides, some researchers code the governing equation in the form of an explicit finite difference equation and solve it by way of dynamic relaxation. The method solves the velocity and movement through the movement equation by assigning a damping value close to the critical damping. The strain rate is then obtained from velocity and used to solve the new stress increment. The process continues until the unbalanced forces are in equilibrium or the system reaches a steady state. The main theory on which the finite difference method is based is not the same as that of the FEM. However, other theories, such as constitutive laws of soil, drained or undrained behaviors, determination of soil parameters, simulations of excavation, etc., are identical with the FEM. If readers adopt the finite difference method to analyze deep excavation, this chapter can also be used for reference.

DOI: 10.1201/9780367853853-8

8.2   FRAMEWORK AND PRINCIPLES

This section only succinctly elucidates the basic principles of the FEM. The objective is to have analysts with a basic understanding on the theory if they are not equipped with full finite element theory but have to perform finite element analysis (FEA) of deep excavation problems.

8.2.1   Linear elastic behavior

As shown in Figure 8.1, set the domain of the influence by construction and boundary condition, and then divide the whole domain into many small elements, called finite elements. The boundaries include external force and/or displacement (Figure 8.1). Depending on the material properties of the elements, establish the stress–strain relation, which is called the constitutive law. In general, the constitutive law for the elastic isotropic material can be expressed as follows:

$$\{\sigma\} = [D]\{\varepsilon\} \tag{8.1}$$

where $\{\sigma\}$ = the stress matrix; the sign $\{\ \}$ refers to a column matrix. In a three-dimensional (3D) space, the stress matrix contains six components such as σ_x, σ_y, σ_z, τ_{xy}, τ_{yz}, and τ_{zx}. In a plane strain or stress condition, it contains only 3 components.

$\{\varepsilon\}$ = the strain matrix. In 3D spaces, the strain matrix contains six components such as ε_x, ε_y, ε_z, γ_{xy}, γ_{yz}, and γ_{zx}. In plane strain or stress conditions, it contains only 3 components.

$[D]$ = the stress–strain or constitutive matrix. The main entries are the Young's modulus (E) and Poisson's ratio (μ), which are also called the deformation parameters.

As shown in Figure 8.2, the relation between the displacement $\{u\}$ at any point within the element and the displacement $\{q\}$ at the nodal points of the element can be expressed as follows:

$$\{u\} = [N]\{q\} \tag{8.2}$$

where $[N]$ is the displacement shape function.

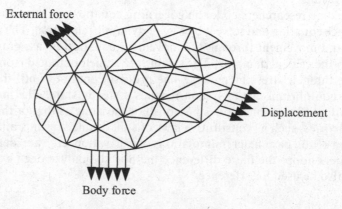

Figure 8.1 Finite element mesh and boundary conditions.

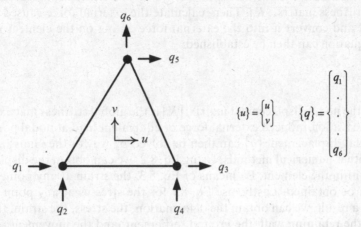

Figure 8.2 Three-node element.

In accordance with the theory of elasticity, the strain and displacement at a point within the element have the following relation:

$$\{\varepsilon\} = [L]\{u\} = [L][N]\{q\} = [B]\{q\} \tag{8.3}$$

where $[L]$ = the linear partial differential operator, such as $\partial/\partial x$, $\partial/\partial y$, etc.;

$[B] = [L][N]$ is the relational matrix between the strain and the nodal displacement.

In accordance with the principle of virtual work, we can derive the work done by the internal force (left side of the equation) and the external force (right side of the equation) in an element as shown in Eq. 8.4, which further can be simplified as the relation between the internal force (the left side of the equation) and external force (the right side of the equation) as expressed by Eq. 8.5.

$$\int_{vol} \{\delta\varepsilon\}^T [\sigma]\, d(vol) = \int_{vol} \{\delta u\}^T \{G\} d(vol) + \int_{area} \{\delta u\}^T [T]\, d(area) \tag{8.4}$$

$$\int_{vol} [B]^T [\sigma]\, d(vol) = \int_{vol} [N]^T \{G\}\, d(vol) + \int_{area} [N]^T [T]\, d(area) \tag{8.5}$$

where $[\delta\varepsilon]$ is the strain increment, $[\sigma]$ is the state of the stress at the current stage, $[\delta q]$ is the displacement increment, $[G]$ is the body force, and $[T]$ is the traction, which is the surface force acting on the element.

The left side of Eq. 8.5 represents equivalent nodal forces produced by internal forces, whereas the right side denotes equivalent nodal forces generated by external forces. The stiffness matrix, $[K_E]$, of the element can be derived by further simplifying Eq. 8.5 as below:

$$[K_E] = \int_{vol} [B]^T [D][B] dV \tag{8.6}$$

After establishing the stiffness matrices for all the elements, assemble them into the global stiffness matrix, $[K]$. Then, calculate the external force caused by excavation or load and convert it into the external force acting on the element nodes. The following equation can then be established:

$$[K]\{q\} = \{F\} \tag{8.7}$$

where $[q]$ is the nodal displacement matrix; $[K]$ is the global stiffness matrix; $\{F\}$ is the matrix of excavation-induced external force or equivalent load at nodal points.

The nodal displacement $\{q\}$ can then be solved by way of the Gauss elimination method or other numerical methods. Using Eq. 8.2, we can obtain the displacement at any point within the element. By means of Eq. 8.3, the strain at any point within the element can be obtained. Lastly, use Eq. 8.1 for the stresses at any point within the element. As a result, we can obtain the deformation, the stress, the strain, the bending moment of the retaining wall, the ground settlement, and the movement of the excavation bottom.

In accordance with Eq. 8.3, when the displacement shape function $[N]$ is quadratic, differentiated by the partial differential operator $[L]$, the matrix $[B]$ becomes linear, which shows that the strain within the element changes linearly. The element within which the strain changes linearly is called the low-order element. Otherwise, elements whose strain changes nonlinearly are called high-order elements.

As the order of the shape functions of high-order elements is higher than those of the low-order elements, the number of nodes of high-order elements is larger than that of low-order elements. High-order elements are more capable of coping with the rapid change of the stress or strain within an element and are thereby more accurate than low-order elements.

The commonly used deformation parameters, in addition to E and μ in $[D]$ in Eq. 8.1, can also be represented by the shear modulus G, the bulk modulus K, and the constrained modulus M. Their definitions are explained in Section 2.8.1.

8.2.2   Plastic behavior consideration

Figure 2.26 shows the soil subject to plastic strain when the state of stress at the yield surface B′B″. One of the purposes of constitutive soil models developed is to find a suitable yield surface function.

Figure 8.3 shows an elastoplastic material or soil subject to a stress increment ($\Delta\sigma$). The total strain increment ($\Delta\varepsilon$), elastic strain increment ($\Delta\varepsilon^e$) and plastic strain increment ($\Delta\varepsilon^p$) have the following relation:

$$\Delta\varepsilon = \Delta\varepsilon^e + \Delta\varepsilon^p \tag{8.8}$$

In the two-dimensional (2D) or 3D condition, the above equation should be expressed in terms of matrix form as

$$\{\Delta\varepsilon\} = \{\Delta\varepsilon^e\} + \{\Delta\varepsilon^p\} \tag{8.9}$$

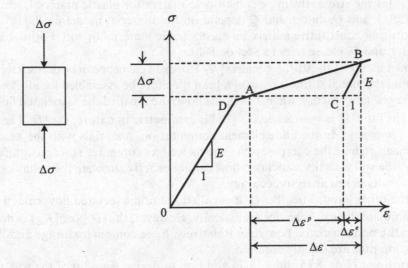

Figure 8.3 Total strain increment, elastic strain increment, and plastic strain increment.

As the stress $\{\Delta\sigma\}$ in the unloading/reloading state, AC as shown in Figure 8.3, produces only the elastic strain $\{\Delta\varepsilon^e\}$, the stress and strain should have the following relation:

$$\{\Delta\sigma\} = [D^e]\{\Delta\varepsilon^e\} \tag{8.10}$$

Then,

$$\{\Delta\varepsilon^e\} = [D^e]^{-1}\{\Delta\sigma\} \tag{8.11}$$

where $[D^e]$ is the elastic constitutive matrix as shown in Eq. 8.1. Substituting $\{\Delta\varepsilon^e\}$ in the above equation into Eq. 8.9, we have the following:

$$\{\Delta\varepsilon\} = \{\Delta\varepsilon^e\} + \{\Delta\varepsilon^p\} = [D^e]^{-1}\{\Delta\sigma\} + \{\Delta\varepsilon^p\} \tag{8.12}$$

Therefore,

$$[D^e]\{\Delta\varepsilon\} = \{\Delta\sigma\} + [D^e]\{\Delta\varepsilon^p\} \tag{8.13}$$

$$\{\Delta\sigma\} = [D^e]\{\Delta\varepsilon\} - [D^e]\{\Delta\varepsilon^p\} \tag{8.14}$$

Assuming that Y represents the yield function of the material or soil and Q represents the plastic potential function, then in accordance with the flow rule, Eq. 8.14 can be derived as follows:

$$\{\Delta\sigma\} = \left([D^e] - [D^p]\right)\{\Delta\varepsilon\} = [D^{ep}]\{\Delta\varepsilon\} \tag{8.15}$$

$$[D^{ep}] = [D^e] - [D^p] \tag{8.16}$$

where $[D^p]$ is the stress–strain or constitutive matrix for plastic material, which is a function of Y and Q, and Y and Q depend on the material model used. $[D^{ep}]$ is the stress–strain or constitutive matrix for elastoplastic material or soil. For more explanation of Y and Q, please refer to Section 8.4.1.

Known the $[D^{ep}]$, the stiffness matrix $[K_E]$ for each element can be derived following Eq. 8.6, and the global stiffness matrix $[K]$ can therefore be assembled for all elements.

In the plasticity theory, when $Y = Q$, the flow rule is called the associated flow rule, and $[D^p]$ in Eq. 8.15 is symmetric. As $[D^e]$ is symmetric in nature, the $[D^{ep}]$ and $[K_E]$ are also symmetric. In the finite element computation, materials with the associated flow rule can reduce the computation time as well as computer storage significantly. However, the soil with the associated flow cannot exactly simulate the behavior of the soil and results in less analysis accuracy.

On the other hand, when $Y \neq Q$, it is called the nonassociated flow rule, which is able to simulate the soil behavior realistically. However, the $[D^{ep}]$ or $[K_E]$ is not symmetric in the nonassociated flow rule. Relatively, huge computer storage and lengthy computation time are usually required.

According to Eqs. 8.15 and 8.16, in addition to derive E and μ of the soil, it is required to have a suitable soil model, that is the yield function Y and plastic potential function Q. Therefore, the required input parameters are E, μ, and the parameters of forming Y and Q, e.g., c' and ϕ'.

8.2.3　Nonlinear behavior consideration

Owing to the nonlinear behavior (nonlinear elastic or nonlinear plastic) of the soil, the global stiffness matrix $[K]$ in Eq. 8.7 is not constant, which would vary with the state of the stress of the soil. Therefore, to solve displacement at each node from Eq. 8.7, it is necessary to perform the nonlinear analysis. Many nonlinear analysis methods available in the literature, such as the tangent stiffness method, viscoplastic method, and modified Newton–Raphson method, each of which is with different strength. The following will summarize the nonlinear analysis procedure, which is divided into 11 steps, with the modified Newton–Raphson method and substepping algorithm stress integration (Figure 8.4):

1. Divide the external load into many small increments $\{\Delta F_{\text{ext}}\} = \Delta \lambda_i \{F_{\text{ext}}\}$, where F_{ext} is the applied external load and $\Delta \lambda_i$ is an incremental ratio.
2. Establish the element stiffness matrices $[K_E]$ for all elements and assembly them into the global stiffness matrix $[K]$ based on the present state of the stress. If the soil model is with elastoplastic behavior, then $[K_E]$ and $[K]$ should consider the yield function Y and plastic potential function Q, that is, adopt $[D^{ep}]$ as shown in Eq. 8.16.
3. Compute the displacement increment at each node $\{\Delta q^k\}$ using the standard FEM procedure where the upper index "k" denotes the iteration number.
4. Compute the strain increment $\{\Delta \varepsilon\}$ following Eq. 8.3.
5. Evaluate the $[D_i^{ep}]$.
6. Divide the strain increment $\{\Delta \varepsilon\}$ into many (e.g., m increments) very small strain increments or strain subincrements, $\{\Delta \varepsilon\}/m$, and then compute the corresponding

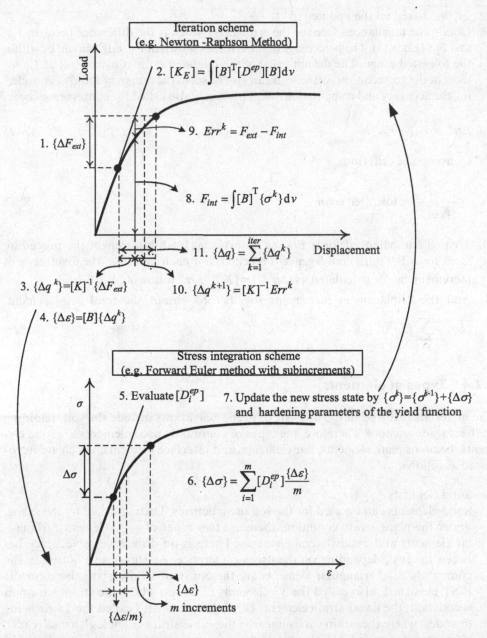

Iteration scheme
(e.g. Newton -Raphson Method)

$2.\ [K_E] = \int [B]^T [D^{ep}][B]\,\mathrm{d}v$

$9.\ Err^k = F_{ext} - F_{int}$

$8.\ F_{int} = \int [B]^T \{\sigma^k\}\,\mathrm{d}v$

Load

$1.\ \{\Delta F_{ext}\}$

$11.\ \{\Delta q\} = \sum_{k=1}^{iter} \{\Delta q^k\}$ Displacement

$3.\ \{\Delta q^k\} = [K]^{-1}\{\Delta F_{ext}\}$ $10.\ \{\Delta q^{k+1}\} = [K]^{-1} Err^k$

$4.\ \{\Delta \varepsilon\} = [B]\{\Delta q^k\}$

Stress integration scheme
(e.g. Forward Euler method with subincrements)

$5.\ \text{Evaluate}\ [D_i^{ep}]$ $7.\ \text{Update the new stress state by}\ \{\sigma^k\} = \{\sigma^{k-1}\} + \{\Delta \sigma\}$
and hardening parameters of the yield function

σ

$\Delta \sigma$

$6.\ \{\Delta \sigma\} = \sum_{i=1}^{m} [D_i^{ep}]\dfrac{\{\Delta \varepsilon\}}{m}$

ε

$\{\Delta \varepsilon\}$
m increments

$\{\Delta \varepsilon/m\}$

Figure 8.4 Procedure for nonlinear finite element computation. (Redraw from Yang (2014).)

small stress increment using the forward Euler method (or other methods, such as the

Runge–Kutta method), and then add them together, that is, $\{\Delta \sigma\} = \displaystyle\sum_{i=1}^{m} \left[D_i^{ep} \right]\dfrac{\{\Delta \varepsilon\}}{m}$

7. Update the new stress state using the following: $\{\sigma^k\} = \{\sigma^{k-1}\} + \{\Delta \sigma\}$ and hardening parameters of the yield function.

8. Obtain the equivalent internal force, F_{int}, using the integration of the left side in Eq. 8.5 based on the updated $\{\sigma^k\}$.

9. Obtain the unbalanced force or the error Err^k, which is the difference between F_{ext} and F_{int} (Eq. 8.17). The percentage of the error, as shown in Eq. 8.18, should be within the tolerated range. The default value of the tolerated error is usually set at 1% for most of the geotechnical software. It can also be set in the ranges of 3%–5%, considering the accuracy and computation rate. This step is also called the convergence check.

$$Err^k = F_{ext} - F_{int} \tag{8.17}$$

Convergence criterion:

$$\frac{\left| F_{ext} - F_{int} \right|}{\left| F_{ext} \right|} < \text{tolerated error} \tag{8.18}$$

10. Treat the unbalanced force Err^k as an external load, and repeat the procedure from steps 1–9 until convergence is reached. For each iteration, the displacement increment can be calculated as $\{\Delta q^{k+1}\} = [K]^{-1} Err^k$ following Eq. 8.7.

11. Add the displacement increments together to obtain the total displacement,

$$\{\Delta q\} = \sum_{k=1}^{\text{iter}} \{\Delta q^k\}.$$

8.2.4 Types of elements

The materials involved in the analysis of deep excavations include the soil, retaining wall, strut, or anchors. Therefore, the types of commonly used elements are solid elements, beam or plate elements, bar elements, and interface elements, which are introduced as follows:

1. Solid elements

    Solid elements can be used for the soil or structures. Distinguished by the shape, under the plane strain condition, there are two types of solid elements: triangular elements and quadrilateral elements. There is no difference in accuracy between the two, depending on the user's preference. As shown in Figure 8.5, the commonly used triangular elements are the constant strain triangular elements (CST elements), also called the T3 elements (3 nodes) where the strain variation is constant; the linear strain elements (LST elements), also called the T6 elements (6 nodes) where the strain varies linearly; the cubic strain triangle elements (CuST elements, also called T15 elements) where the strain changes cubically. Both T3 and T6 elements are low-order elements, and CuST elements (15 nodes) are of a high order. The commonly used quadrilateral elements are Q4 elements and Q8 elements. The former consists of 4 nodes, and strains change linearly, thereby called low-order elements. The latter have 8 nodes, and the strain changes nonlinearly, thus belonging to high-order elements, as shown in Figure 8.6.

    High-order plane strain elements, for example, a Q8 element, are more recommended, considering the accuracy of the analysis. Usually, the accuracy

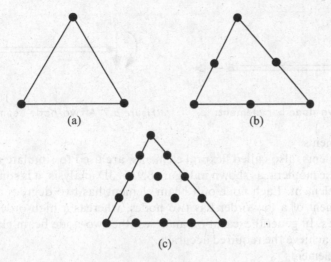

Figure 8.5 Triangular elements (a) CST element (b) LST element (c) CuST element.

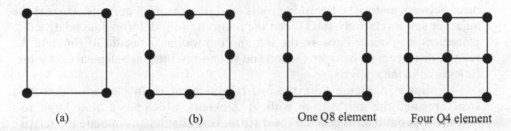

Figure 8.6 Quadrilateral elements (a) Q4 element (b) Q8 element.

Figure 8.7 Comparison of accuracy between a Q8 element and four Q4 elements.

of a Q8 element is better than that of four Q4 elements (Figure 8.7). However, a high-order element has more nodes than a low-order one, therefore require greater computation time and computer storage although use of high-order elements can achieve a higher accuracy. In practical application, one can adopt low-order elements initially to perform calibration and then use high-order elements for formal analysis.

Similarly, 3D analysis certainly requires 3D elements. Although 3D elements can achieve a higher accuracy, it is with too many nodes and therefore requires a lengthy computation time and a huge computer storage. Therefore, most of the geotechnical software only provides low-order 3D elements.

2. Bar elements

Bar elements, also called truss elements or anchor elements, are used to simulate struts, anchors, or other members bearing only the axial stress of structural members, as shown in Figure 8.8. Each node of a bar element has only one degree of freedom. A bar element of a low order has two nodes, whereas a high-order element has three nodes. In general excavation analyses, two node bar elements can achieve good accuracy.

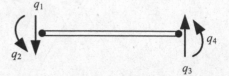

Figure 8.8 A two-node bar element.   Figure 8.9 A two-node beam element.

3. Beam elements

    Beam elements, also called flexural elements, are used to simulate members sub-
    jected to the moment, as shown in Figure 8.9. In 3D analysis, it is sometimes called
    the plate element. Each node of a beam element has two degrees of freedom. A
    beam element of a low order has two nodes, whereas a high-order element has
    three nodes. In general excavation analyses, the two-node beam element is good
    enough to achieve the required accuracy.

4. Interface elements

    The FEM is based on continuum mechanics and is incapable of effectively eval-
    uating the loading and displacement conditions induced by the relative displace-
    ment between materials. Retaining walls used in excavation are stiff, whereas the
    adjacent soil is relatively soft. When the retaining wall deforms, the relative dis-
    placement may occur between the soil and the wall as a result. To simulate the
    relative displacement between the soil and structures, interface elements are some-
    times used in the analysis.

    As shown in Figure 8.10, the interface element is an element connecting
    structures and the soil, with or without thickness, which has a quite large nor-
    mal stiffness but relatively small shear stiffness so that it can simulate the relative
    displacement between the soil and structures. Concerning the studies of interface
    elements, readers can refer to the literature (for example, Goodman et al., 1968;
    Sharma and Desai, 1992).

    Although interface elements can rationally simulate the relative displacement
    between the soil and structures, extra parameters, which are not easily obtained
    from conventional soil tests, will be introduced. If the interface elements are not

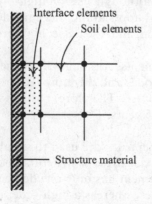

Figure 8.10 An interface element.

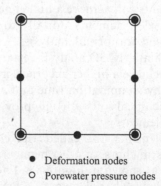

● Deformation nodes
○ Porewater pressure nodes

Figure 8.11 An element for the coupled analysis
(eight deformation nodes and four
porewater pressure nodes).

adopted, the soil in the vicinity of the structure can be considered to divide into fine elements. When the retaining wall is deformed, these fine soil elements can easily attain the plastic state, which will then produce larger deformation. Therefore, a rational analysis result is also attainable.

8.3 EFFECTIVE STRESS ANALYSIS AND TOTAL STRESS ANALYSIS

As mentioned in Section 2.9.3.1, the effective stress analysis treats the target soil as a two-phase material, i.e., the soil and pore water, and the corresponding stress and strain of the soil and pore water are computed, respectively. For the undrained condition, one should establish the constitutive matrix (Eq. 8.1) for the soil and pore water, respectively, which is called the effective stress undrained analysis (Section 2.9.3.1).

Considering a soil element subject to loading, the total stress increment and strain increment produced are $\{\Delta\sigma\}$ and $\{\Delta\varepsilon\}$, respectively. Under undrained conditions, the pore water pressure generated is $\{\Delta u\}$. Based on the principle of the effective stress, the following can be obtained

$$\{\Delta\sigma\} = \{\Delta\sigma'\} + \{\Delta u\} \tag{8.19}$$

In the undrained condition, the soil and pore water will deform simultaneously, so the strain of the soil and pore water should be the same. The effective stress acting on the soil and the pore water pressure of the pore water can be obtained as follows:

$$\{\Delta\sigma'\} = [D']\{\Delta\varepsilon\} \tag{8.20}$$

$$\{\Delta u\} = [D_w]\{\Delta\varepsilon\} \tag{8.21}$$

where $[D']$ and $[D_w]$ represent the constitutive matrix for the soil and pore water, respectively.

In accordance with the effective Young's modulus and effective Poisson's ratio of the soil, $[D']$ can be constructed; similarly, in accordance with the bulk modulus of pore water, $[D_w]$ can be obtained. Substituting Eqs. 8.20 and 8.21 into Eq. 8.19, then we can obtain the following:

$$[D] = [D'] + [D_w] \tag{8.22}$$

The procedure of the effective stress undrained analysis is explained as follows:

1. Input effective soil parameters such as c', ϕ', E', and μ', and construct the soil constitutive matrix, $[D']$.
2. Input the stiffness of pore water, such as the bulk modulus, and construct the pore water constitutive matrix, $[D_w]$.
3. Establish the total constitutive matrix as $[D] = [D'] + [D_w]$.
4. Compute the displacement increment, $\{\Delta q\}$, at each node using the standard finite element computation procedure (Eq. 8.7), i.e., $\{\Delta q\} = [K^{-1}]\{\Delta F\}$.
5. Compute the strain increment, $\{\Delta\varepsilon\}$, in each element, i.e., $\{\Delta\varepsilon\} = [B]\{\Delta q\}$.
6. Compute the stress increment in each element using Eq. 8.1 or Eq. 8.10, i.e., $\{\Delta\sigma'\} = [D']\{\Delta\varepsilon\}$.
7. Compute the pore water pressure increment, i.e., $\{\Delta u\} = [D_w]\{\Delta\varepsilon\}$.

The total stress analysis treats the soil and pore water as a one-phase material. No pore water pressure is generated in this material. Therefore, in the undrained condition, the total stress analysis method does not consider the excess pore water pressure caused by the soil subject to loading. However, the pore water pressure affects the effective stress, thus affecting the strength and stiffness of the soil. Moreover, the excess pore water pressure generated is closely related to the stress path of the soil subject to loading. Therefore, to obtain a reasonable analysis result, the strength parameters (s_u, $\phi_u = 0$) and deformation parameters (E_u, μ_u) should take into account the stress path of the soil or be estimated based on empirical formulas. The procedure of the total stress undrained analysis can be explained as follows:

1. Input the total stress parameters of soil such as s_u, ϕ_u, E_u, and μ_u, and then construct the constitutive matrix, $[D]$, for the one-phase material.
2. Compute the displacement increment $\{\Delta q\}$ at each node using the standard finite element computation procedure (Eq. 8.7), i.e., $\{\Delta q\} = [K]^{-1}\{\Delta F\}$.
3. Compute the strain increment $\{\Delta \varepsilon\}$ in each element, i.e., $\{\Delta \varepsilon\} = [B]\{\Delta q\}$.
4. Compute the total stress increment in each element using Eq. 8.1, i.e., $\{\Delta \sigma\} = [D]\{\Delta \varepsilon\}$.

In the effective stress analysis (drained or undrained), the parameters of the strength and stiffness are soil basic properties, which can be derived theoretically or tested experimentally, and the ground water level should be set following the actual condition. The effective stress undrained analysis can yield the effective stress and pore water pressure, of which two terms are added together to obtain the total stress. In the total stress (undrained) analysis, the parameters of the soil and stiffness are not the basic soil properties, and hence, the parameters are obtained mostly from empirical formulas. No matter where the actual groundwater level is located, the groundwater level in an analysis must be set at the bottom of the finite element mesh to exclude the influence of analysis results. The saturated unit weight should be adopted for the soil below the original ground water level. In the total stress undrained analysis, only the total stress can be obtained, but not the effective stress and pore water pressure.

The forgoing undrained analysis is based on the assumption that the excess pore water pressure is not dissipated at all. However, real soil behavior is often time related, with the pore water pressure response and rate of loading. To account for such behavior, it is necessary to couple the continuity of the equation or general consolidation equation with the constitutive and equilibrium equations. The method is then called the coupled analysis, which is subsumed to the effective stress analysis, that is, the effective stress parameters are required in the analysis. In addition to the parameters of soil models, the coupled analysis also requires the coefficient of the permeability and loading time.

The coupled method uses the displacement and pore water pressure as unknowns and therefore results in both displacement and pore water pressure degrees of freedom at element nodes. As a result, it can compute the displacements and stresses of a soil element as well as the pore water pressures based on the effective stress. Figure 8.11 shows a quadrilateral element used for the coupled analysis where the element has eight displacement nodes, four of which are simultaneously pore water pressure nodes.

If using the coupled method to analyze the completely undrained behavior, we could simply set the time of loading at a very short time. For example, assume there is a site to be excavated 4 m deep. The coupled method simply sets the 4 m excavation to be finished within a very short time when considering the undrained condition. If considering the drained condition, the excavation time can be set with a very long period.

8.4 COMMONLY USED SOIL MODELS AND RELATED PARAMETERS

Figure 8.12 shows a typical soil stress–strain relation where the soil has the characteristics of nonlinear, plastic, and confining pressure-related behaviors. Based on the principle of the effective stress, the stress–strain relation of the soil relates to the stress path, and sometimes even to the time. The soil model in the excavation analysis should fully consider these soil properties. This section will introduce the Mohr–Coulomb (MC) model, Duncan–Chang (DC) model , the modified Cam-clay (MCC) model, the hardening soil (HS) model, and small strain soil models.

8.4.1 The MC model: a linear elastoplastic model

The Mohr-Coulomb (MC) model assumes the soil is linear elastic before the yield where the yield function (Y) is defined by the MC failure criterion, as stated in Section 2.8.3. As shown in Figure 8.13, if the plastic potential function (Q) is the same as the yield function, then the constitutive law is called the associated flow rule and the plastic strain increment, $\{\Delta \varepsilon^p\}$, is perpendicular to the yield surface (Line AB in Figure 8.13). Under such a condition, when the soil undergoes shear strain, it will dilate, and its dilation angle, BAC as shown in Figure 8.13, should be equal to the friction angle, ϕ, which is unreasonably large and does not conform the actual soil behavior. To solve this problem, a plastic potential function is usually adopted, such as Q with slope ψ (Figure 8.13), and hence, the nonassociated flow rule is used. From the geometric relation in Figure 8.13, we can know the dilation angle, DAC, is equal to ψ. As long as a reasonable ψ is selected, the model is capable of simulating the actual dilation behavior of the soil.

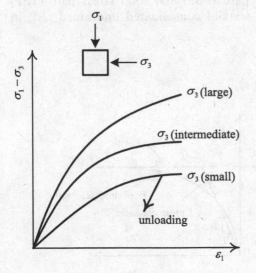

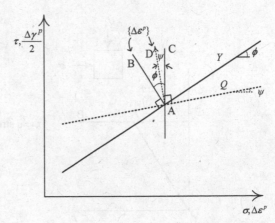

Figure 8.12 Typical stress–strain relations of soils.

Figure 8.13 The yield function and plastic potential function for the MC model.

Figure 8.14a shows a representative stress–strain behavior with the MC model where Point A denotes the yield stress, and the OA segment represents linear elastic. When the stress of the soil exceeds point A, only the plastic strain occurs but not the elastic strain. Therefore, the AB segment is called a perfectly plastic behavior. The MC model is a type of linear elastic perfectly plastic model. The overconsolidated clay in both primary loading and unloading states is of elastic behavior, and general soil in the unloading/reloading state is also elastic. Since the OA segment in Figure 8.14a is linear elastic, the MC model is suitable for both primary loading and unloading behaviors for overconsolidated clay and also suitable for the unloading/reloading behavior for general soil.

Figure 8.14b shows a typical stress–strain behavior of soil subject to the triaxial compression. If the MC model is adopted, the secant Young's modulus (E_{50}) is usually adopted to simulate the soil behavior. The Young's modulus for other loading stress paths may not be the same as that of the axial compression, and hence, the secant Young's modulus (E_{50}) for a certain stress path should be adopted. In other words, the parameters used in the MC model are closely related to the stress path.

To illustrate the use of the MC model in the simulation of the shear behavior of the soil under drained and undrained conditions, the stress acting on an elastic material is divided into the average principal stress (p) and shear stress (q) where $p = (\sigma_1 + \sigma_2 + \sigma_3)/3$ and $q = (\sigma_1 - \sigma_3)$. In the case of drained condition, p will only cause the volume change (compression or dilation) of the elastic material but will not produce shear strain (change its shape), whereas q only causes the shear strain (change the shape) without changing its volume. The behavior of the saturated elastic material under undrained shear conditions is more complicated. Under the action of p, the volume of the elastic material (before yielding) does not change, which leads to an increase in the pore water pressure. The amount of the increase is equal to p, so the average effective stress (p') remains unchanged. When q is applied, the volume of the material is also unchanged, and the pore water pressure is also unchanged because under drained conditions, the increase in q will not cause the volume of the elastic material to change.

Figure 8.15 shows the effective stress path (ESP) and total stress path (TSP) of the saturated elastic clay subject to a triaxial consolidated undrained test, in

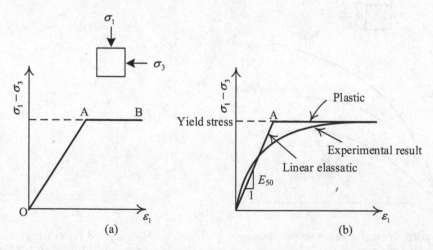

Figure 8.14 (a) Stress–strain behavior of soil derived from the MC model (b) Modeling a real stress–strain model of soil with the MC model.

which the consolidation stress is 100kPa ($p' = 100$kPa). In the undrained shear stage, when the axial stress (σ_1) increases to 130kPa, both σ_2 and σ_3 still maintain at 100kPa. Because the p rises to 110kPa from 100kPa, the pore water pressure increment is 10kPa, and the average effective stress (p') is still 100kPa, and $q = 30$kPa. From the above values, it can be seen that the average effective principal stress on the elastic material is unchanged during the undrained shear process, thus making the ESP move upward vertically.

The stress–strain behavior of the real soil is nonlinear and plastic, so its undrained ESP and TSP should be as shown in Figure 8.16a. If the MC model is used for conducting the effective stress undrained analysis, its input parameters are c', ϕ', E', μ', and ψ, which are exactly the same as those in the effective (drained) analysis. The analysis method is called the MC undrained A model. The E' and μ' are the entries of $[D^e]$ in Eq. 8.15, implying E' and μ' are the *elastic* effective deformation parameters. However, as shown in Figure 8.16b, the MC undrained A model will overestimate the undrained shear strength of the real soil because the soil behavior before the yield is simulated as the elastic model, and its ESP and TSP should be similar to those in Figure 8.15.

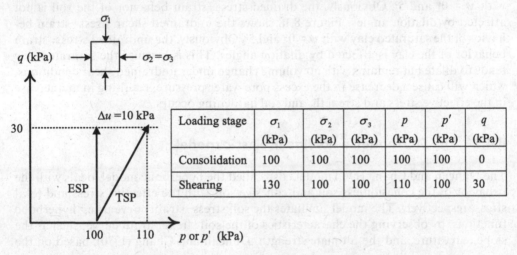

Loading stage	σ_1 (kPa)	σ_2 (kPa)	σ_3 (kPa)	p (kPa)	p' (kPa)	q (kPa)
Consolidation	100	100	100	100	100	0
Shearing	130	100	100	110	100	30

Figure 8.15 The ESP and TSP of an elastic material in triaxial consolidated undrained tests

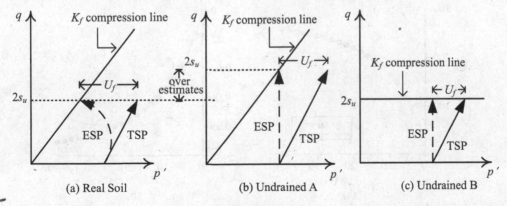

(a) Real Soil   (b) Undrained A   (c) Undrained B

Figure 8.16 The stress path of a real soil and MC model (a) real soil (b) stress path for MC undrained A model (b) stress path from MC undrained B model.

To obtain the behavior close to real soil, it is necessary to force the soil being failed or yielded at the actual undrained shear strength (s_u). Therefore, it is necessary to have input parameters s_u, $\phi_u = 0$, E', μ', and ψ. Similar to the MC undrained A model, the E' and μ' are also *elastic* effective parameters. The analysis method is called the MC undrained B model.

On the other hand, the undrained behavior of the saturated clay can also be analyzed by the total stress analysis. As stated in Section 2.9.3.2, the total stress undrained analysis method regards the soil and its pore water as a mixture of the soil and water, which is a one-phase material. The soil parameters used in the analysis are all expressed in the undrained condition, that is, s_u, $\phi_u = 0$, E_u, and μ_u. The E_u and μ_u are undrained deformation moduli. The analysis method is called the MC undrained C model.

The dilation angle ψ can be directly calculated from the volumetric strain–axial strain relationship of a drained test, or it can be obtained by using the finite element effective stress analysis method to simulate the actual test result. If there is no relevant test data, $\psi = \phi - 30°$ can be assumed (Bolton, 1986).

Figure 8.17 shows the different drained shear stress–strain behavior of the clay with $\psi = 0°$ and $5°$. Obviously, the drained stress–strain behavior of the soil is not affected by dilation angles. Figure 8.18 shows the undrained shear stress–strain behavior of the saturated clay with $\psi = 0°$ and $5°$. Obviously, the undrained stress–strain behavior of the clay is affected by dilation angles. This is because the saturated clay tends to dilate but remains with no volume change under undrained shear conditions, which will cause a decrease in the excess pore water pressure, resulting in an increase in the effective stress and strength, and soil hardening occurs.

8.4.2   The DC model: a nonlinear elastic model

The Duncan and Chang (DC) model, also called the hyperbolic model, deals with the behavior of the cohesionless soil and cohesive soil with the effective stress and total stress, respectively. The model simulates the soil stress–strain curve using hyperbolic functions, by observing the characteristics of the soil stress–strain curve, such as the slope, curvature, and the ultimate strength. Duncan and Chang (1970), based on the

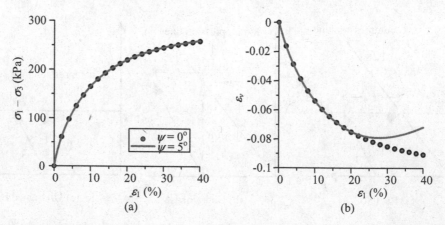

Figure 8.17 Stress–strain behavior of saturated clay computed from the MC model in drained shear conditions.

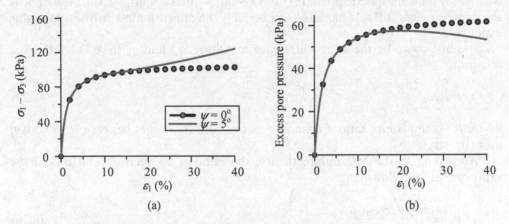

(a) (b)

Figure 8.18 Stress–strain behavior of saturated clay computed from the MC model in undrained shear conditions.

Konder's study (1963), proposed that the soil stress–strain behaviors be expressed by a hyperbolic function, as shown in Figure 8.19, as follows:

$$q = \frac{\varepsilon_1}{\dfrac{1}{E_i} + \dfrac{\varepsilon_1}{q_a}} \tag{8.23}$$

where q is the deviator stress, $q = \sigma_1 - \sigma_3$; σ_1 and σ_3 are major and minor principal stresses, respectively; ε_1 is the strain in the direction of the major principal stress; E_i is the initial tangent modulus; $q_a = (\sigma_1 - \sigma_3)_{\text{ult}}$ is the asymptote of the stress–strain curve, representing the ultimate strength.

Under drained conditions, the stress–strain relation and the confining pressure are related. Janbu (1963) thus proposed the relation between E_i and σ_3 as follows (see Figure 8.20):

$$E_i = KP_a \left(\frac{\sigma_3}{P_a} \right)^n \tag{8.24}$$

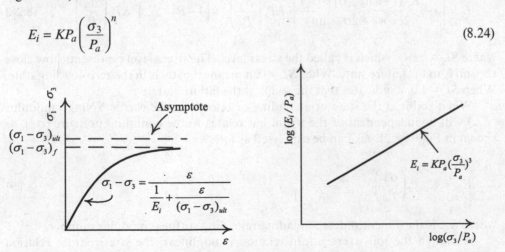

Figure 8.19 The DC model.

Figure 8.20 The relation between the initial Young's modulus and confining pressure.

where P_a is the atmospheric pressure ($1.033 \text{ kg/cm}^2 = 101.3 \text{ kN/m}^2 = 2{,}116.2 \text{ lb/ft}^2$); K is the dimensionless stiffness modulus number; n is the dimensionless stiffness modulus exponent.

Assume q_f to be the stress difference at failure. q_f and q_a have the following relation:

$$q_f = R_f q_a \tag{8.25}$$

where R_f is the failure ratio. For most types of soil, R_f ranges between 0.5 and 0.9; $q_f = (\sigma_1 - \sigma_3)_f$.

According to the MC failure theory, the relation between q_f and σ_3 can be expressed as the following:

$$q_f = \frac{2c \cos\phi + 2\sigma_3 \sin\phi}{1 - \sin\phi} \tag{8.26}$$

where c and ϕ are the strength parameters of the soil.

For cohesionless soils, the friction angle decreases in proportion to the logarithm of the confining pressure. That is, the friction angle of the soil under any confining pressure can be expressed as follows:

$$\phi = \phi_0 - \Delta\phi \log_{10}\left(\frac{\sigma_3}{P_a}\right) \tag{8.27}$$

where ϕ_0 is the friction angle of the soil under the confining pressure of one atmospheric pressure; $\Delta\phi$ is the slope of the $\phi - \log_{10}(\sigma_3/P_a)$ curve

Substitute Eqs. 8.24–8.26 in Eq. 8.23, and differentiate with respect to the strain, and then, we can obtain the tangent elastic modulus E_t for any stress state as follows:

$$E_t = \left[1 - \frac{R_f(1-\sin\phi)(\sigma_1 - \sigma_3)}{2c\cos\phi + 2\sigma_3\sin\phi}\right]^2 KP_a\left(\frac{\sigma_3}{P_a}\right)^n = \left[1 - R_f \cdot SL\right]^2 KP_a\left(\frac{\sigma_3}{P_a}\right)^n \tag{8.28}$$

where $SL = q / q_f$, which is called the stress level. The stress level represents how close the soil is to the failure state. When $SL = 0$, it means the soil is in the zero-loading state. When $SL = 1.0$, it indicates that the soil is at the failure state.

When soil is at the state of unloading or reloading, the elastic Young's modulus (E_{ur}) will be independent of the strain but relates to the confining pressure (σ_3), as shown in Figure 8.21. E_{ur} can be expressed as follows:

$$E_{ur} = K_{ur} P_a \left(\frac{\sigma_3}{P_a}\right)^n \tag{8.29}$$

where K_{ur} is the dimensionless unloading/reloading stiffness modulus number.

Although the soil stress–strain relation is nonlinear, the stress–strain relation at each load stage can be simulated as linear elastic if loading is divided into many stages, and the Young's modulus for the load stage can be expressed by E_t, as shown

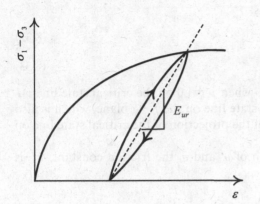

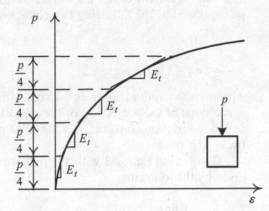

Figure 8.21 Unloading-reloading Young's modulus.

Figure 8.22 Variation of the tangent Young's modulus with strain.

in Figure 8.22. Although the soil stress–strain relation is inelastic (i.e., plastic), the permanent deformation can be obtained by the model with the introduction of E_{ur}. Thus, the hyperbolic model can simulate the characteristics of the nonlinearity and plasticity of soil.

In accordance with Eqs. 8.1 and 8.6, to establish the stiffness matrix of an element, the required parameters are E (representing E_t or E_{ur}) and μ. In accordance with Eqs. 8.27–8.29, if the DC model is used to describe the stiffness matrix at a stress level or in a stress state, the required parameters are c, ϕ_0, $\Delta\phi$, K, n, K_{ur}, R_f, and μ.

In the analysis with the DC model, the external load has to be divided into several increments. For each increment, E_t (or E_{ur}) and μ will be adopted to enter into $[D]$ in Eq. 8.1. The analysis procedure is based on the theory of elasticity without involving the plasticity theory. The DC model is therefore categorized as an elastic model.

8.4.3   The MCC model: a critical state model

The modified Cam-clay (MCC) model is an effective stress model, in which all the stresses are expressed in terms of the effective stress. The strength parameters and deformation moduli are also effective terms. The model is derived in accordance with critical state soil mechanics. First, define the mean effective stress (p') and the deviator stress (q), on the basis of the test results of laboratory triaxial tests:

$$p' = \frac{1}{3}(\sigma_1' + \sigma_2' + \sigma_3') \tag{8.30}$$

$$q = \sigma_1' - \sigma_3' \tag{8.31}$$

where σ_1', σ_2', and σ_3' are the major, intermediate, and minor principal stresses, respectively.

The critical state of the soil refers to the state where the volumetric strain is not further produced with the increase of the shear strain at the large shear strain.

The state is usually either the failure state or the ultimate state. The critical state can be expressed by the following two equations:

$$q = Mp' \tag{8.32}$$

$$e = e_{cs} - \lambda \ln p' \tag{8.33}$$

where e is the void ratio; e_{cs} is the void ratio when $p' = 1.0$ on the critical state line; M is the slope of the projection of the critical state line on the $p'-q$ plane, which is also called the friction constant; λ is the slope of the projection of the critical state line on the $e-\ln p'$ plane.

Combining Eq. 8.32 with the definition of p' and q, the friction constant, M, is given by the following:

$$M = \frac{6\sin\phi'}{3 - \sin\phi'} \tag{8.34}$$

Assume the soil consolidation is in the isotropic consolidation state ($\sigma_1 = \sigma_2 = \sigma_3$). The $e-p'$ curve obtained from the isotropic consolidation is called the virgin isotropic consolidation line, as shown in Figure 8.23b. It can be expressed by the following equation:

$$e = e_a - \lambda \ln p' \tag{8.35}$$

where e_a is the void ratio when $p' = $ one unit; λ is the slope. In the $e-p'$ coordinate, the critical state line, the virgin one-dimensional consolidation line, and the isotropic virgin consolidation line are parallel to one another.

The unloading/reloading line represents the soil in the overconsolidated state. No plastic strain would be generated in the soil in this state, that is, the soil would be in the elastic condition. If we designate the slope as κ, the unloading/reloading equation can be expressed as follows:

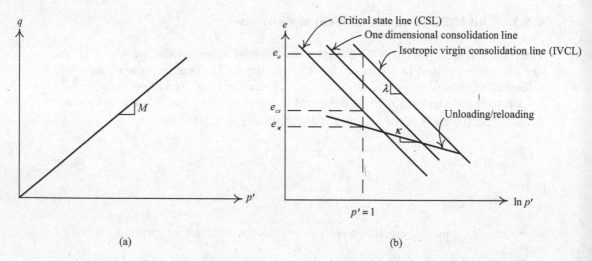

(a) (b)

Figure 8.23 Definitions of various parameters in critical soil mechanics (a) $p'-q$ relation in the critical state (b) critical state line and the consolidation line.

$$e = e_\kappa - \kappa \ln p' \tag{8.36}$$

where e_k is the void ratio when $p' =$ one unit; κ is the slope of the unloading/reloading line.

Critical state soil mechanics assumes that there exists a single state boundary surface for normally consolidated or lightly oveconsolidated soils in the $e-p'-q$ space, as shown in Figure 8.24. The state boundary surface intersects the plane $q = 0$ along the line YY, which is called the virgin isotropic consolidated line. The border line of the state boundary surface is the line XX. The state boundary surface represents the stress states in the various ultimate conditions (the coordinates of the stress states are e, p', and q), that is, the stress states of the soil exist only on or below the state boundary surface.

If the stress state of the soil is on the swelling line (EF), only the elastic deformation will be produced, i.e., no plastic deformation is produced. Furthermore, all paths that remain on the curved vertical plane above the swelling line EF, but below the state boundary surface, will only induce elastic deformation. Thus, this curve plane is called the elastic wall. The line EX, the state boundary surface intersecting with the elastic wall, is then projected onto the plane of $e = 0$, which forms the line E'X'. E'X' is generally defined as a yielding surface. As discussed above, if the state boundary surface is known, we can then derive the yielding function.

The MCC model derives the state boundary surface and the yielding function on the basis of energy dissipation of the soil during shearing. With a rational assumption of energy dissipation during the shear, the MCC model has become one of the most widely adopted analysis model in geotechnical engineering. According to Roscoe and Burland (1968), the equations for the state boundary surface with the MCC model are given as the below:

$$\frac{p'}{p_e} = \left(\frac{M^2}{M^2 + q^2 / p'^2} \right)^{(1-\kappa/\lambda)} \tag{8.37}$$

$$p_e = \exp\left(\frac{e_a - e}{\lambda} \right) \tag{8.38}$$

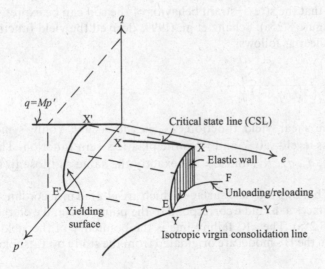

Figure 8.24 State boundary surface (Borja, 1990).

The yielding equation is given as follows:

$$p' = p'_0 \left(\frac{M^2}{M^2 + q^2 / p'^2} \right) \tag{8.39}$$

where p'_0 is the p'-value when $q = 0$.

With the yielding equation given above, we can then obtain the constitutive law for the MCC elastoplastic model in accordance with Eqs. 8.15 and 8.16. As discussed above, the parameters related to the state boundary surface or the yielding surface are M (or ϕ'), λ, and κ. Thus, if the MCC model is adopted as the yielding function, the required input parameters are M (or ϕ'), λ, and κ, plus elastic constants E'_{ur} and μ'_{ur} that form the elastic constitutive matrix $[D^e]$ in Eq. 8.15.

The MCC model can simulate the elastoplastic behavior of the soil during the shear. They thereby also are elastoplastic models. Although the derivation of the MCC model is theoretically rigorous, having the theoretical basis and having it confirmed by soil tests, it is still limited in application because it does not consider the anisotropic, dilatant, and overconsolidated behaviors of the soil. Therefore, some high-order elastoplastic models have been developed, for example, the hardening soil (HS) model, etc. These high-order models can analyze the soil behavior more accurately and rationally.

8.4.4   The HS model: a nonlinear elastoplastic model

The hardening soil (HS) model (Schanz et al., 1999) was derived on the basis of the DC model, the MC model, the MCC model, and the Rowe's theory (refer to Section 2.5). The HS model is an effective stress model, in which all the stresses are expressed in terms of the effective stress. The strength parameters and deformation moduli are also effective terms. This section summarizes the main part of the HS model. Interested readers are advised to refer the literatures (e.g., Schanz et al., 1999).

Assuming that the stress–strain behavior of the soil can be expressed by a hyperbolic curve (Figure 8.25a), Schanz et al. (1999) derived the yield function for the soil subject to the shear as follows:

$$Y_s = \frac{2 - R_f}{E'_{50}} \frac{q}{1 - q / q_a} - \frac{2q}{E'_{ur}} - \gamma^p \tag{8.40}$$

where Y_s is the shear yield function, E'_{50} is the secant Young's modulus corresponding stress level = 50%, and γ^p is the plastic strain function. The rest parameters such as q, q_a, q_f, R_f, and E'_{ur} are exactly the same as those in the DC model (Section 8.4.2).

Figure 8.25b shows the yield surfaces which are plotted in accordance with Eq. 8.40 where yield surfaces **a**, **b**, and **c** correspond to the points **a**, **b** and **c** on the stress–strain curve (Figure 8.25a). The MC failure line is the boundary of the yield surfaces. The yield surfaces in the HS model are originated from the study by Tatsuoka and Ishihara (1974).

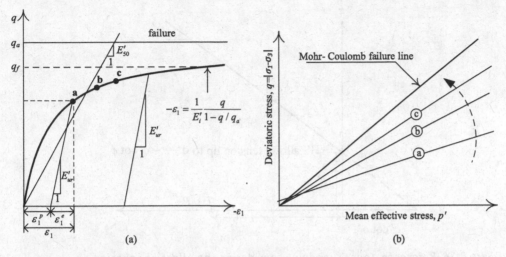

Figure 8.25 The HS model (a) the stress–strain curve by a hyperbola (b) the shear yield surface (tensile strain is in positive).

Similar to the DC model, the E'_{50} is related to the confining pressure σ'_3. The HS model uses the following expression to relate E'_{50} to σ'_3 as follows:

$$E'_{50} = E'^{\text{ref}}_{50} \left(\frac{c' \cot\phi' + \sigma'_3}{c' \cot\phi' + p'^{\text{ref}}} \right)^m = E'^{\text{ref}}_{50} \left(\frac{c' \cos\phi' + \sigma'_3 \sin\phi'}{c' \cos\phi' + p'^{\text{ref}} \sin\phi'} \right)^m \tag{8.41}$$

where p'^{ref} is the reference pressure, which is similar to the atmospheric pressure P_a in the DC model; usually sets $p'^{\text{ref}} = 100$ stress unit (when the unit is in kPa, $p'^{\text{ref}} = 100\,\text{kPa}$). E'^{ref}_{50} is the reference stiffness (Young's) modulus corresponding to p'^{ref}. m is the stiffness modulus exponent.

Eq. 8.41 is similar to Eq. 8.24. However, to take into account the effect of cohesion, both σ'_3 and p'^{ref} add $c' \cot\phi'$ together, as illustrated in Figure 8.26. Similar to Eq. 8.41, E'_{ur} is also related to σ'_3 and should consider the effect of cohesion, which is expressed as the below:

$$E'_{ur} = E'^{\text{ref}}_{ur} \left(\frac{c' \cot\phi' + \sigma'_3}{c' \cot\phi' + p'^{\text{ref}}} \right)^m = E'^{\text{ref}}_{ur} \left(\frac{c' \cos\phi' + \sigma'_3 \sin\phi'}{c' \cos\phi' + p'^{\text{ref}} \sin\phi'} \right)^m \tag{8.42}$$

where E'^{ref}_{ur} is the reference unloading/reloading Young's modulus.

As the yield surfaces, as illustrated in Figure 8.25b, do not take into account the plastic volumetric strain due to the isotropic compression, Schanz et al. (1999) adopted the yield surface similar to that in the MCC model, as shown by E'X' on the $q - p'$ space in Figure 8.24. The complete yield surface of the HS model is as illustrated in Figure 8.27 where Y_c is called the cap yield surface, which is a function of the tangent

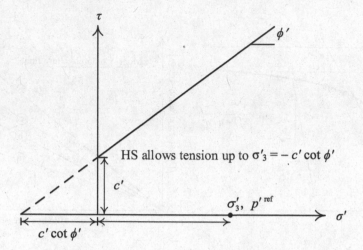

Figure 8.26 Reference Young's modulus considering the effect of cohesion.

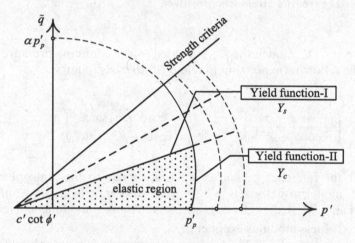

Figure 8.27 Yield surfaces used in the HS model.

oedometer Young's modulus, E'_{oed}; the coefficient of the at-rest earth pressure of the soil under normally consolidated state, $K_{0,NC}$; and the preconsolidation pressure, p'_c.

The definition of the E'_{oed} is shown in Figure 8.28. Similar to Eqs. 8.41 and 8.45, the E'_{oed} can be obtained from the following expression:

$$E'_{oed} = E'^{ref}_{oed} \left(\frac{c' \cot\phi' + \sigma'_1}{c' \cot\phi' + p'^{ref}} \right)^m = E'^{ref}_{oed} \left(\frac{c' \cos\phi' + \sigma'_3 \sin\phi' / K_{0,NC}}{c' \cos\phi' + p'^{ref} \sin\phi'} \right)^m \qquad (8.43)$$

where E'^{ref}_{oed} is the reference tangent oedometer Young's modulus.

The HS model adopts a plastic potential function, Q, with a slope, ψ_m, similar to the yield surfaces **a**, **b**, and **c**, as shown in Figure 8.25b where ψ_m is the dilation angle at the current stress state. For examples, at the stress state, σ'_1 and σ'_3 (or q and p'), the corresponding yield surface is the yield surface **a**; if the corresponding dilation angle

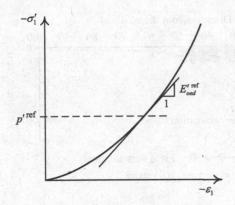

Figure 8.28 Reference oedometer Young's modulus considering the effect of cohesion.

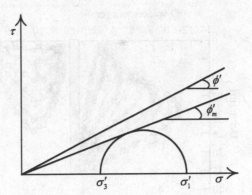

Figure 8.29 Physical implication of mobilized friction angle (ϕ'_m).

is ψ_m, then the corresponding plastic potential function, Q, is a line with the slope, ψ_m, that passes through the current stress state, σ'_1 and σ'_3, similar to the yield surface function. ψ_m is not constant but changes with the dilation behavior at different stress states. The dilation behavior of the soil simulated by the HS model follows the Rowe's theory (refer to Section 2.5).

As shown in Figure 2.8 for dense sand, before the shear stress reaches the critical stress (point A in Figure 2.8), the soil undergoes volumetric compression. After the shear stress exceeds point A, the soil begins to dilate; as the shear stress keeps increasing, the dilation rate gradually increases, that is, the dilation angle gradually increases. When the shear stress reaches the peak strength (corresponding to ϕ'), the volumetric dilation rate reaches the maximum, and the dilation angle reaches the maximum; after that, as the shear stress continues to increase, the dilation rate gradually decreases, i.e., the dilation angle gradually decreases. When the shear stress drops to the critical stress, both the dilation rate and dilation angle are equal to 0. To accurately represent the shear stress and dilation behavior of the soil at various stages (points A–C, etc. in Figure 2.8), the mobilized friction angle, ϕ'_m, and the mobilized dilation angle, ψ_m, are used. When the soil is at the stress state, σ'_1 and σ'_3, its corresponding mobilized friction angle, ϕ'_m, can be obtained from the Mohr's circle shown in Figure 8.29 as follows:

$$\sin\phi'_m = \frac{\sigma'_1 - \sigma'_3}{\sigma'_1 + \sigma'_3 - 2c'\cot\phi'} \tag{8.44}$$

At the same stress state, the mobilized dilation angle, ψ_m, is given as follows:

$$\sin\psi_m = \frac{\sin\phi'_m - \sin\phi'_{cv}}{1 - \sin\phi'_m \sin\phi'_{cv}} \tag{8.45}$$

where ϕ'_{cv} is the friction angle at the critical state or at the state of the constant volume.

When the soil is at failure, $\phi'_m = \phi'$ and $\psi_m = \psi$. Substitute those values into Eq. 8.45, we can obtain the following:

$$\sin\psi = \frac{\sin\phi' - \sin\phi'_{cv}}{1 - \sin\phi' \sin\phi'_{cv}} \tag{8.46}$$

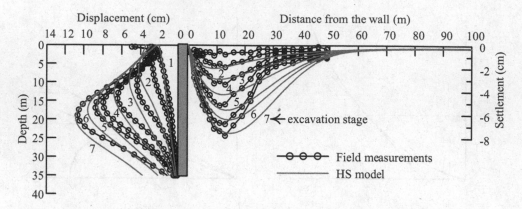

Figure 8.30 Comparison of wall deflections and surface settlements computed from the HS model for the TNEC case with field measurements.

Move the term ϕ'_{cv} to the left-hand side (Eq. 8.46), and make a necessary simplification; we can obtain the following:

$$\sin\phi'_{cv} = \frac{\sin\phi' - \sin\psi}{1 - \sin\phi'\sin\psi} \tag{8.47}$$

Therefore, as long as the friction angle and dilation angle are known, the friction angle ϕ'_{cv} at the critical state can be obtained from Eq. 8.47, and the mobilized friction angle ϕ'_m and mobilized dilation angle ψ_m can be obtained from Eqs. 8.44 and 8.45, respectively.

It can be found from the description above, in addition to the elastic parameters E'_{ur} (or E''^{ref}_{ur}) and μ'_{ur}, the HS model requires c', ϕ', R_f, p'^{ref}, E'_{ur} (or E''^{ref}_{ur}), E'_{50} (or E''^{ref}_{50}), E'_{oed} (or E''^{ref}_{oed}), $K_{0,NC}$, and p'_c to form the yield function and ψ to construct the plastic potential function. Figure 8.30 shows the comparison of wall deflections and surface settlements of the TNEC case (see Section 3.6 and Appendix B) computed from the HS model and field measurements. Except for the surface settlement, the computed wall deflection is close to field measurements.

8.4.5 Small strain model

Recent studies have found that the FEM with traditional soil models can reasonably predict the deformation of the retaining wall, but it cannot reasonably predict the ground settlement. The results of the surface settlement from finite element analysis still have a gap with the actual observation results, as shown in Figure 8.31.

According to Smith et al. (1992), inside the initial yielding surface (Y3) of clayey soils should have two subyielding surfaces, Y1 and Y2, which divide the stress space into four areas, as shown in Figure 8.32. Area 1, which is the innermost one of the four, is very small in range. The soil behavior is linear elastic, and the strain is completely recoverable. The strain is usually in the range of 10^{-5} % to 10^{-3}%. When the stress state is in Area 2, the soil behavior changes to nonlinear elastic, and the

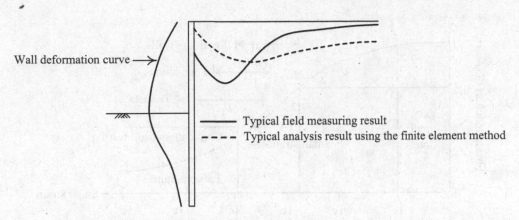

Figure 8.31 Wall deflection and surface settlement computed from a typical conventional FEM.

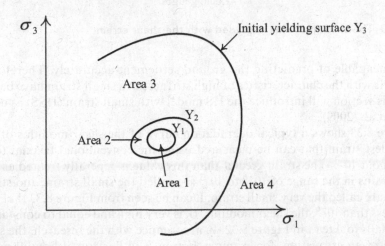

Figure 8.32 Yield surfaces at different strains of clay soils.

stiffness of the soil decreases rapidly although the strain is still completely recoverable. In general, the strain is basically smaller than $10^{-2}\%$. When the stress state is in Area 3, if the loading is decreased, the strain will no longer be recoverable, and a plastic strain is generated, and the soil behavior is inelastic. When exceeding the yielding surface, Y3, and entering Area 4, the soil deformations and plastic strain are all largely increased, and a large quantity of deformation is generated. Compared with that in Area 4, the plastic strain occurring in Area 3 is relatively much smaller. The Smith's study also discovers that the soil stiffness in Area 1 (with small strain) is very high.

Because of general elastoplastic models such as the Cam-clay model, all assume the state of the soil within the initial yielding surface (Y3) is elastic and do not take into consideration the high stiffness of the soil in Area 1. As the soil outside of the influence range is not much influenced by excavation and thus has very small strain, that is, the soil is very possibly at the linear (or nonlinear) elastic stage, the FEM is

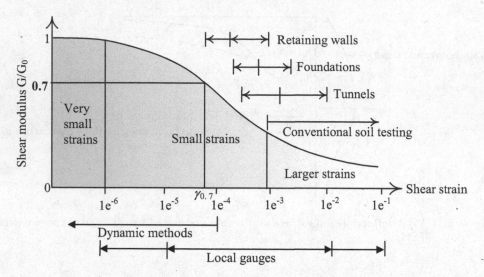

Figure 8.33 The shear modulus degraded with the shear strain.

thereby incapable of predicting the ground settlement accurately. Therefore, some soil models with the characteristics of high stiffness at a small strain have been developed. This section will introduce the HS model with small strain (HSS) as developed by Benz et al. (2009).

Figure 8.33 shows a typical degradation curve of the shear modulus of the clay. The smallest strain that can be measured using the conventional triaxial test apparatus is about 10^{-3}. The strain greater than this value is generally treated as the large strain. Strains in the range of 10^{-6} to 10^{-3} are called the small strain, and strains less than 10^{-6} are called the very small strain. It can be seen from Figure 8.33 that when the strain is less than 10^{-6}, the shear modulus, G, is very high and equal to constant, which basically falls to Area I in Figure 8.32. In accordance with the research, the soil in the boundary of an excavation, for example, bottom and horizontal boundaries, is less affected by excavation, so the strain is very small, quite close to the small strain. On the basis of the HS model, Benz et al. (2009) developed the hardening soil model with small strain (HSS), which considers the characteristics of high stiffness at small strain. In addition to the parameters in the HS model, HSS requires two more parameters, i.e., G_0 and $\gamma_{0.7}$, where G_0 is the shear modulus at the small strain and $\gamma_{0.7}$ is the shear strain corresponding to $G/G_0 = 0.7$, as shown in Figure 8.33. G_0 can be obtained from the bender element test, seismic survey, a small strain triaxial test, or from empirical formulas. Interested readers are advised to refer to relevant literatures to know about how to obtain the related parameters (Benz et al., 2009; Teng et al., 2014).

Figure 8.34 shows the computed wall deflections and surface settlements by using the HSS model to analyze the TNEC case, along with field measurements. In addition to G_0 and $\gamma_{0.7}$, the parameters used in the analysis with the HSS model are exactly the same as those in Figure 8.30 using the HS model. Compared with Figure 8.30, it is found that the difference in the analysis results using the HSS model are very small. Although the HSS model takes into account the small strain characteristics, the improvement of the accuracy of the ground settlement analysis is quite limited. According to Vermeer

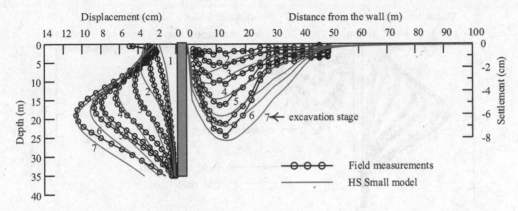

Figure 8.34 The comparison of wall deflections and surface settlements computed from using the HSS model for the TNEC case with field measurements.

and Wehnert (2005), the small strain zone normally occurs in the range of $(a - H_e)$ to a below the excavation bottom where a is the minimum depth of the boundary, which is equal to $2H_e$ (Figure 8.43). In the case of the TNEC, the small strain should occur in the depth of 20–40 m below the excavation bottom. The soil in this range is dense sand and gravel soil which already have a quite high stiffness. Therefore, the HSS model has a limited improvement in the surface settlement for TNEC case.

Figure 8.35 shows the analyzed wall deflection and surface settlement for a 20 m deep excavation in 80 m thick soft clay. It is found from the figure that the HSS model do improve the accuracy of the surface settlement analysis. If the soil stiffness in a small strain area, as indicated in Figure 8.43, is raised by 5 times in both the HS model and MC undrained B model, that is, $5E'_{ur}$, both models can obtain the wall deflection

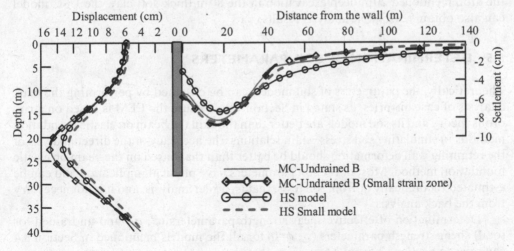

Figure 8.35 The effect of small strain on the analysis results (a) zone of small strain in an excavation (b) computed wall deflections and surface settlements computed from the HS, HSS, and MC undrained B models at the final stage for the 80 m deep soft clay.

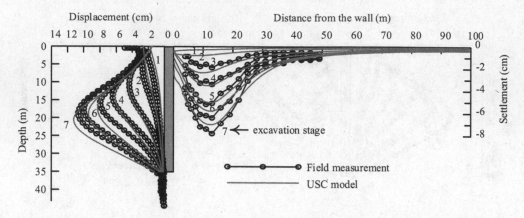

Figure 8.36 The comparison of wall deflections and surface settlements computed from using the USC model for the TNEC case with field measurements.

and surface settlement close to those from the HSS model. The results are also shown in Figure 8.35.

Conventional total stress models cannot take into account the changes in the pore water pressure induced by different stress paths of the soil at the various locations in an excavation area and therefore cannot take into account the variation of the strength and stiffness due to the different stress paths in the excavation. The undrained soft clay (USC) model, as developed by Hsieh and Ou (2012), can consider the strength and stiffness for different stress paths and the undrained Young's modulus at different stress levels, and therefore, it can simulate the stress–strain behavior more reasonably. Compared with the field measurement, the USC model can result in more rational wall deflection and surface settlement (Figure 8.36), even better than the HSS model. For the aforementioned 20 m deep excavation in the 80 m thick soft clay, the USC model can also obtain reasonable analysis results.

8.5 DETERMINATION OF SOIL PARAMETERS

Theoretically, the parameters of soil models can be obtained by performing the back analysis of case histories, as stated in Section 7.8. Because the FEM is based on a rigorous theory and its soil models are better than those in the beam on elastic foundation methods in simulating soil stress–stain relations, the accuracy of the direct analysis of the retaining wall deformation should be better than that based on the beam on elastic foundation method. Most of the soil parameters have physical implication and can be estimated theoretically or from soil tests, that is, direct analysis, and it is not necessary from the back analysis.

Determination of effective shear strength parameters (c', ϕ') and undrained (or total) shear strength parameters (s_u, $\phi=0$) for all the models mentioned in Section 8.4 can refer to Sections 2.5 and 2.6.

The MCC model is only applicable to the clay. In addition to M (or ϕ'), the required parameters are λ, κ, E'_{ur}, and μ'_{ur}. λ and κ are defined in terms of $\ln p'$, whereas in conventional soil mechanics, the C_c and C_s are expressed in terms of $\log p'$. In accordance with the chain rule, the above two have the following relations:

$$\lambda = \frac{C_c}{2.303} \tag{8.48}$$

$$\kappa = \frac{C_s}{2.303} \tag{8.49}$$

The C_c and C_s can be derived from either the one-dimensional consolidation test or the index test. In general, the result of the consolidation test is little affected by the disturbance of the sampling process, and the obtained C_c and C_s are therefore highly reliable. C_c and C_s have the following relationship,

$$C_s = \left(\frac{1}{10} \text{ to } \frac{1}{5}\right) C_c \tag{8.50}$$

In most cases, lower values are for clays with low plasticity. If values of the C_s are not available, one can use the above relation to estimate the C_s. As the MCC model is symmetric to the isotropic consolidation line, C_s used for the analysis may need to raise slightly, which can be expressed as follows (Lim et al., 2010):

$$C_s = \left(\frac{1}{5} \text{ to } \frac{1}{4}\right) C_c \tag{8.51}$$

Differentiating Eq. 8.36 at both sides, we have the following:

$$de = -\kappa \frac{dp'}{p'} \tag{8.52}$$

In accordance with the definition of the bulk modulus, we can derive the effective (or drained) bulk modulus as follows:

$$K'_{ur} = -\frac{dp'}{d\varepsilon_v} = -\frac{dp'}{de/(1+e)} = \frac{(1+e)p'}{\kappa} = \frac{2.303(1+e)p'}{C_s} \tag{8.53}$$

The unloading/reloading or elastic Young's modulus can then be calculated based on the formula in Table 2.13 as follows:

$$E'_{ur} = 3K'_{ur}(1-2\mu'_{ur}) = \frac{6.909(1+e)(1-2\mu'_{ur})p'}{C_s} \tag{8.54}$$

The unloading/reloading or elastic, $\mu'_{ur,}$ usually falls into a narrow range, which can be reasonably assumed ($\mu'_{ur} \approx 0.2$). With given the C_s value, the elastic E'_{ur} can thus be derived from Eq. 8.54.

When applying the HS model in clay, in addition to c' and ϕ', the required parameters are R_f, p^{ref}, E'_{ur} (or E'^{ref}_{ur}), E'_{50} (or E'^{ref}_{50}), E'_{oed} (or E'^{ref}_{oed}), μ'_{ur}, and ψ. Basically, all the parameters can be obtained by the simulation of triaxial compression tests with unloading/reloading cycles and oedometer tests using the FEM with the HS model. Besides, the parameters can also be obtained from correlations (Schanz et al., 1999; Calvello and Finno, 2004; Schweiger, 2010). Similar to the MCC model, E'_{ur} (or E'^{ref}_{ur})

and μ'_{ur} represent the unloading/reloading Young's modulus and Poisson's ratio, which are elastic deformation parameters. Still $\mu'_{ur} \approx 0.2$. E'_{ur} can be determined by using Eq. 8.54, but real C_s values should be used rather than the adjusted one following Eq. 8.50. After the determination of E'_{ur}, $E'^{ref}_{50} = E'^{ref}_{ur}/(3{\sim}5)$ for the normally consolidated clay; $E'^{ref}_{50} = E'^{ref}_{ur}/(2{\sim}3)$ for the overconsolidated clay; the default value in some software such as PLAXIS, $E'^{ref}_{50} = E'^{ref}_{ur}/3$. Then, $E'^{ref}_{oed} \approx (0.7-1.0)E'^{ref}_{50}$. As observed for the clay, m can be taken equal to 1.0. $R_f = 0.9$ is a suitable assumption. The dilation angle can be estimated by the following empirical formula (Bolton, 1986):

$$\psi = \phi' - 30° \tag{8.55}$$

When applying the HS model in granular soil such as sand, the parameters may not be determined from the stress–strain test data unless undisturbed soil samples are used for testing. Alternatively, the parameters can also be obtained from correlations. The unloading/reloading Young's modulus (E'_{ur}) for excavation problems can be estimated in accordance with the following formula (Khoiri and Ou, 2013):

$$E'_{ur} = (2,000 \text{ to } 3,000)N \tag{8.56}$$

where E'_{ur} is in kN/m^2 and N is the SPT-N value.

As the E'_{ur} is known, the E'^{ref}_{ur} can be obtained from Eq. 8.42. Again, the $\mu'_{ur} \approx 0.2$ is also an appropriate assumption for granular soil such as sand. $E'^{ref}_{50} = E'^{ref}_{ur}/(3{\sim}5)$, and $E'^{ref}_{50} = E'^{ref}_{ur}/3$ is the default value for some software. $E'^{ref}_{oed} \approx E'^{ref}_{50}$. $m = 0.4$ to 0.6. According to Wong and Broms (1989), the failure strain of sandy or gravelly soils is rather small, and the failure ratio (R_f) is about 0.5 to 0.6. Similar to clay, the dilation angle can be estimated using Eq. 8.55.

The MC undrained B model can be used for clay. In addition to strength parameters (s_u, $\phi_u = 0$), the required parameters for the MC undrained B model are E', μ', and ψ. As most of the soil in the excavation area is subject to unloading, the E' and μ' should be estimated from the unloading stress path, that is, the unloading/reloading (or elastic) effective Young's modulus (E'_{ur}) and Poisson's ratio (μ'_{ur}) should be used, which have exactly the same meaning as the E'_{ur} and μ'_{ur} in the HS model. Therefore, the $E' = E'_{ur}$, which can be estimated from Eq. 8.54, and $\mu' = \mu'_{ur} \approx 0.2$ can be used for analysis of excavation problems. The dilation angle (ψ) can be estimated using Eq. 8.55.

When applying the MC model in sand, in addition to c' and ϕ', the required parameters are E', μ' and ψ. The dilation angle (ψ) can be estimated using Eq. 8.55. The magnitude of parameters E' and μ' is related to the stress or loading path. Because the soil in front of the wall is subject to unloading stress path, the E' can be determined following Eq. 8.56, and μ' can be set equal to 0.2. For the soil subject to the loading stress path such as the soil behind the wall, the E' is closer to E'_{50} (refer to Figure 8.14b). Therefore, the relationship of E'_{50} and E'_{ur} as used in the HS model can be applied. The μ' should be in the range of 0.3–0.4.

The MC undrained C model can also be used for clay. In addition to undrained strength parameters (s_u, $\phi_u = 0$), the required parameters for the MC undrained C model are undrained E_u and μ_u. The saturated clay has no volume change under undrained conditions, the undrained Poisson's ratio $\mu_u = 0.5$ (refer to Section 2.8.2).

However, to maintain numerical stability during computation, $\mu_u = 0.495$ is usually assumed. Considering the fact that the pore water in soil cannot bear the shear stress, the effective shear modulus (G') should be equal to undrained shear modulus (G_u), that is, $G' = G_u$. By equating Eq. 8.57 to Eq. 8.58, the undrained Young's modulus can be derived as shown in Eq. 8.59.

$$G' = \frac{E'}{2(1+\mu')} \tag{8.57}$$

$$G_u = \frac{E_u}{2(1+\mu_u)} \tag{8.58}$$

$$E_u = \frac{E'(1+\mu_u)}{(1+\mu')} = \frac{E'_{ur}(1+\mu_u)}{(1+\mu'_{ur})} \tag{8.59}$$

For saturated clay, the undrained Poisson's ratio $\mu_u = 0.5$ (see Section 2.8.2). Moreover, the effective Poisson's ratio for saturated clay can be reasonably assumed, $\mu' = 0.3$ (refer to Table 2.15). Substituting those values into Eq. 8.59, $E_u \approx 1.15E'$ can be found. For excavation problems, the effective unloading/reloading Young's modulus and Poisson's ratio can be used. Therefore, the undrained Young's modulus (E_u) can be derived, that is, $E_u \approx 1.25E'_{ur}$. Alternatively, the E_u can also be determined from empirical formulas or from the back analysis of well-documented case histories. As shown in Figure 8.37, the ratio of the undrained Young's modulus to the undrained shear strength (E_u/s_u) can also be estimated based on the plasticity index (PI) and OCR of the soil. Based on the author's experience, if $E_u = (500 \text{ to } 700)s_u$ is assumed for normally consolidated clay with low plasticity, reasonable excavation analysis results with the MC undrained C model can be obtained. As a matter of fact, this range is close to the lower bound in Figure 8.37.

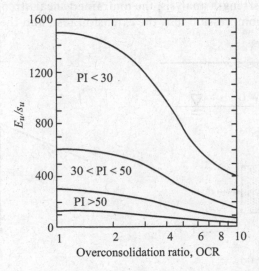

Figure 8.37 The relation of E_u/s_u with OCR and PI. (Redraw from Duncan and Buchignani (1976)

8.6 DETERMINATION OF INITIAL STRESSES

There are two methods for the computation of initial stresses: the direct input method and the gravity generation method.

8.6.1 Direct input method

The direct input method is also called the K_0-procedure. After the calculation of initial stresses, they can be input directly to the FEA software.

In the effective stress analysis, the initial effective vertical (σ'_y) and horizontal stresses (σ'_x) at a point in the soil in the free-field horizontal surface (Figure 8.38a) can be calculated following the principle of the effective stress in soil mechanics or those in Section 4.2. The initial stresses in the sloping ground with no groundwater (Figure 8.38b) can be calculated as $\sigma'_y = \gamma_m h$, $\sigma'_x = K_0\sigma'_y$, and $\tau_{xy} = \gamma_m h\sin\alpha / 2$ where K_0 is the coefficient of the earth pressure at rest. If there is groundwater seepage, the pore water pressure can be estimated using the flow net method, and then, σ'_y and σ'_x can be determined accordingly.

In the total stress analysis, the coefficient of the earth pressure at rest should be estimated with the consideration of the isotropic strength or anisotropic strength analysis. As shown in Figure 8.39, after the soil is consolidated at the K_0 state, the stress–strain curves obtained by compression and extension tests are significantly different. At the initial state (consolidated state), its major principal stress and minor principal stress are in the vertical and horizontal directions, respectively. After being subjected to extension pressure to failure, the directions of the major principal stress and the minor principal stress change to horizontal and vertical, respectively. The pore water pressure generated by principal stress rotation is much greater than that without rotation (i.e., compression test), so the undrained shear strength (s_{ue}) obtained by the extension test is lower than that by the compression test (s_{uc}).

In the isotropic strength analysis, the undrained shear strength used is the average value from the compression test, the extension test, and the direct simple shear

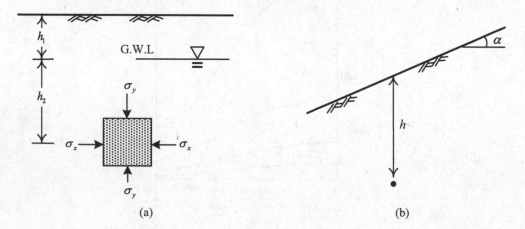

(a)   (b)

Figure 8.38 Initial stresses (a) in the free field (b) in the sloping ground.

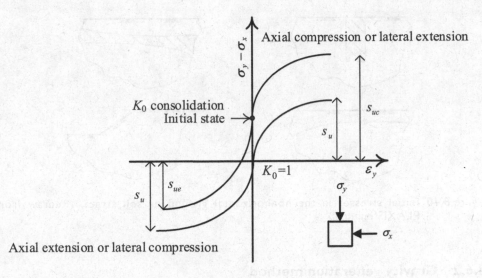

Figure 8.39 Calculation of initial stresses with the total stress.

test. If the results of the simple shear test are not available, the average value of the undrained shear strength from extension and compression tests can also be directly used:

$$s_u = \frac{s_{uc} + s_{ue}}{2} \tag{8.60}$$

where s_u is the undrained shear strength used in the isotropic strength analysis. In addition to Eq. 8.60, the average undrained shear strength can also be obtained from the vane shear test, or the triaxial unconsolidated undrained (UU) test, or the unconfined compression (UC) test (Section 2.6.3).

As shown in Figure 8.39, the isotropic strength analysis is equivalent to moving the initial stresses at the K_0-consolidated state to the origin, equivalent to the $K_0 = 1.0$ state. It seems to be plausible to substitute $\phi = 0$ into the Jaky's equation (Eq. 4.3), which seems to be unreasonable because the effective friction angle (ϕ') should be used in the Jaky's equation rather than the total stress friction angle (ϕ).

For some total stress anisotropic soil models, the undrained shear strength (s_u) is the largest when the major principal stress is vertical (axial compression) and the smallest when the major principal stress is horizontal. The s_u changes with the angle of the principal stress rotation. At this time, the coefficient of the initial stress cannot be estimated using Eq. 4.3 or using $K_0 = 1.0$. Instead, the horizontal total stress (σ_h) and vertical total stress (σ_v) should be calculated separately, and then, $\overline{K_0} = \sigma_h / \sigma_v$ is taken to estimate the initial total stresses.

The direct input method is only applicable to the horizontal ground surface or infinite long slope (Figure 8.38). For nonhorizontal strata or soil layers, only the gravity generation method can be used (Figure 8.40).

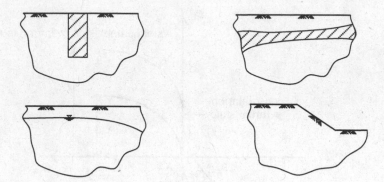

Figure 8.40 Initial stresses in the nonhorizontal ground or soil strata. (Redraw from PLAXIS manual.)

8.6.2 Gravity generation method

Change the boundaries of excavations to be rollers (this is a temporary measure for computing initial stresses. When starting the formal analysis, it should be set to what they should be). Assign a suitable μ value to each element, the magnitude of which strongly influences the resulting initial stresses. It is important to select values of μ that gives realistic values of initial stresses or K_0 values. For an example, the lateral strain of each element in the initial state is 0 in one-dimensional consolidated layers; in accordance with the theories of elasticity, we can infer that $\mu = K_0 (1 + K_0)$. Then, apply the gravity body force over the whole area, and use the standard finite element computation procedure to solve the initial stresses. These data sets should be changed back to what they should be in the formal analysis. It is noted that when undrained materials are used in the effective stress analysis, the effect of undrained behavior should be ignored to ensure that reasonable initial effective stresses are obtained.

The initial stresses computed by the gravity generation method still needs to be examined for correctness. The method of examination is to select several representative elements from the region, and use the methods introduced in Section 8.6.1 to check the results computed by the gravity generation method. Finally, reset all the induced displacements equal to 0 after the determination of initial stresses.

8.7 STRUCTURAL MATERIAL MODELS AND RELATED PARAMETERS

The yielding stress of the retaining wall and struts is usually very high and can be analyzed using linear elastic models, that is, the Young's modulus and Poisson's ratio are assumed to be constant. In the analysis, the inertia of moment (I) and stiffness (EI) of a retaining wall such as sheet piles, soldier piles, row piles, and diaphragm walls should be reduced rationally. The axial stiffness (AE / L) of a strut or floor slab should also be reduced. The reasons and how much should be reduced are the same as used with the beam on the elastic foundation method. Readers can refer to Section 7.7. They will not be repeated here.

8.8 MESH GENERATION

The shape of the element used in an analysis can strongly affect the results obtained. The following are the commonly adopted rules for mesh generation:

8.8.1 Shape of the element

Elements used in the FEA should avoid irregular shapes. It is better to be as regular as possible because elements in irregular shapes will cause numerical instability or inaccuracy of the numerical analysis. Whether an element is in a good shape can be evaluated using its aspect ratio. The aspect ratio is the ratio of the length to the width of an element (L / B), as shown in Figure 8.41. The closer the aspect ratio is to 1.0, the better is the shape, that is, the square or an equilateral triangle is the best choice. Elements with angles of 90° (quadrangles) or 60° (triangles) are also good elements. Because neither squares nor equilateral triangle are easily found, elements with an aspect ratio within the range $1.0 \leq L / B \leq 2.0 \sim 2.5$ can be viewed as good ones.

The shape of an element will influence the analytical accuracy of the element and the surrounding elements. It is therefore necessary to place good elements in interested areas. In less interested areas, some elements not so good can be placed. For example, if the retaining wall is an important object of the analysis, good elements should be placed in its surroundings. On the other hand, the boundary areas are not important areas, and some elements not that good can be placed there.

8.8.2 Density of mesh

In principle, the mesh in the areas of the stress concentration, of rapid strain changing, of the crucial areas, and of the object zones should be finer. The retaining wall is a rigid structure, and the soil is comparatively a soft material. The mesh in the transition zone between the wall and the surrounding soils should, therefore, be as fine as possible because a larger stress gradient will be generated there. The farther from the wall, the lower can be the density of the mesh.

As the unloading forces caused by excavation act directly on the excavation bottom in the excavation zone, the density of the mesh in the excavation zone will affect the analysis results. Therefore, the density of the mesh in the excavation zone should be finer than that outside the excavation zone.

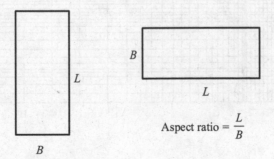

Aspect ratio $= \dfrac{L}{B}$

Figure 8.41 Definition of the aspect ratio.

8.8.3 Boundary condition

If considering the symmetry of an excavation and taking a half for the analysis as shown in Figure 8.42a, the symmetric boundary (line B'–B' in Figure 8.42b) should be equipped with rollers to restrain the lateral displacement and allow vertical displacement. Moreover, Ou and Shiau (1998) found that to analyze movement in an excavation, rollers will produce better results than hinges placed on the line C'–C' (Figures 8.42b). Also, the rollers should be placed at a distance more than three times excavation depths ($3H_e$) and four times excavation depths ($4H_e$) from the retaining wall for the analysis of the wall deformation and surface settlement, respectively. In principle, the farther the boundary (C'–C' for example) is from the wall, the better the analysis results, although it takes more computation time. On the base, either rollers or hinges can be placed. In general, the hinges or rollers should be placed on hard soils or a certain distance below the retaining wall bottom.

The location of the boundary of the mesh can also be determined from the convergence study from the FEA. As shown in Figure 8.42b, assume the boundary is line G'–G' first and then carry out the analysis. Extend the boundary to line H'–H' and perform the analysis again. If the two analyses come out similar in stress, strain, or displacement along, for example, X'–X', it means that the boundary can be set at G'–G' or H'–H'. Otherwise, the boundary should be moved to C'–C'.

Figure 8.43 shows the suggested boundaries for stability analysis, structural force analysis, and deformation analysis.

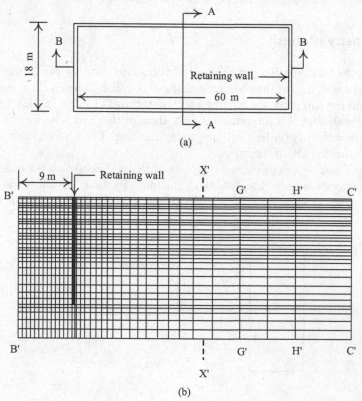

Figure 8.42 Plane strain analysis of an excavation (a) plan (b) finite element mesh.

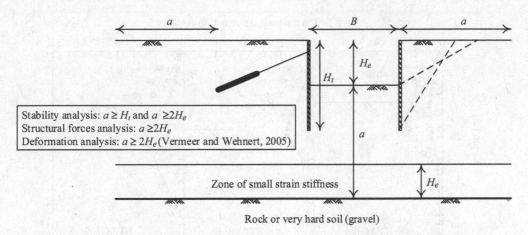

Stability analysis: $a \geq H_t$ and $a \geq 2H_e$
Structural forces analysis: $a \geq 2H_e$
Deformation analysis: $a \geq 2H_e$ (Vermeer and Wehnert, 2005)

Figure 8.43 Distance of the finite element mesh boundary (Waterman, 2009).

8.9   PLANE STRAIN ANALYSIS AND 3D ANALYSIS

Although in general engineering practice, the plane strain analysis is capable of obtaining a rational result, the wall deformation, and surface settlement on the section of the shorter side (B-B section in Figure 8.42) or that near the corner are 3D behaviors, and the plane strain analysis would overestimate the deformation or settlement. The 3D analysis can solve the problems.

In the 3D analysis, the stress state at a point has six components: σ_{xx}, σ_{yy}, σ_{zz}, τ_{xy}, τ_{yz}, and τ_{zx}. The theories of the 3D FEA, soil models, and analysis procedure are all similar to those for the plane strain analysis. In practice, the 8-noded low-order hexahedron element (H8) (Figure 8.44a), also called the brick element or solid element, is usually used for soil. Alternatively, the 10-noded tetrahedron element can be used (Figure 8.44b). The high-order 3D element, although more accurate, is usually not adopted because it requires too much computer memory and is very time consuming in computation.

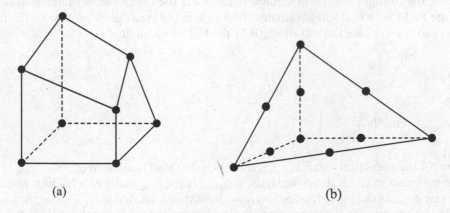

(a)   (b)

Figure 8.44 3D elements.

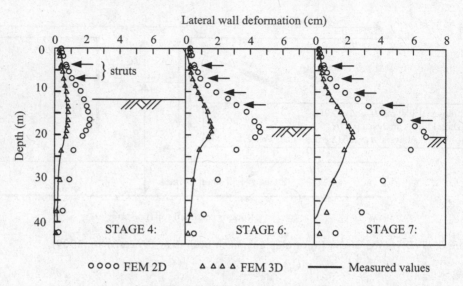

Figure 8.45 Comparisons of the wall deflection from the plane strain analysis, 3D analysis and field measurement, respectively, in a corner of the Haihaw Financial Center excavation.

Figure 8.45 shows wall deformations at the corner of the Haihaw Financial Center excavation computed from the plane strain analysis, 3D analysis, and field measurement (Ou et al., 1996). As shown in the figure, the plane strain analysis will overestimate, whereas the 3D analysis will obtain rational results. Besides, in excavations that use soil improvement, cross walls or buttress walls to reduce deformation (see Sections 11.6–11.8), wall deformation and ground settlement are with 3D behaviors. In such cases, only the 3D analysis can obtain rational results.

8.10   FINITE ELEMENT STABILITY ANALYSIS

In addition to predicting the movement of the retaining wall and soil, the FEM can also use the strength reduction method to calculate the safety factor of an excavation. Use the FEM with real soil parameters to simulate the excavation process to the final stage, and then reduce the soil strength by the following methods:

$$c_{input} = \frac{c_{original}}{SR} \tag{8.61}$$

$$\phi_{input} = \tan^{-1}\left(\frac{\phi_{original}}{SR}\right) \tag{8.62}$$

where SR is a reduction value, $c_{original}$ is the original (real) cohesion of soil, c_{input} is the input cohesion in the stability analysis, $\phi_{original}$ is the original (real) friction angle of soil, and ϕ_{input} is the input friction angle in the stability analysis.

In the analysis, first assume an SR value greater than 1.0, for example, $SR = 1.1$; then reduce the values of cohesion and friction angle with Eqs. 8.61 and 8.62; and then perform the analysis following the standard nonlinear finite element procedure, steps 1 to 11, as described in Section 8.2.3. For a certain *SR* value, the displacement increment at each node $\{\Delta q\}$ can be obtained in Step 3, and the equivalent nodal force (F_{int}) can be solved in Step 8. If the percentage of the error as calculated by using Eq. 8.18 is less than the tolerated error, for example, 1%, it can be called the computation convergence. Under such a condition, the excavation is stable rather than in or near failure because the displacements in the excavation are not sufficiently large. Then, assume a larger *SR* value than the previous one, for example, 1.15, and repeat the above computation procedure. By continuously increasing the *SR* value until the percentage of the error is greater than the tolerated error, which is called the computation divergence. When the computation reaches divergence, the displacements obtained in Step 3 are too large, leading the F_{int} no longer being balanced with the F_{ext}. An extraordinarily large displacement in an excavation implies the excavation in or near the state of failure. The safety factor of the excavation is equal to the *SR* value.

The authors used the strength reduction method to perform the stability analysis of a failure excavation. In the analysis, if the retaining wall and struts are assumed to be elastic, it means that the retaining wall and support will never fail or yield, and the failure or yield only occurs in the soil. However, if in the analysis, the retaining wall and struts are assumed to be elastoplastic, it means that in addition to the soil, the retaining wall and struts may also fail or yield. If the retaining wall or struts fail or yield first, then the entire excavation system will collapse, and the soil fail or yield accordingly. Based on the author's experiences, the factor of safety obtained by this approach is closer to the real factor of safety. For the relevant theories and computations of using the strength reduction method to compute the factor of safety, please refer to the relevant literature (Do et al., 2016). Figure 8.46 shows the variation of displacement at a point in the soil with the *SR* value. As the *SR*

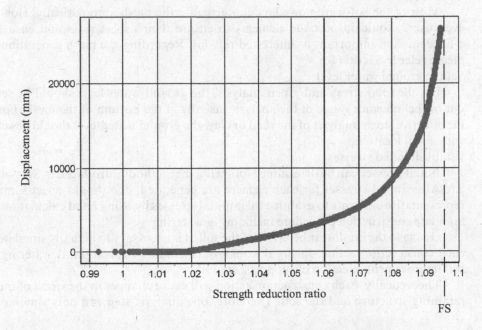

Figure 8.46 Variation of wall deflections with strength reduction ratio.

value is equal to 1.095, the displacement is sufficiently large so that the computation diverges, and the excavation can be regarded as in a state of failure. The factor of safety of the excavation is thus equal to the SR value when the computation diverges, that is, $F_b = 1.095$.

8.11   FEA PROCEDURE

The following are the procedure for the FEA.

1. Define the problem dimension

    For the target geotechnical structure, determine the sections to be analyzed and then perform 2D or 3D analysis based on the selected sections. The 3D analysis is certainly of longer time of computation and data processing than the 2D analysis. The 3D analysis usually gives more realistic results, whereas the result from the 2D analysis is conservative. Determination of the 2D or 3D analysis is mostly based on the engineering judgement (Section 8.9).

2. Set the target geometry and boundaries

    Determine the geometry of the target for the analysis. Assign appropriate boundary conditions for the region in accordance with the contents in Section 8.8.3.

3. Select the material model and evaluate their input parameters

    For the target with various stress paths, a high level constitutive model such as the HS or HSS model would be a good choice. For simple problems, monotonic loading or unloading, for example, the MC model with the parameters obtained from the designated stress path can be selected. The introduction to various constitutive models can be referred to in Section 8.4. Estimation of the relevant parameters can be referred to in Section 8.5.

4. Generate the mesh

    Most of the software nowadays can generate the mesh automatically. However, users should inspect the elements to ensure their aspect ratio and ensure a finer mesh in important or interested regions. Regarding the mesh generation, please refer to Section 8.8.

5. Set the ground water level

    For the total stress undrained analysis, the ground water level should be set out of the influence range of the analysis, usually at the bottom of the mesh. For the effective stress analysis of the sand or clay, the ground water level should be set with a real location.

6. Establish initial stresses

    Initial stresses can be determined following the methods introduced in Section 8.6. After initial stresses for each element are generated, one should select some representative elements to examine their initial stresses by using hand calculation.

7. Simulate construction procedure including dewatering

    Owing to the nonlinear behavior of the soil, it is necessary to actually simulate each construction action during the analysis, including excavation, dewatering, soil improvement, erection of supports, etc.

    Theoretically, each construction action will cause changes in the stress of the retaining structure and the soil. Therefore, one analysis step can only simulate one construction action. However, sometimes to simplify the analysis procedure, several construction actions are often combined into one analysis stage; for example, dewatering and excavation are combined into one analysis stage. The analysis results of dewatering and excavation combined into one analysis stage are almost identical to the results of each analysis stage of dewatering and excavation.

However, some analysts further combine the installation of support, preloading, dewatering, and excavation into one analysis stage to further simplify the analysis process. However, the analysis results found that the deformation of the retaining wall obtained by combining those steps is greater than that without combination (see Figure 7.8). This simplified analysis result is obviously unreasonable.

The combination of several steps into one step needs to calibrate before conducting the formal analysis.

8. Check convergence

Each stage of computation during analysis has to achieve convergence. The convergence criterion can follow Eq. 8.18. The relevant theory can refer to Section 8.2.3.

9. Validate the analysis results

As the FEA procedure is complicated and any small neglect is likely to lead to wrong results, the results from the FEA should be examined with case histories or other methods, simplified methods (Chapter 6) for example, to ensure the reasonability of the results.

PROBLEMS

8.1   Explain why when the Poisson's ratio of soil is 0.5, the volume strain would be 0.

8.2   Explain why the perfectly plastic model can only produce plastic strain without elastic strain.

8.3   Explain why the drained shear modulus is the same as the undrained shear modulus.

8.4   Same as Example 6.1, the excavation of building P. The excavation profile and geological conditions are as shown in Figure 6.43. The observations of deformations in all the excavation stages are as shown by Figure 6.44. If we analyze the deformation and stress of the retaining wall using the FEM (or the finite difference method) and the HS model, estimate the required parameters for each soil layer.

8.5   Same as above. Use the FEM (or the finite difference method) and the parameters determined in the previous problem to analyze the deformation of the retaining wall at each stage, and compare the results with the observations.

8.6   Same as Problem 8.4. If we adopt the MC elastoplastic model to simulate the stress–strain behaviors of sand and clay, the latter of which is analyzed with the MC undrained B model, determine the required parameters for each soil layer.

8.7   Same as above. Use the FEM (or the finite difference method) and the parameters obtained in the previous problem to analyze the deformation at each stage. Compare the results with the observations.

8.8   Same as Problem 8.6. If we use the MC elastoplastic model to analyze the excavation-induced deformation and stress of the retaining wall, except that the

MC undrained C model is to be adopted for the clayey layer, determine the required input parameters for each layer of the soil.

8.9   Same as above. Use the FEM (or the finite differential method) and the parameters obtained in the previous problem to analyze the deformation at each stage. Compare the results with the observations.

8.10   Same as Problem 8.4. If we adopt the MC elastoplastic model to simulate the stress–strain behavior of sandy soils and the MCC elastoplastic model for the stress–strain behaviors of clayey soils, determine the required parameters for each layer of the soil.

8.11   Same as above. Use the FEM (or the finite difference method) and the parameters obtained in the previous problem to analyze the deformation at each stage. Compare the results with the observations.

8.12   Same as Problem 6.11. The excavation profile and geological profile of Building S are as shown in Figure P6.11. The observations of deformations at all the stages are as shown in Figure P7.6. If we use the FEM (or the finite difference method) and the HS model to analyze the excavation-induced wall deformation and stress, estimate the input parameters for each layer of the soil.

8.13   Same as Problem 8.12. Use the FEM (or the finite difference method) and the parameters obtained in the previous problem to analyze the deformation at each stage. Compare the results with the observations (Figure P7.6).

8.14   Same as Problem 8.12. If we use the FEM (or the finite difference method) to analyze the excavation-induced deformation and stress of the retaining wall and adopt the MC elastoplastic model to simulate the stress–strain behaviors of sand and clay, the latter of which the MC undrained B is applied, determine the input parameters required for each layer.

8.15   Same as above. Use the FEM (or the finite difference method) and the obtained parameters from the previous problem to analyze the deformation at each stage. Compare the results with the observations.

8.16   Same as Problem 8.14. If we use the MC elastoplastic model to analyze the excavation-induced wall deformation and stress, except that the MC undrained C is adopted for the clayey layers, determine the input parameters required for each layer.

8.17   Same as above. Use the FEM (or the finite difference method) and the parameters obtained in the previous problem to analyze the deformation at each stage. Compare the results with the observations.

8.18   Same as Problem 8.12. If we adopt the MC elastoplastic model to simulate the stress–strain behaviors of sand and the MCC elastoplastic model for the clay, determine the input parameters required for each soil layer.

8.19   Same as above. Use the FEM (or the finite difference method) and the parameters obtained in the previous problem to analyze the deformation at each stage. Compare the results with the observations.

8.20   Same as Problem 6.4. Figure P6.4 shows the excavation profile and geological profile of Building R in Taipei. The measured deformation of the retaining wall at each stage is as shown in Figure P7.7. If we use the FEM (or the finite difference method) with the HS model to analyze the excavation-induced wall deformation and stress, estimate the input parameters required for each layer of the soil.

8.21 Same as above. Use the FEM (or the finite difference method) and the parameters obtained in the previous problem to analyze the deformation at each stage and compare the results with the observations.

8.22 Same as Problem 8.20. If we use the FEM (or the finite difference method) to analyze the wall deformation and stress induced by excavation and adopt the MC elastoplastic model for the simulation of the stress-strain relation for both sand and clay, for the latter of which the MC undrained B is to be performed, determine the input parameters required for each soil layer.

8.23 Same as above. Use the FEM (or the finite difference method) and the parameters obtained in the previous problem to analyze the deformation at each stage and compare the results with the observations.

8.24 Same as Problem 8.22. If we use the MC elastoplastic model to analyze the excavation-induced wall deformation and stress, except that the MC undrained C for clay, determine the input parameters required for each soil layer.

8.25 Same as above. Use the FEM (or the finite difference method) and the parameters obtained in the previous problem to analyze the deformation at each stage and compare the results with the observations.

8.26 Same as Problem 8.20. If we adopt the MC elastoplastic model to simulate the stress–strain behavior of sand and the MCC elastoplastic model for clay, determine the input parameters required for each soil layer.

8.27 Same as above. Use the FEM (or the finite difference method) and the parameters obtained above to analyze the deformation at each stage and compare the results with the observations.

Chapter 9

Dewatering in excavations

9.1 INTRODUCTION

Based on investigations, most problems encountered in a deep excavation have direct or indirect relations with groundwater. Therefore, whether groundwater has been properly dealt with is a crucial point for the success of an excavation. Groundwater-induced problems in an excavation may arise from insufficient investigation of groundwater or geological conditions that lead to the inability to fully control the groundwater. It may also arise from misunderstanding of the influence of groundwater so that a wrong excavation method is adopted. Thus, it is necessary to perform detailed investigations of groundwater and its influences on soils or structures during excavation.

The permeability of clay is lower than 10^{-6} cm/s, and the flow velocity of groundwater in clay is rather slow. As a result, it is not necessary to deal with groundwater problems in clay during an excavation. The permeability of sand or gravel is usually greater than 10^{-3} cm/s, and the flow velocity of groundwater in sand or gravel is rather high. The groundwater may leak into the excavation zone during excavation and cause much trouble. In the worst cases, it may bring about the loosening of soils, sand boiling, or upheaval failure. To avoid such conditions, it is necessary to design comprehensive dewatering schemes before or during excavation. On the other hand, with the higher flow velocity of sand or gravel, simple methods such as pumping are usually sufficient to lower the groundwater levels.

This chapter will explain the goals of dewatering in excavations and introduce some commonly used methods of dewatering, including their basic theories and design methods.

9.2 GOALS OF DEWATERING

As explained above, dewatering methods are the methods of lowering the groundwater level in sandy or gravelly grounds rather than in clayey grounds. Generally speaking, the goals of lowering the groundwater level are to keep the excavation bottom dry, to prevent leakage of water or sand, to avoid sand boiling or upheaval failure, and to forestall the occurrence of floating basements, as explicated in the following:

1. To keep the excavation bottom dry
    With a higher flow velocity of groundwater in sand or gravel, groundwater may flow well into the excavation zone, which may cause inconvenience for construction.

DOI: 10.1201/9780367853853-9

To keep the excavation bottom dry, the groundwater level is generally lowered to 0.5–1.0 m below the excavation bottom, as shown in Figure 9.1. Groundwater flows so slowly in clayey soils that flowing of groundwater into the excavation zone will not occur. There is no need to lower the groundwater level in clay.

2.   To prevent leakage of groundwater or soils

To excavate in sandy or gravelly soil with a high groundwater level, using either soldier piles or sheet piles which are not satisfactorily watertight, or diaphragm walls or bored piles with joints that may have defects, will risk the possibility of the leaking of groundwater into the excavation zone through the retaining wall. The leakage of groundwater and soils may lead to disastrous results and bring about collapses and failures of excavations in the worst cases, when the leaking is great enough to enlarge the holes in the retaining wall. For a more detailed content, please see Section 11.11.1.

3.   To avoid sand boiling

To keep the excavation bottom dry when excavating in sandy or gravelly soils requires lowering the groundwater level within the excavation zone to 0.5–1.0 m below the excavation bottom at least. While excavation proceeds, the difference between the groundwater levels within and outside the excavation zone grows larger. When the hydraulic gradient around the excavation bottom grows larger or equals the critical hydraulic gradient of soils, sand boiling will occur. Many methods are available to avoid sand boiling. One of them is to lower the groundwater level outside the excavation zone. However, the possibility of ground settlement outside the excavation zone has to be considered (see Figure 9.2).

4.   To forestall the upheaval failure

As shown in Figure 9.3, there exists a permeable layer (such as sand or gravel) underlying the clayey layer. The water pressure from the permeable layer will generate an upheaving force against the clayey layer. When the water pressure acting on the bottom of the clayey layer is larger than the total weights of the above soil layers, upheaval failure will occur. One of the methods to prevent the occurrence of upheaval failure is to lower the water pressure in the permeable layer by pumping.

5.   To keep the basement from floating

Basement construction will start right after the completion of the excavation. In sandy soils, the phenomenon of the floating of the basement is likely to happen

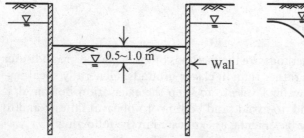

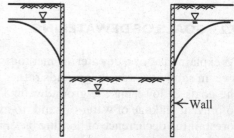

Figure 9.1 Keeping the excavation bottom dry by lowering the groundwater level beneath the excavation bottom.

Figure 9.2 Lowering the groundwater level outside of the excavation zone to avoid sand boiling.

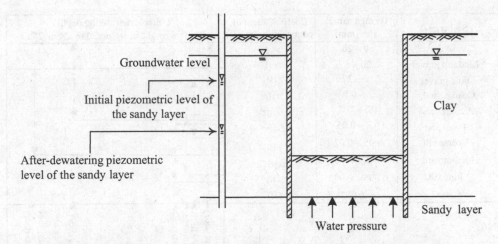

Figure 9.3 Lowering the piezometric level in sandy soils to avoid upheaval failure.

during the stage of basement construction if the weight of the basement structure is smaller than the water pressure acting on the foundation base. Once the floating phenomenon happens, with the differential heaves at different locations of the foundation, the floated basement will not necessarily sink back to the original elevation while building construction proceeds, which may lead to the damage of the structures. In the worst condition, the basement or foundation may need to be demolished or reconstructed. Therefore, dewatering is usually required at the stage of basement construction to keep the upheaving force on foundation bottoms smaller than the weight of structures during construction. Dewatering is to be continued until the upheaving force is smaller than the weight of the structure or basement during construction.

9.3 METHODS OF DEWATERING

The open sump or ditch method, the deep well method, and the well point method are commonly used dewatering methods in excavations. Figure 9.4 diagrams the application ranges of these methods. The following will explain their basic principles and implementation.

9.3.1 Open sumps or ditches

The open sump method is to construct an open sump on the excavation bottom to collect the groundwater seeping into the sump by gravity or other natural means and then pump the collected water out. If the excavation area is very large or the base has a long narrow shape, several sumps may be placed along the longer side or simply use a long narrow sump, which is called an open ditch. Both the open sump and the ditch are gravity draining methods.

As shown in Figure 9.5, open sumps are usually placed near the retaining wall at the lowest excavation bottom. The depth of an open sump is generally 0.6–1.0 m.

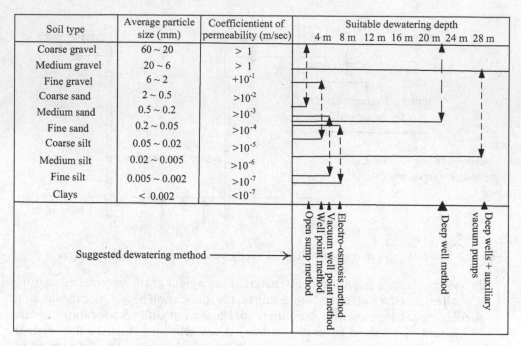

Soil type	Average particle size (mm)	Coefficientient of permeability (m/sec)	Suitable dewatering depth 4 m 8 m 12 m 16 m 20 m 24 m 28 m
Coarse gravel	60 ~ 20	> 1	
Medium gravel	20 ~ 6	> 1	
Fine gravel	6 ~ 2	$+10^{-1}$	
Coarse sand	2 ~ 0.5	$>10^{-2}$	
Medium sand	0.5 ~ 0.2	$>10^{-3}$	
Fine sand	0.2 ~ 0.05	$>10^{-4}$	
Coarse silt	0.05 ~ 0.02	$>10^{-5}$	
Medium silt	0.02 ~ 0.005	$>10^{-6}$	
Fine silt	0.005 ~ 0.002	$>10^{-7}$	
Clays	< 0.002	$<10^{-7}$	

Suggested dewatering method ———→

Open sump method
Well point method
Vacuum well point method
Electro-osmosis method
Deep well method
Deep wells + auxiliary vacuum pumps

Figure 9.4 Application conditions for various dewatering methods.

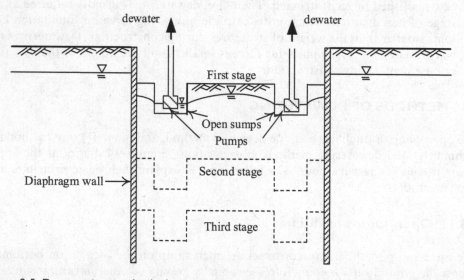

Figure 9.5 Dewatering method: the open sump method.

The side walls of the open sump sometimes need to be protected to prevent them from collapse. Some common protective measures are slope side walls, wood planks, iron barrels, and concrete culverts. Figure 9.6 is a photo of protecting the open sump by a concrete culvert.

Figure 9.6 The photo of a concrete culvert protecting the side wall of a sump.

The open sump method is the most economical method of dewatering. However, its application is confined to permeable layers such as sandy and gravelly soils. Because the bottom of the open sump is lower than the excavation bottom, it will shorten the seepage path along which groundwater from outside seeps into the excavation zone, and as a result, the exit gradient on the sump bottom will be larger than that on the excavation bottom. This fact may bring about sand boiling on the sump bottom, which has to be prevented. Sand boiling has occurred in many excavation case histories (please refer to Section 5.11).

9.3.2 . Deep wells

As shown in Figure 9.7, the deep well method is to drill a well near the excavation zone and pump water out of it to make groundwater around the well flow into it under the influence of gravity. As a result, the groundwater level in the vicinity of the well will decline. The deep well method is another gravity draining method.

The diameter of a deep well is about 150–500 mm. If the goal of pumping is confined to the lowering of the groundwater level to keep the excavation bottom dry, the depth of the well could be set around 2.0–5.0 m below the excavation bottom and lower than the bottom of the retaining wall. The wells are to be located around the vicinity of the excavation zone. The pump used in a deep well can be a submerged pump or centrifugal pump. Concerning the design, functions, and application limits of various pumps, please consult the related documents. Based on the type and arrangement of pumps used in a deep well, the dewatering depth can reach more than 30 m.

Figure 9.8 diagrams the construction of a deep well. As illustrated in the figure, drill the well to the designed depth, install a case (a PVC pipe for example) and a screen, and place filters between the well sides and the case. A well screen is also a

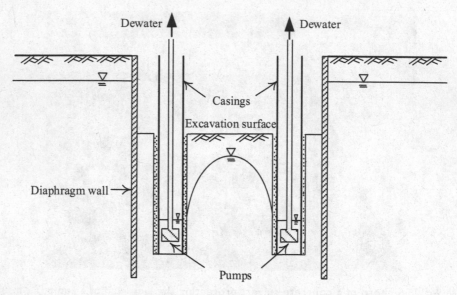

Figure 9.7 Dewatering method: the deep well method.

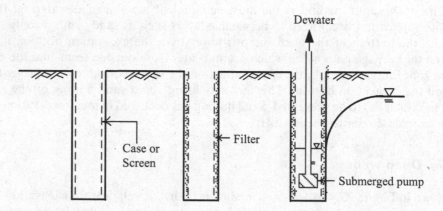

Figure 9.8 Construction of deep wells.

filtering device that serves as the intake portion of the well. A screen permits groundwater to enter the well from the aquifer and prevents fine particles from entering the well. After the filter is placed, the well has to be flushed to wash away the muddy soils generated during drilling, so as to avoid mud clogs in the filters, which will reduce the efficiency of the well.

The filter has to be made of a material which is fine enough to prevent soil particles from flowing into the filter. On the other hand, the filter should have voids large enough to keep good permeability and to avoid flowing into the well screen. The Japanese Society of Architecture (JSA, 1988) proposes the filter should meet the following criteria:

The criterion that the in situ soil particles do not flow into the filter:

$$\frac{D_{15}(\text{filter})}{D_{85}(\text{ground})} < 5$$

The criterion that the filter material keeps good permeability:

$$\frac{D_{15}(\text{filter})}{D_{15}(\text{ground})} > 5$$

The criterion that the filter material does not flow into the well screen:

$$\frac{D_{85}(\text{filter})}{D(\text{screen diameter})} > 2$$

Figure 9.9a shows the layout of deep wells during the construction of the Daqiao station of the Taipei metro system, and Figure 9.9b shows the profile of the section m-m of the dewatering system, retaining wall, and geological formation. The deep wells were installed inside the diaphragm wall, passing through the bottom of the diaphragm wall to the aquifer. To lower the piezometric level in the gravel located 50 m below the surface to forestall the upheaval failure, 21 deep wells, with the diameter of 0.4–0.5 m, were allocated. In addition, many observation wells for monitoring the

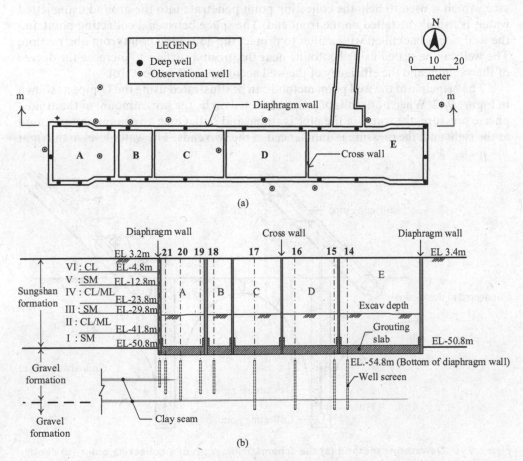

Figure 9.9 Dewatering system at the Daqiao station of the Taipei metro system (a) layout of the deep wells and excavation plan (b) profile of the section m-m of the deep wells, diaphragm wall, and geological formation.

water pressure in gravel formation were installed to ensure the successful dewatering during excavation. More details of the dewatering plan can be referred to in a study by Ou and Chen (2010).

9.3.3　Well points

The well point method is also called the vacuum well point method. The method is to place collecting points connected to the pumping pipe inside a small-diameter well and have them arranged in a line or in a rectangle. The collecting points are usually separated by a distance of 0.8–2.0 m. The top of the pumping pipe is connected to the common collecting pipe, which is then pumped out with vacuum, drawing out pore water from the soil and lowering the groundwater level accordingly. Figure 9.10 is a schematic diagram illustrating dewatering by way of the well point method.

The collecting pipes are usually arranged around the vicinity of an excavation site. Collecting points are the main structures of the well point method. As shown in Figure 9.11a, a collecting point is usually 100 cm long and has an external diameter of 5–7 cm, and many small holes are bored along the side to collect groundwater. A spurting device, which is used to help the collecting point penetrate into the ground using jetted water, is usually installed on the front end. The space between a collecting point and the well side is backfilled with a filter to protect the collecting point from obstruction. The well can be sealed using bentonite near the ground surface to increase the degree of the vacuum and the efficiency of the well accordingly (Figure 9.11b).

The principle of the well point method can be illustrated using the U-pipe as shown in Figure 9.12. When both ends of a pipe are acted on by the same amount of the atmospheric pressure, the water in the pipe is stable and still. Once a vacuum pump is used at the right end, the pressure is unbalanced at the two ends. The water level in the right

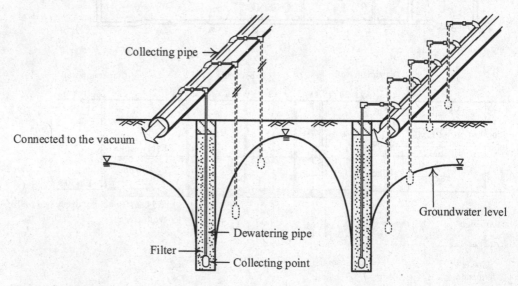

Figure 9.10 Dewatering method (a) the schematic diagram of a collecting point (b) configuration of the well point method.

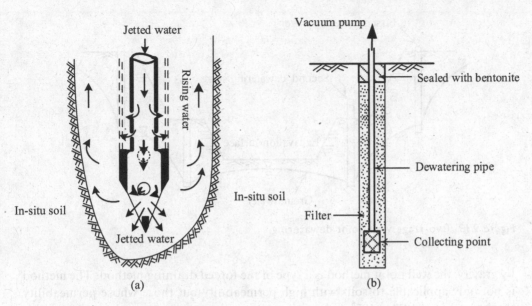

Figure 9.11 A well point dewatering system.

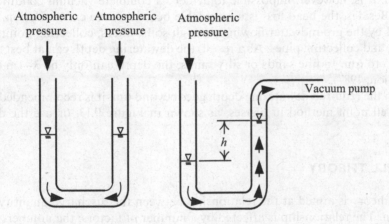

Figure 9.12 U-pipes acted on by a vacuum pump.

end of the U-pipe will rise, and a height difference (h) is thus generated. When the right end of the U-pipe reaches the complete vacuum state, h will be 1 atmospheric pressure, or 103.3 kPa, which equals a 10.33 m high water column. Obviously, the height to which the water table rises relates to the degree of the vacuum at the right end.

The principle of the well point method is similar to the above-mentioned principle of the U-pipe. Acted on by a vacuum pump, the pressure inside a collecting point is far lower than the pressure of groundwater which is exposed to the atmospheric pressure that leads to the groundwater being sucked into the collecting point. As the ground-water is sucked into the well out of the vacuum and drained instead of flowing into it

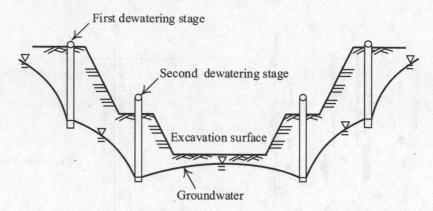

First dewatering stage

Second dewatering stage

Excavation surface

Groundwater

Figure 9.13 Two-stage well point dewatering.

by gravity, the well point method is a type of the forced draining method. The method is not only applicable to soils with high permeability but those whose permeability reaches as low as 10^{-4} to 10^{-5} cm/s, such as silts.

Theoretically, the dewatering depth of the well point method can reach as deep as 10.33 m. It is, however, impossible to achieve a complete vacuum condition inside the well. Besides, the head loss is unavoidable because there exists friction, which is generated by the groundwater flowing through soils, filters, collecting points, pumping pipes, and collecting pipes. As a result, the dewatering depth can at best reach the depth of 5 to 6 m. In fine sands or silty sands, the depth can only be 3–4 m (Quinion and Quinion, 1987).

When the required dewatering depth goes beyond 6 m, it is recommended to carry out the well point method in phases, as shown in Figure 9.13, or use the deep well method.

9.4 WELL THEORY

The well theory is aimed at the relationship between the discharge quantity and the drawdown. The relationship is affected by a number of factors: the numbers of wells, their structures, the geological conditions, and pumping time. The relationship between the discharge quantity and drawdown can be either solved using mathematical well formulas or numerical groundwater modeling. The former are inferred by assuming ideal conditions, whereas the latter is applicable to any geological or pumping condition and is therefore widely adopted. There are a great number of well formulas, and it is impossible to introduce all of them here. Only the most widely used well formulas are introduced here for application. As for the method of numerical groundwater modeling, please refer to the referent books (Powers, 1992 for example). Table 9.1 summarizes the well formulas for a single well under various conditions. The theories will be explicated as follows.

Table 9.1 Summary of Commonly Used Well Formulas

	Confined Aquifer		Free Aquifer	
	Full Penetration Well	Partial Penetration Well	Full Penetration Well	Partial Penetration Well
Non-equilibrium equation	Theis' equation: $s = \frac{Q}{4\pi T}\{-0.5772 - \ln u$ $- \sum (-1)^n \dfrac{u^n}{n \times n!}\}$ Jacob's modified equation: $s = \frac{Q}{4\pi T}(-0.5772 - \ln u)$ (Symbols are as shown in Fig. 9.16)		If the drawdown induced by dewatering is far smaller than the thickness of the aquifer, Theis' and Jacob's equations can be used instead.	
Equilibrium equation	Thiem's equation: $Q = \dfrac{2\pi k D(s_1 - s_2)}{\ln(r_2/r_1)}$ (Symbols are as shown in Fig. 9.16)	Kozeny: $Q = \dfrac{2\pi T(H - h_w)}{\ln(R/r_w)}\,\mu$ $\mu = \dfrac{D_1}{D}\left(1 + 7\sqrt{\dfrac{r_w}{2D_1}}\cos\dfrac{\pi D_1}{2D}\right)$ (Symbols are as shown in Fig. 9.19)	Dupuit-Thiem's equation: $Q = \dfrac{\pi k(h_2^2 - h_1^2)}{\ln(r_2/r_1)}$ (Symbols are as shown in Fig. 9.20)	Hausman's suggestion: $Q = \dfrac{\pi k[H^2 - h_w^2]\alpha}{\ln(R/r_w)}$ $\alpha = \sqrt{\dfrac{H - D_1}{H}}\sqrt[4]{\dfrac{H + D_1}{H}}$ (Symbols are as shown in Fig. 9.21)

9.4.1 Confined aquifers

Permeable layers are also called aquifers. When both above and below an aquifer, with the piezometric level higher than the top of the aquifer, are impermeable layers, the aquifer is called a confined aquifer.

When pumping an aquifer with a single well, the piezometric level will decline and a virtual drawdown cone will be formed, centered at the well, as shown in Figure 9.14. From the beginning of pumping, the area of the virtual drawdown cone will grow with the continuation of pumping. After a long period, the expansion rate of the area of the virtual drawdown cone will decline or stop expanding. This section is to introduce the pumping-induced drawdown in accordance with the penetration depth of the well into an aquifer and the expansion conditions of the virtual drawdown cone.

1. Full penetration wells

   A full penetration well is one that fully penetrates through the aquifer, and its flow direction is thus horizontal (Figure 9.15a). Theis (1935) derived an equation of the drawdown curve with regard to time based on the following assumptions:

   - The aquifer has a uniform thickness and is a homogeneous and isotropic confined aquifer.
   - The well has to be one that fully penetrates through the aquifer.
   - The well is 100% efficient. That is to say, no drop exists between the water table in the well and the drawdown curve.
   - Meets the Dupuit–Thiem assumption. That is, the hydraulic gradient of any point on the drawdown curve is the slope at the point.
   - No recharge water within the influence range.
   - The radius of the well is small enough that the amount of retained water can be ignored.

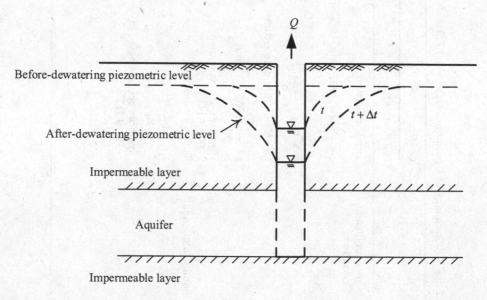

Figure 9.14 Dewatering in a confined aquifer.

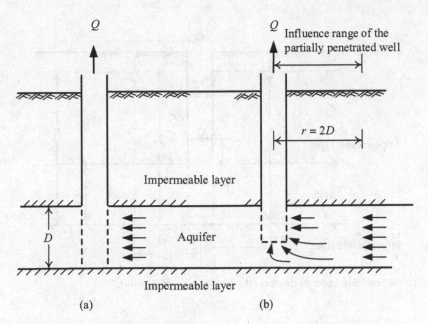

Figure 9.15 Flow directions around a well in a confined aquifer (a) full penetration well (b) partial penetration well.

Theis' drawdown equation is written as follows (Figure 9.16):

$$s = \frac{Q}{4\pi kD}W(u) = \frac{Q}{4\pi T}W(u) \tag{9.1}$$

$$W(u) = \int_u^\infty \frac{e^{-u}}{u}du = -0.5772 - \ln u - \sum_{n=1}^\infty (-1)^n \frac{u^n}{n \times n!} \tag{9.2}$$

$$u = \frac{r^2 S}{4Tt} \tag{9.3}$$

where
  s = drawdown,
  Q = the discharge quantity of the well,
  k = the coefficient of permeability or permeability,
  D = thickness of the aquifer,
  $T = kD$ = the coefficient of transmissivity,
  $W(u)$ = the well function,
  u = the parameter of the well function,
  r = the distance to the center of the well.
  S = the coefficient of storage or storativity
  t = time since pumping started

Eq. 9.1 is the Theis' nonequilibrium equation. Table 9.2 lists the values of well functions calculated from Eq. 9.2.

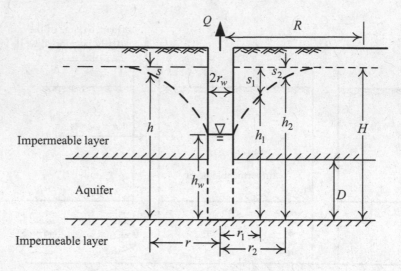

Figure 9.16 Notations used in dewatering in a confined aquifer.

Table 9.2 Well function $W(u)$ (Ferris et al., 1962)

N	$u = N \times 1$	$u = N \times 10^{-1}$	$u = N \times 10^{-2}$	$u < 10^{-2}$
1.0	0.2194	1.8229	4.0379	$W(u) = -0.577 - \ln u$
1.1	0.1860	1.7371	3.9436	
1.2	0.1584	1.6595	3.8576	
1.3	0.1355	1.5889	3.7785	
1.4	0.1162	1.5241	3.7054	
1.5	0.1000	1.4645	3.6374	
1.6	0.08631	1.4092	3.5739	
1.7	0.07465	1.3578	3.5143	
1.8	0.06471	1.3098	3.4581	
1.9	0.05620	1.2649	3.4050	
2.0	0.04890	1.2227	3.3547	
2.1	0.04261	1.1829	3.3069	
2.2	0.03719	1.1454	3.2614	
2.3	0.03250	1.1099	3.2179	
2.4	0.02844	1.0762	3.1763	
2.5	0.02491	1.0443	3.1365	
2.6	0.02185	1.0139	3.0983	
2.7	0.01918	0.9849	3.0615	
2.8	0.01686	0.9573	3.0261	
2.9	0.01482	0.9309	2.9920	
3.0	0.01305	0.9057	2.9591	
3.1	0.01149	0.8815	2.9273	
3.2	0.01013	0.8583	2.8965	
3.3	0.008939	0.8361	2.8668	
3.4	0.007891	0.8147	2.8379	
3.5	0.006970	0.7942	2.8099	
3.6	0.006160	0.7745	2.7827	
3.7	0.005448	0.7554	2.7563	
3.8	0.004820	0.7371	2.7306	
3.9	0.004267	0.7194	2.7056	

(Continued)

Table 9.2 (Continued) Well function $W(u)$ (Ferris et al., 1962)

N	$u = N \times 1$	$u = N \times 10^{-1}$	$u = N \times 10^{-2}$	$u < 10^{-2}$
4.0	0.003779	0.7024	2.6813	$W(u) = -0.577 - \ln u$
4.1	0.003349	0.6859	2.6576	
4.2	0.002969	0.6700	2.6344	
4.3	0.002633	0.6546	2.6119	
4.4	0.002336	0.6397	2.5899	
4.5	0.002073	0.6253	2.5684	
4.6	0.001841	0.6114	2.5474	
4.7	0.001635	0.5979	2.5268	
4.8	0.001453	0.5848	2.5068	
4.9	0.001291	0.5721	2.4871	
5.0	0.001148	0.5598	2.4679	
5.1	0.001021	0.5478	2.4491	
5.2	0.0009086	0.5362	2.4306	
5.3	0.0008086	0.5250	2.4126	
5.4	0.0007198	0.5140	2.3948	
5.5	0.0006409	0.5034	2.3775	
5.6	0.0005708	0.4930	2.3604	
5.7	0.0005085	0.4830	3.3437	
5.8	0.0004532	0.4732	2.3273	
5.9	0.0004039	0.4637	2.3111	
6.0	0.0003601	0.4544	2.2953	
6.1	0.0003211	0.4454	2.2797	
6.2	0.0002864	0.4366	2.2645	
6.3	0.0002555	0.4280	2.2494	
6.4	0.0002279	0.4197	2.2346	
6.5	0.0002034	0.4115	2.2201	
6.6	0.0001816	0.4036	2.2058	
6.7	0.0001621	0.3959	2.1917	
6.8	0.0001448	0.3883	2.1779	
6.9	0.0001293	0.3810	2.1643	
7.0	0.0001155	0.3738	2.1508	
7.1	0.0001032	0.3668	2.1376	
7.2	0.00009219	0.3599	2.1246	
7.3	0.00008239	0.3532	2.1118	
7.4	0.00007364	0.3467	2.0991	
7.5	0.00006583	0.3403	2.0867	
7.6	0.00005886	0.3341	2.0744	
7.7	0.00005263	0.3280	2.0623	
7.8	0.00004707	0.3221	2.0503	
7.9	0.00004210	0.3163	2.0386	
8.0	0.00003767	0.3106	2.0269	
8.1	0.00003370	0.3050	2.0155	
8.2	0.00003015	0.2996	2.0042	
8.3	0.00002699	0.2943	1.9930	
8.4	0.00002415	0.2891	1.9820	
8.5	0.00002162	0.2840	1.9711	
8.6	0.00001936	0.2790	1.9604	
8.7	0.00001733	0.2742	1.9488	
8.8	0.00001552	0.2694	1.9393	
8.9	0.00001390	0.2647	1.9290	
9.0	0.00001245	0.2602	1.9187	
9.1	0.00001115	0.2557	1.9087	
9.2	0.00000998	0.2513	1.8987	
9.3	0.000008948	0.2470	1.8888	
9.4	0.000008018	0.2429	1.8791	
9.5	0.000007185	0.2387	1.8695	
9.6	0.000006439	0.2347	1.8599	
9.7	0.000005771	0.2308	1.8505	
9.8	0.000005173	0.2269	1.8412	
9.9	0.000004637	0.2231	1.8320	

Habitually, the coefficients of permeability, transmissivity, and storage are all called hydraulic parameters. The coefficient of permeability is an intrinsic parameter of a soil, which is defined and can be tested as introduced in Section 2.2.4. The coefficient of transmissivity is the coefficient of permeability multiplied by the thickness of the aquifer. As the thickness of a confined aquifer is a constant, the coefficient of transmissivity of a confined aquifer is also a constant.

The coefficient of storage is defined as the drained volume of the pore water due to lowering a unit pressure head per unit surface area of an aquifer (Figure 9.17). As soil in a confined aquifer is always in the saturated state, the drainage of pore water during dewatering should be caused by the decrease of the thickness of the aquifer. Therefore, the coefficient of storage of a confined aquifer, which generally ranges between 0.0005 and 0.0001, is similar to the compressibility of soils (Powers, 1992).

When pumping time (t) is rather long or the distance (r) is quite short, the parameter, u, will be very small, as listed in Table 9.2. Thus, when $u \le 0.05$, the high order of the multinomial at the right side of the equation can be ignored. Eq. 9.1 can then be rewritten as follows:

$$s = \frac{Q}{4\pi T}(-0.5772 - \ln u) \tag{9.4}$$

which can also be written as

$$s = \frac{0.183Q}{T} \log \frac{2.25Tt}{r^2 S} \tag{9.5}$$

Eq. 9.4 or 9.5 is called the Jacob's modified nonequilibrium equation (Jacob, 1940), which is only applied when $u \le 0.05$.

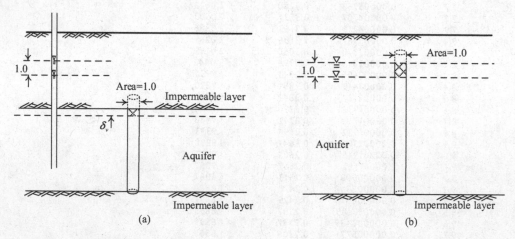

Figure 9.17 Definition of coefficient of storage: S = water drained from a volume of $\delta_v \times 1.0$ (a) confined aquifer (b) free aquifer.

Assuming the influence range of pumping is defined as the distance where the drawdown just declines to 0, the influence range of pumping-induced drawdown (R) can be calculated from Eq. 9.5 as follows:

$$R = \sqrt{\frac{2.25Tt}{S}} \tag{9.6}$$

In addition to the above equation, many people have proposed different formulas to estimate influence ranges of pumping-induced drawdown. One of them is the Kozeny's formula (1953):

$$R = 1.5\frac{\sqrt{Hkt}}{n} \tag{9.7}$$

where
  n = porosity of the soil,
  H = the groundwater level before pumping,
  t = pumping time,
  k = the coefficient of permeability.

The Sichart's formula (1928) can be given as the below:

$$R = 3,000s_w\sqrt{k} \tag{9.8}$$

Eq. 9.8 is an empirical formula where R represents the influence range measured in meter; s_w is the drawdown in the well, also measured in meter; k is the coefficient of permeability, and its unit is m/sec.

Assume a hypothetical dewatering case: $Q = 1.0$ m^3 / min, $T = 1.0$ m^2 / min, and $S = 0.0001$. Compute the relationship between the drawdown and pumping time in terms of the logarithm scale using separately Theis' and Jacob's equations, as shown in Figure 9.18a. As illustrated in the figure, when t is considerably large, causing $u \leq 0.05$, the relationship between the drawdown and pumping time in terms of the logarithm value becomes linear, which is consistent with the Jacob's modified nonequilibrium equation.

Similarly, the relationship between the drawdown and distance on the logarithm scale, as shown in Figure 9.18b, can be computed following Theis' and Jacob's equations, respectively. We can see when r is rather small, it also leads to $u \leq 0.05$, which is to say the drawdown and distance on the logarithm scale have a linear relationship. The influence range of pumping in Eq. 9.6 is expressed by the intersection point where the extension of the line meets the line $s = 0$.

Assume the drawdown curve achieves equilibrium after pumping (which follows that the drawdown curve will not expand with regard to pumping time). The Thiem's equilibrium equation (Thiem, 1906) can thus be derived from the differential equation of the drawdown curve (see Figure 9.16)

$$Q = \frac{2\pi kD(s_1 - s_2)}{\ln(r_2/r_1)} \tag{9.9}$$

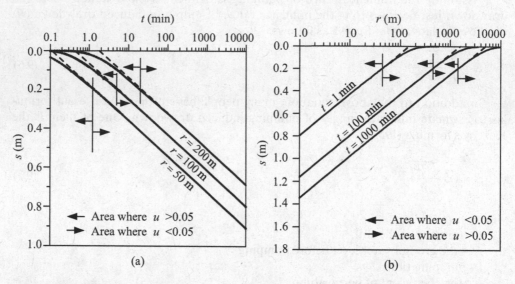

Figure 9.18 Variations of drawdowns in a confined aquifer (a) the relation between the drawdown and time for different distances (b) the relation between drawdown and distance for different time.

where r_1 and r_2 are the distances of the first observation well and the second observation well to the well center, respectively; s_1 and s_2 are drawdowns in the first and the second observation well, respectively.

Eq. 9.9 is the most common equation of pumping introduced in the soil mechanics textbooks. Let R be the influence range and substitute for r in the above equation. We can have the drawdown (s) at any distance. Thus, the Thiem's equilibrium equation can be rewritten as follows:

$$s = \frac{Q\ln(R/r)}{2\pi kD} \tag{9.10}$$

Let the drawdown in the pumping well be $(H - h_w)$ and the radius of the well be r_w; substitute them into Eq. 9.9. We then have the Thiem' equilibrium equation rewritten as follows:

$$Q = \frac{2\pi kD(H - h_w)}{\ln(R / r_w)} \tag{9.11}$$

where
$H =$ the piezometric level before pumping,
$h_w =$ the groundwater level in the pumping well after pumping,
$r_w =$ the radius of the pumping well.

The Thiem's equilibrium equation, Eq. 9.9, is derived from the differential equation of the drawdown curve, assuming that the drawdown after pumping is in equilibrium and that the drawdown curve does not expand with the pumping time. Basically, the Thiem's equilibrium equation is a special case of the Jacob's modified nonequilibrium equation. That is to say, when the drawdown curve does not expand any more, the pumping time coming to t and the influence range extending to R, the Jacob's modified nonequilibrium equation will be identical to the Thiem's equilibrium equation.

After pumping for a period of time, turn off the pump, and the drawdown curve will gradually return to the original water table. The relationship between the recovered drawdown curve and time can be used to derive the coefficient of transmissivity or permeability, as is designated as the recovery method. According to Theis (1935), the relationship between the residual drawdown and time can be expressed as follows:

$$s' = \frac{Q}{4\pi T}\ln\frac{t}{t'} = \frac{2.3Q}{4\pi T}\log\frac{t}{t'} \qquad (9.12)$$

where

s' = the residual drawdown, that is, the distance between the water level in the pumping well and the original groundwater level;
Q = the recovered quantity, equivalent to discharge quantity;
T = the coefficient of transmissivity;
t = time since pump started;
t' = time since pump is stopped.

The recovery method can be used to examine the results of pumping tests. What's more, during a long period of pumping and dewatering, the activities of pumping may be suspended temporarily for certain reasons, and the drawdown curve will recover gradually. The recovery method can be used under such conditions. The results can not only offer extra data but also be compared with those from pumping tests.

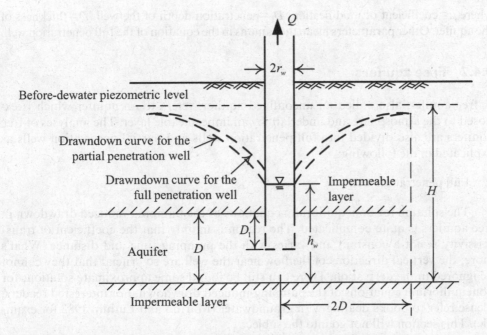

Figure 9.19 Drawdown for the partial penetration well in a confined aquifer.

2. Partial penetration well

In engineering practice, the partial penetration well is more widely used than the full penetration well. As shown in Figure 9.15b, within the range of $r \leq 2D$, the flow lines in the vicinity of the well are not necessarily horizontal. Instead, some are vertical. For most types of soils, the vertical coefficients of permeability are about one-tenth of the horizontal one. Thus, the well formulas concerning partial penetration wells need modifying.

There have been many nonequilibrium theories for partial penetration wells, and most of them are very complicated. This chapter will leave them aside. For interested readers, please see the related literature (Kruseman and de Ridder, 1990; Hantush, 1962).

There are also many equilibrium theories available. They are not, nevertheless, so complicated as nonequilibrium theories. This section will introduce a formula derived by Kozeny (1953).

As shown in Figure 9.19, when $r > 2D$, the effects of a partial penetration well can be ignored because its drawdown curve can be fully solved as that of full penetration wells. When $r \leq 2D$, however, the amount of the drawdown for a partial penetration well is larger than that for a full penetration well. In accordance with the Kozeny's derivation (1953), the quantity of water to be pumped to achieve the designed drawdown can be expressed as follows:

$$Q = \frac{2\pi T(H - h_w)}{\ln(R/r_w)} \mu \tag{9.13}$$

$$\mu = \frac{D_1}{D}\left(1 + 7\sqrt{\frac{r_w}{2D_1}} \cos\frac{\pi D_1}{2D}\right) \tag{9.14}$$

where μ = coefficient of modification; D_1 = penetration depth of the well; D = thickness of the aquifer. Other parameters mean the same as in the equation of the full penetration well.

9.4.2 Free aquifers

A free aquifer, also called an unconfined aquifer, refers to an aquifer which is exposed to the atmosphere and underlain by an impermeable layer. The analyses of free aquifers are also divided into full penetration wells and partial penetration wells as explicated in the following:

1. Full penetration well

The solution for nonequilibrium equations of the pumping-induced drawdown in free aquifers is quite complicated. The reason is mainly that the coefficient of transmissivity is not a constant and varies with the pumping time and distance. What's more, the vertical directions of the flow near the well are so crucial that they cannot be ignored in the derivation. There can still be found some approximate solutions for nonequilibrium equations of the pumping-induced drawdown. For interested readers, please refer to books dealing with groundwater (Marino and Luthin, 1982 for example). This section will not go into the subject.

On the other hand, if the drawdown is much smaller than the thickness of aquifers, the thickness of aquifers can thus be assumed to be a constant during dewatering. As the coefficient of transmissivity is the product of the coefficient of permeability and the thickness of the aquifer, the coefficient of transmissivity is also a constant. The Theis' nonequilibrium equation or Jacob's modified nonequilibrium equation can also be applied to pumping in free aquifers accordingly. The required hydraulic parameters are also the coefficients of transmissivity (T) and storage (S).

As in Section 9.4.1, the coefficient of storage is given as the drained volume of the pore water due to lowering a unit pressure head per unit surface area of an aquifer. The drainage per unit surface area of a free aquifer is caused by the decline of the groundwater level (Figure 9.17b). Therefore, in a free aquifer, the coefficient of storage is the same as the ratio of free water in the soil to the soil volume (Powers, 1992). Its value is either smaller than or the same as the porosity of soil. For example, if the porosity of soil is 30% and two-thirds of pore water is drained as a result of the lowering of the groundwater level by pumping, the coefficient of storage will then be 20% or 0.2. For most free aquifers, the coefficient of storage would be around 0.2–0.3.

As shown in Figure 9.20, under the Dupuit–Thiem assumption (Dupuit, 1863; Thiem, 1906), the equilibrium equation of drawdown of a full penetration well in a free aquifer is as follows:

$$Q = \frac{\pi k \left(h_2^2 - h_1^2\right)}{\ln\left(r_2 / r_1\right)} = \frac{\pi k \left(h_2^2 - h_1^2\right)}{2.3 \log\left(r_2 / r_1\right)} \tag{9.15}$$

where h_1 and h_2 are the heights of the water levels at the distances of r_1 and r_2, respectively.

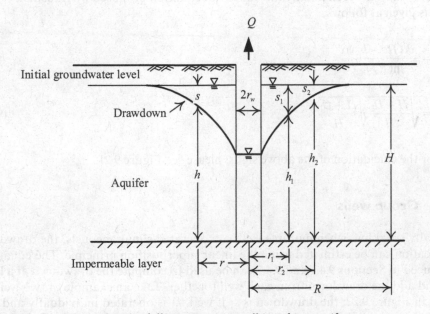

Figure 9.20 Drawdown for the full penetration well in a free aquifer.

Eq. 9.15 is the generally adopted equation in books on soil mechanics. Substituting the influence range (R) and the radius of the well (r_w) for r_2 and r_1, respectively, in the above equation, we have the following:

$$Q = \frac{\pi k (H^2 - h_w^2)}{\ln(R / r_w)} \tag{9.16}$$

Eq. 9.15 can also be rewritten to obtain the groundwater level or drawdown at any distance (r) as follows:

$$H^2 - h^2 = \frac{Q \ln(R / r)}{\pi k} \tag{9.17}$$

or

$$h^2 - h_w^2 = \frac{Q \ln(r / r_w)}{\pi k} \tag{9.18}$$

where h is the groundwater level at the distance of r.

2. Partial penetration well

The same as the full penetration well, the nonequilibrium equation of the drawdown of a partial penetration well is rather complicated and is not to be discussed in this book. For interested readers, please refer to related literatures (for example, Marino and Luthin, 1982). This section only introduces an equation proposed by Hausman (1990), which is given as follows:

$$Q = \frac{\pi k (H^2 - h_w^2) \alpha}{\ln(R / r_w)} \tag{9.19}$$

$$\alpha = \sqrt{\frac{H - D_1}{H}} \sqrt[4]{\frac{H + D_1}{H}} \tag{9.20}$$

For the elucidation of the above signs, please see Figure 9.21.

9.4.3 Group wells

Generally speaking, when multiple wells are operated simultaneously, the drawdown at a location can be estimated using the linear superposition principle. The equations introduced in Sections 9.4.1 and 9.4.2 can be used to compute the drawdown at a location and add the drawdown from each well together. Take an example of two wells as shown in Figure 9.22; the drawdown is s_1 if well #1 is operated individually and s_2 if well #2 is operated individually. Then, the total drawdown (s_t) should be the sum of s_1 and s_2 when wells #1 and #2 are operated simultaneously.

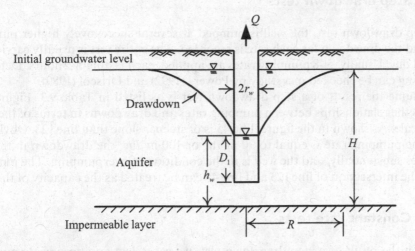

Figure 9.21 Drawdown for the partial penetration well in a free aquifer.

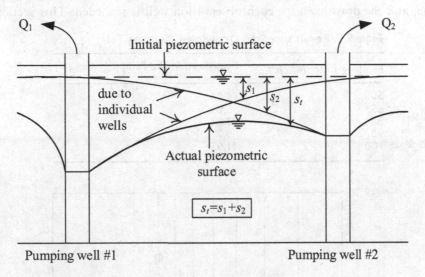

Figure 9.22 Calculation of the drawdown for group wells.

9.5 PUMPING TESTS

Two types of pumping tests, i.e., the step drawdown and constant rate tests, are usually carried out before dewatering. The objective of a step drawdown test is to determine the capacity of a well, whereas the goal of a constant rate test is to obtain the hydraulic parameters, such as the coefficients of permeability, transmissivity, and storage. Although there are full penetration well and partial penetration well theories, as introduced in Section 9.4, the test well for a pumping test should fully penetrate into the aquifer in order for its results to be analyzed using the well theories of the full penetration well, considering that the equations for a partial penetration well are too complicated and some of them are empirical formulas.

9.5.1   Step drawdown tests

In a step drawdown test, the well is pumped at several successively higher pumping rates and the drawdown for each rate is recorded. The entire test is usually carried out within 1 day. Usually, 5–8 pumping rates are applied, each lasting 1–2 hours. Details of the testing can be checked in a study by Powers (1992) and Driscol (1989).

Assume the result of a step drawdown test is as listed in Table 9.3. Figure 9.23 displays the relationships between pumping rates and drawdowns in terms of the logarithm scale. As shown in the figure, line $\overline{45}$ is of steeper slope than line $\overline{123}$. Obviously, when the pumping rate is equal to 90 m^3/hr or 100 m^3/hr, the drawdown in the well increases substantially, and the well is in the condition of overpumping. The pumping rate at the intersection of line $\overline{123}$ and line $\overline{45}$ can be treated as the capacity of the well.

9.5.2   Constant rate tests

Based on the result of a step drawdown test, we can select an appropriate pumping rate, slightly smaller than the capacity of the well, to pump the water for a long period of time, and the drawdown for each observation well is recorded. This section will

Table 9.3 Result of a Step Drawdown Pumping Test

Pumping Rate (m³ / hr)	Drawdown in the Pumping Well (m)
50	4.06
65	5.81
80	7.81
90	9.7
100	11.95

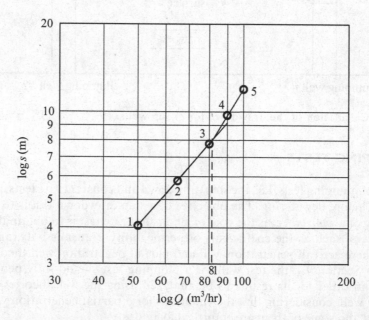

Figure 9.23 The relation between drawdowns and pumping rates for a step drawdown test.

introduce the analytical methods for constant rate pumping tests in confined and free aquifers, respectively.

1. Confined aquifers

Given the results of pumping tests of a confined aquifer, the coefficients of transmissivity and storage of the aquifer can be obtained using the nonequilibrium and the equilibrium equations. As the computing of Theis' nonequilibrium equation is too complicated, it is rarely adopted, and therefore, the Jacob's nonequilibrium equation is most widely used.

Refer Eqs. 9.1–9.3; when t is greater or r is small enough so that $u \leq 0.05$, the Jacob's nonequilibrium equation can be used. We can depict the relations between s and $\log t/r^2$, whose values are obtained from a certain observation well (i.e., r is a constant; t is a variable). If there is more than one well, the $s \sim t/r^2$ relation at a specific time (i.e., t is a constant; r is a variable) can also be adopted. Based on the $s \sim \log t/r^2$ curve and Eq. 9.5, we have the following:

$$T = 0.183 \frac{Q}{(\Delta s)_J} \tag{9.21}$$

where $(\Delta s)_J =$ the slope of the $s \sim \log t/r^2$ curve, the value of which equals the decreased amount of s when t/r^2 increases by a factor of ten.

The coefficient of transmissivity can thus be derived using the above equation. If we extend the line segment of $s \sim \log t/r^2$, the corresponding value of t/r^2 is $(t/r^2)_{s=0}$ where the extended line intersects with $s = 0$ at $(t/r^2)_{s=0}$ (see Figure 9.18). In accordance with Eq. 9.5, we can compute the coefficient of storage using the following equation:

$$S = 2.25T \left(\frac{t}{r^2} \right)_{s=0} \tag{9.22}$$

where $(t / r^2)_{s=0}$ is the corresponding value of t/r^2 when $s = 0$.

The detailed procedure of deriving hydraulic parameters using the Jacob's nonequilibrium equation is as illustrated in Example 9.1.

If the drawdown curve of a pumping test reaches equilibrium or changes little, the Thiem's equilibrium equation can then be adopted for the calculation of hydraulic parameters. In accordance with Eq. 9.9, the coefficient of permeability can be obtained as follows:

$$k = \frac{Q \ln(r_2 / r_1)}{2\pi D(s_1 - s_2)} = \frac{0.366Q}{D(s_1 - s_2)} \log \frac{r_2}{r_1} \tag{9.23}$$

and the coefficient of transmissivity is given as follows:

$$T = kD = \frac{Q \ln(r_2 / r_1)}{2\pi(s_1 - s_2)} \tag{9.24}$$

If there are more than 2 observation wells, first establish the relations between the drawdown (s) and the distance on the logarithm scale ($\log r$) for each well, which will be similar to Figure 9.18b. Then, the coefficient of transmissivity can be derived from Eq. 9.23 as follows:

$$T = \frac{0.366Q}{(\Delta s)_T} \tag{9.25}$$

where $(\Delta s)_T$ is the slope of the $s \sim \log r$ curve and its value equals the decreased amount of s when r increases ten times.

Similarly, if we extend the line segment of the $s \sim \log r$ curve, the extended line will intersect with $s = 0$ at R (as shown in Figure 9.18b). The same as earlier discussions, R represents the influence range of the drawdown. Based on the influence range in Eq. 9.6, the coefficient of storage can be derived as follows:

$$S = \frac{2.25Tt}{R^2} \tag{9.26}$$

Similarly, the detailed computing procedure of Thiem's is as illustrated in Example 9.1.

The coefficient of transmissivity can also be derived from the recovery method. Similarly, depict the $s' \sim \log(t/t')$ relation based on the data in observation wells. Using Eq. 9.12, we can derive the following:

$$T = \frac{0.183Q}{(\Delta s')_r} \tag{9.27}$$

where $(\Delta s')_r$ = the slope of the $s' \sim \log(t/t')$ curve and its value is the decreased amount of s' when t / t' increases ten times.

As for the detailed computing procedure of the recovery method, please see Example 9.1.

2.   Free aquifers

If the pumping-induced drawdown is much smaller than the thickness of the free aquifer, the Jacob's modified nonequilibrium equation for confined aquifers is applicable, and the required hydraulic parameters are also the coefficients of transmissivity and storage. If the drawdown is not much smaller than the thickness of the aquifer, the Jacob's modified nonequilibrium equation for confined aquifers is not applicable to the free aquifer. More complicated equations or numerical simulations of groundwater are to be used. Please refer to related books on groundwater for the subject.

If there are two observation wells, let the distances between the pumping well and the two observation wells be r_1 and r_2, respectively, and the water levels in the two observation wells are separately h_1 and h_2. Based on Eq. 9.15, the coefficient of permeability will be as follows:

$$k = \frac{2.3Q}{\pi(h_2^2 - h_1^2)} \log\frac{r_2}{r_1} \tag{9.28}$$

9.6   DEWATERING PLAN FOR AN EXCAVATION

A dewatering plan includes the selection of dewatering methods, the estimation of the hydraulic parameters or the pumping tests, the computation of the pumping rate for a single well, determination of the number of wells, the computation of drawdown and the influence range, etc. This section will introduce these procedures step by step.

9.6.1   Selection of dewatering methods

An appropriate dewatering method can be selected based on soil types and dewatering depths as shown in Figure 9.4. With the advance in technology, pumps have improved in their performance, and the dewatering depths have also been raised significantly. In practice, one or more methods can be adopted in a single phase or many phases (refer to Figure 9.13).

9.6.2   Determination of hydraulic parameters

Initial estimation of the coefficients of permeability can be referred to in Tables 2.5 and 2.6. Besides, as discussed in Sections 9.4.1 and 9.4.2, the coefficient of storage of a confined aquifer, which ranges from 0.0005 to 0.001, is about the same as the compressibility of the soil. The coefficient of storage of a free aquifer is slightly smaller than the porosity of a soil and ranges from 0.2 to 0.3.

   More accurate estimation of the coefficients of permeability can be obtained with laboratory tests, empirical formulas, or pumping tests. However, because of sample disturbance during sampling and complicated soil stratum conditions, the results of laboratory tests are not reliable for a design. The most reliable method of obtaining the coefficient of permeability is the in situ pumping test, which is introduced in Section 9.5.

9.6.3   Determination of the capacity of wells

The efficiency of a well cannot reach 100% because there exists friction in the well and between soil particles and the water flow. As a result, the well capacity (Q_w) has to be estimated. As elucidated in Section 9.5.1, the most reliable method of estimating Q_w is to conduct a step drawdown pumping test.

   In addition, the Q_w may be estimated using empirical formulas. Take an example of pumping in a free aquifer as shown in Figure 9.24; the quantity of groundwater flows into a deep well (Q) can be expressed as follows:

$$Q = 2\pi r_w h_w k i_e \tag{9.29}$$

where
   i_e = the entry hydraulic gradient of groundwater flowing into a well,
   h_w = the groundwater level at which groundwater flows into the well.

   According to Sichart and Kyrieleis (1930), the entry hydraulic gradient cannot be larger than the following:

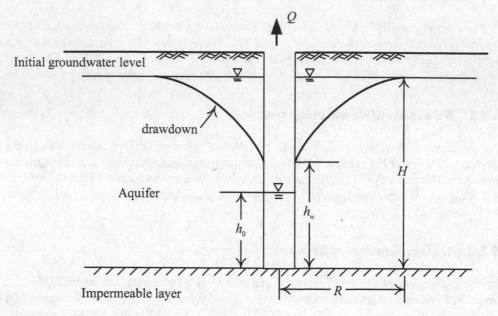

Figure 9.24 Estimation of the well capacity.

$$i_{e\,max} = \frac{1}{15\sqrt{k}} \tag{9.30}$$

where k is the coefficient of permeability in m/sec.

If one substitutes $i_{e\,max}$ for i_e in Eq. 9.29, the capacity of a deep well (Q_w) will be as follows:

$$Q_w = 2\pi r_w h_w \frac{\sqrt{k}}{15} \tag{9.31}$$

The above equation is an empirical formula where the r_w and h_w are in meter, the k in m/sec, and the Q_w in m^3 / sec.

We can also use a similar procedure to derive the well capacity in a confined aquifer as follows:

$$Q_w = Av = (2\pi r_w \ell_w)(ki_e) = 2\pi r_w \ell_w i_e \frac{\sqrt{k}}{15} \tag{9.32}$$

where ℓ_w is the length of the well screen in the confined aquifer.

Because the value of h_w is difficult to estimate, assume it to be about the groundwater level in the deep well at the preliminary estimation, that is, $h_w = h_0$. Then, examine it using pumping tests.

The capacity of the well point method can also be evaluated based on the above procedure. In addition, Tables 9.4 and 9.5 list some suggested values of the capacities of the well point method (JSA, 1988).

Table 9.4 Relationship Between
Coefficient of Permeability and
Q_w for Well Points (JSA, 1988)

k (cm/sec)	Q_w
1.0×10^{-3}	$1 \sim 5$
5.0×10^{-3}	$5 \sim 10$
1.0×10^{-2}	$10 \sim 20$
5.0×10^{-2}	$40 \sim$

Table 9.5 Relationship Between Soil Type and
Q_w for Well Points (JSA, 1988)

Soil type	Q_w
Gravel	$50 \sim 70$
Coarse gravel	$30 \sim 50$
Coarse sand	$20 \sim 25$
Sand	Around 15
Fine sand	$8 \sim 10$

9.6.4 Estimation of the number of wells

Enough pumping wells have to be placed at an excavation site to lower the groundwater level to a depth below the excavation bottom. Generally speaking, the groundwater level has to be at least 0.5 to 1.0 m below the excavation bottom. The design of deep wells is elucidated as follows:

1. Compute the total quantity of water to be pumped in the excavation area

   Assume the excavation site is an imaginary well, with a radius of R_w, (see Figure 9.25) and compute the required total quantity (Q_{tot}) of water to be pumped in the imaginary well using either equilibrium or nonequilibrium equations. The computation of the pumped water is as discussed in Sections 9.4.1 and 9.4.2.

   The radius of the imaginary well (R_w) can be computed by way of the equivalent area or perimeter:

   $$R_w = \sqrt{\frac{a \times b}{\pi}} \text{ or } R_w = \frac{a + b}{\pi} \tag{9.33}$$

2. Compute the pumping capacity of each well

   Select a radius of each deep well, and assume its water level to be h_w, which should be a little lower than the water level, usually 0.5–1.0 m below the excavation bottom, at the excavation center (see Figure 9.7). Then, compute the discharge quantity of each well. The capacity of each well (Q_w) can be estimated following

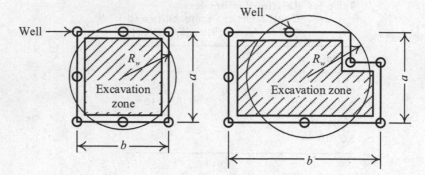

Figure 9.25 The radius of an imaginary well.

the method introduced in Section 9.6.3. The pumping rate (Q_p) should be less than the discharge quantity and capacity of each well.

3. Compute the number of deep wells
    The number of the deep wells (n) is given as follows:

$$n = \frac{Q_{tot}}{Q_w} \tag{9.34}$$

4. Examine the drawdown at the excavation center
    Compute the drawdown at the excavation center and corners, and check if the computed values are smaller than the designed values. As shown in Figure 9.26, there are four full penetration wells in an excavation site. To obtain the drawdown at the excavation center (point A) with the four wells pumping simultaneously, use the well formulas to compute the drawdown of each well at point A and add up those values using the superposition principle to find the total drawdown at point A.

The design procedures of the well point are explicated as follows:

1. Compute the total quantity of water to be pumped in the excavation area
    The method of computing the total quantity of water to be pumped is identical to the deep well method introduced above.

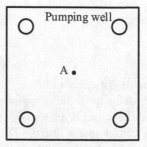

Figure 9.26 An excavation site with pumping wells.

2. Compute the number of well points

    With the general spacing of 0.7 to 2.0 m between well points, assume the spacing as a, and the number of well points can be obtained using the following:

$$n = \frac{L}{a} \qquad\qquad (9.35)$$

where L is the length of the collecting pipes, which are usually arranged around the excavation site.

3. Examine the pumping capacity

    Check whether the pumping capacity of each well point is within the suggested range (Tables 9.4 and 9.5).

4. Examine the drawdown at the excavation center

    The method of examination is identical with that with the deep well method.

9.6.5   Computation of the influence range of the drawdown

In Section 9.4, the influence range of the drawdown is defined as the distances from which the drawdown is none. The influence range can be either obtained by extending the line segment of the $s \sim \log r$ curve to intersect the line $s = 0$ or computed using Eqs. 9.6 to 9.8. The influence range thus obtained, however, is too large to be meaningful in engineering practice. Therefore, the reasonable influence range should be computed on the basis of $s = s_a$ (s_a is the allowable drawdown). For example, if the objective of computing the influence range of the drawdown is to estimate the influence on the supply of water or the amount of settlement, it is feasible to assume a value of s_a ($s_a = 0.5$ m, for example) and compute the influence range. If the objective of computing the influence range is to back analyze hydraulic parameters, it may be still necessary to assume $s = 0$ or use Eqs. 9.6–9.8 to compute the influence range. Note that the drawdown does not necessarily cause settlement and the influence range of the drawdown does not equal the influence range of settlement, either. As for the discussion of the influence range of settlement, please refer to Section 9.7.

Figure 9.27a shows a diaphragm wall penetrating into an impermeable layer. As pumping is confined to the excavation area, it will not have influence behind the wall and the influence range of the drawdown is 0 accordingly. Figure 9.27b shows the pumping carried out behind the wall, and the influence range of the drawdown extends by a distance behind the wall as a result. Figure 9.27c shows the pumping executed in the permeable layer under an impermeable layer. While water is pumped in a confined aquifer, the piezometric level in the confined aquifer will decline. Figure 9.27d is a diaphragm wall located in a permeable layer, and the well depth is less than the depth of the wall. As a result, pumping will not cause the drawdown of the groundwater level behind the wall. Figure 9.27e is a case in which the well depth is larger than the depth of the wall, and the influence range of the drawdown may extend behind the wall by a distance.

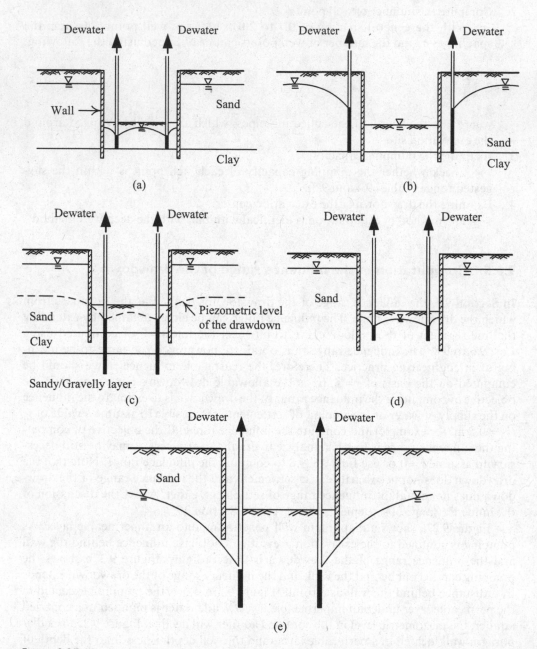

Figure 9.27 Various scenarios for the influence of dewatering in excavations.

9.7   DEWATERING AND GROUND SETTLEMENT

Dewatering will decrease the pore water pressure in soil and increase the effective stress of the soil accordingly. In sandy or gravelly soils, the increase of the effective stress will produce elastic settlement. In clayey soils, not only elastic settlement but also consolidation settlement will be induced. Generally speaking, the amount of elastic settlement is far less than that of consolidation settlement and usually can be neglected. As far as dewatering and pumping are concerned, the consolidation settlement should be considered. Regarding the calculation of consolidation settlement, the readers can refer to the book of soil mechanics, and this section will not introduce it again.

Whether dewatering or pumping will cause settlement relates to the locations of wells, the depths, and the drawdowns. Figure 9.27a shows a case where there is no drawdown behind the wall. Thus, the effective stress in clay behind the wall stays unchanged, and no elastic settlement or consolidation settlement in clay occurs behind the wall as a result. Figure 9.27b shows a case where pumping causes the drawdown in sand behind the wall. The effective stress of clay increases under such a condition, and consolidation settlement will occur behind the wall. Figure 9.27c shows the pumping executed in the permeable layer underlying a clay layer below the diaphragm wall. The pumping certainly will cause the decline of the piezometric level in the confined aquifer and the increase of effective stress in the clayey layer below. Consolidation settlement occurs behind the wall as a result. Figure 9.27d is a case where no drawdown behind the wall is generated, and no consolidation settlement occurs behind the wall, either. Figure 9.27e shows the condition where pumping is carried out at a certain depth below the wall bottom and induces the drawdown of groundwater behind the wall. With sandy soils, however, only elastic settlement is to occur whereas consolidation settlement will not.

We can see from the above elucidation that pumping a soil deposit containing clay layers will generate consolidation settlement behind the wall; if it is a purely sandy ground, pumping behind the wall will basically not cause a severe settlement (consolidation settlement). It does not follow, however, that pumping behind the wall to lower the groundwater level is always feasible. The reason is that in situ geological conditions are usually complicated so that clayey soils may be mingled in. Considering the above possibility, pumping behind the wall is recommended to be carried out, along with the water recharge, to avoid the drawdown of the groundwater level as shown in Figure 9.28.

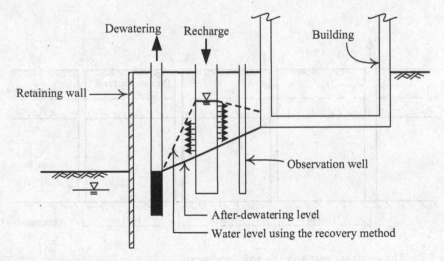

Figure 9.28 The recovery method.

Theoretically, the influence range of dewatering-induced settlement is the distance from which settlement declines to 0. However, the definition has little meaning in engineering practice. As explicated above, the drawdown is the major cause of settlement. If we apply the Theis' or Jacob's nonequilibrium theory to compute the influence range of the drawdown, the range will work out as extremely far. Does it follow that the influence range of settlement is actually extremely far? Certainly not. Thus, the reasonable settlement influence range should be determined on condition that $\delta = \delta_a$ (δ_a is the allowable settlement) or when the angular distortion is small. As for the determination of the allowable settlement and the angular distortion, please refer to Section 11.2.

Example 9.1

Figure 9.29 shows the allocation of pumping wells and observation wells. The thickness of the permeable layer is 10 m. The distances between the pumping well center and the observation wells K-1, K-2, and K-3 are 10, 20, and 30 m, respectively. The radius of the pumping well $(r_w) = 0.5$ m; the pumping rate $(Q) = 2,000$ cm^3 / sec; Table 9.6 lists the drawdowns observed in the pumping well and the three observation wells. Table 9.7 lists the recovered water levels after pumping is stopped. Compute the coefficients of transmissivity and storage using Jacob's, Thiem's, and the recovery methods, respectively.

Solution

1. The Jacob's modified nonequilibrium equation
   Choose one of the four wells (including the pumping well) and take its observation data for analysis. We can use either r (the location of the well) as a variable at a specific time t, or t as a variable at a specific location. Let K-2 be the object of analysis and t be the variable. Compute the $s \sim t / r^2$ relation as shown in Table 9.8. Plot the

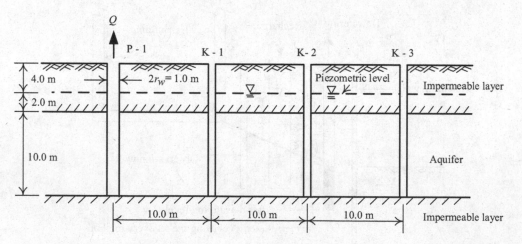

Figure 9.29 The pumping test and the soil condition.

Table 9.6 Results of the Pumping Test

| Time t (sec) | Drawdown s(cm) | | | |
	Pumping Well P-1	Observation Well K-1	Observation Well K-2	Observation Well K-3
0	0	0	0	0
30	10.0	0.3	0	0
60	12.5	0.6	0.1	0
120	15.5	1.1	0.3	0.1
240	18.1	2.1	0.7	0.3
480	20.2	3.5	1.2	0.6
900	22.4	5.0	2.0	1.0
1,800	24.8	6.6	3.4	1.8
3,600	27.5	8.4	4.9	3.1
5,400	29.2	9.4	6.0	4.1
7,200	30.9	10.2	6.7	4.7
10,800	32.0	11.3	7.7	5.6

Table 9.7 Recovered Water Levels in the Pumping Well

Time t(sec)	Time after Pumping Is Stopped t'(sec)	t / t'	Residual Drawdown s'(cm)
10,800	0	—	32.0
10,830	30	361	15.2
10,860	60	181	12.6
10,920	120	91	9.8
11,040	240	46	7.5
11,280	480	23.5	4.9
11,700	900	13	3.3
12,600	1,800	7	2.3
14,400	3,600	4	1.6
18,000	7,200	2.5	1.1

Table 9.8 Computation Based on Time as Variable (Observation Well K-2, $r = 20$ m)

t (sec)	30	60	120	240	480	900
t/r^2	7.5×10^{-6}	1.50×10^{-5}	3.0×10^{-5}	6.0×10^{-5}	1.20×10^{-4}	2.25×10^{-4}
s (cm)	0	0.1	0.3	0.7	1.2	2.0

t(sec)	1,800	3,600	5,400	7,200	10,800
t/r^2 (sec/cm²)	4.5×10^{-4}	9.00×10^{-4}	1.35×10^{-3}	1.8×10^{-3}	2.7×10^{-3}
s (cm)	3.4	4.9	6.0	6.7	7.7

$s \sim \log t / r^2$ relation as shown in Figure 9.30a where we have $(\Delta s)_J = 5.61$ cm and $(t / r^2)_{s=0} = 0.00012 \, \text{sec/cm}^2$. Using Eq. 9.21, we have the following:

$$T = 0.183 \frac{Q}{(\Delta s)_J} = 0.183 \frac{2,000}{5.61} = 65.2 \, \text{cm}^2 / \text{sec}$$

$$k = \frac{T}{D} = \frac{65.2}{1,000} = 6.52 \times 10^{-2} \, \text{cm} / \text{sec}$$

Using Eq. 9.22, we have the following:

$$S = 2.25T(\frac{t}{r^2})_{s=0} = 2.25 \times 65.2 \times 0.00012 = 1.76 \times 10^{-2}$$

On the other hand, it is also feasible to carry out analysis by treating r as a variable. Based on the data in Table 9.9, depict the $s \sim \log t / r^2$ curve as shown in Figure 9.30b. The relations between K-3, K-2, K-1, and P-1 are nonlinear.

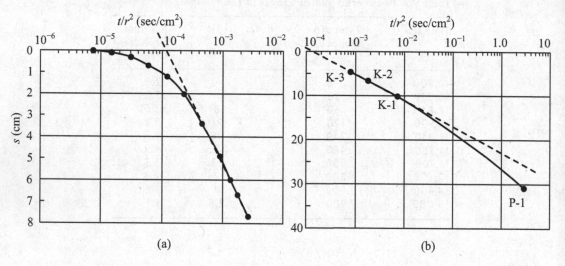

Figure 9.30 The Jacob's method (a) r is constant while t is a variable (b) t is constant while r is a variable.

Table 9.9 Computation Based on Distance as Variable ($t=7200$sec)

Variant	Pumping Well P-1	Observation Well K-1	Observation Well K-2	Observation Well K-3
r (cm)	50	1,000	2,000	3,000
t / r^2 (sec/cm²)	2.88	7.2×10^{-3}	1.8×10^{-3}	8.0×10^{-4}
s (cm)	30.9	10.2	6.7	4.7

Considering the lower accuracy of the data of P-1, only adopt the data of K-3, K-2, and K-1 for regression analyses as shown in the figure where we have $(\Delta s)_J = 6.1\,\text{cm}$ and $(t/r^2)_{s=0} = 0.00012\,\text{sec}/\text{cm}^2$.

Using Eq. 9.21, we then have the following:

$$T = 0.183\frac{Q}{(\Delta s)_J} = 0.183\frac{2,000}{6.1} = 60\ \text{cm}^2/\text{sec}$$

$$k = \frac{T}{D} = \frac{60}{1,000} = 6.0\times10^{-2}\ \text{cm}^2/\text{sec}$$

Using Eq. 9.22, we have the following:

$$S = 2.25T\left(\frac{t}{r^2}\right)_{s=0} = 2.25\times61\times0.00012 = 1.65\times10^{-2}$$

2. The Thiem's equilibrium equation

Assume wells K-1, K-2, and K-3 reach equilibrium when $t = 10,800$ sec. From Table 9.6, depict the $s\sim\log r$ curve as shown in Figure 9.31 where we have $(\Delta s)_T$, the decreased amount of s when r increases ten times, equal to 12.1 cm. That is to say, when $\log r_2/r_1 = 1.0$, $s_1 - s_2 = 12.1$cm. Using Eq. 9.23, we have the following:

$$k = \frac{0.366Q}{D(s_1-s_2)}\log\frac{r_2}{r_1} = \frac{0.366\times2,000}{1,000\times12.1} = 6.05\times10^{-2}\ \text{cm/sec}$$

The coefficient of transmissivity can be derived from $T = kD$, or find R on the figure $R = 87\,\text{m}$ and use Eq. 9.25 to obtain the following:

$$T = \frac{0.366Q}{(\Delta s)_T} = \frac{0.366\times2,000}{12.1} = 60.5\ \text{cm}^2/\text{sec}$$

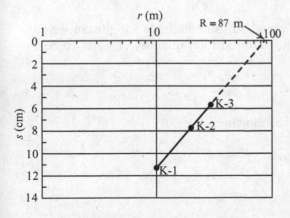

Figure 9.31 The Thiem's method.

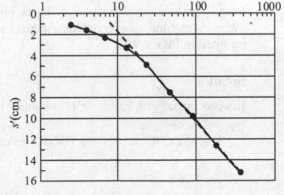

Figure 9.32 The recovering method.

Table 9.10 Hydraulic Parameters from Various Methods

Method	$k(cm/sec)$	$T(cm^2/sec)$	S
Jacob's method (t as variable)	6.52×10^{-2}	65.2	1.76×10^{-2}
Jacob's method (r as variable)	6.00×10^{-2}	60.0	1.65×10^{-2}
Thiem's method	6.05×10^{-2}	60.5	1.94×10^{-2}
Recovering method	4.31×10^{-2}	43.1	—

Using Eq. 9.26, we obtain the following:

$$S = \frac{2.25Tt}{R^2} = \frac{2.25 \times 60.5 \times 10800}{(8700)^2} = 1.94 \times 10^{-2}$$

3. The recovery method
    Based on the data in Table 9.6, plot the $s' \sim \log(t/t^2)$ curve as shown in Figure 9.32, where we can find $(\Delta s')_r = 8.5$ cm. Using Eq. 9.27, we then obtain the following:

$$T = \frac{0.183Q}{(\Delta s')_r} = \frac{0.183 \times 2,000}{8.5} = 43.1 \, cm^2/sec$$

and the coefficient of permeability:

$$k = \frac{T}{D} = \frac{43.1}{1,000} = 4.31 \times 10^{-2} \, cm/sec$$

Table 9.10 lists the results of Jacob's, Thiem's, and the recovery methods. As shown in the table, the Thiem's method cannot obtain the coefficient of storage.

Example 9.2

Based on the analysis results of the Jacob's method listed in Table 9.10 (distance as a constant; time as a variable), estimate the required total quantity of water to be pumped in Figure 9.33.

Solution

In accordance with Example 9.1, we know the following:

$$k = 6.52 \times 10^{-2} \, cm/sec = 3.91 \times 10^{-2} \, m/min$$

$$S = 1.76 \times 10^{-2}$$

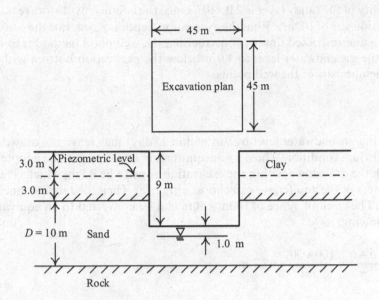

Figure 9.33 The plan and profile of an excavation.

Using the method of equivalent area, compute the radius of the imaginary well as follows:

$$R_w = \sqrt{\frac{a \times b}{\pi}} = \sqrt{\frac{45 \times 45}{\pi}} = 25.4 \text{ m}$$

The excavation depth is 9 m. The groundwater level is to be lowered to 1.0 m below the excavation bottom. As a result, the drawdown $(s) = 7$ m. Assume the pumping time $(t) = 15$ days $= 21,600$ minutes. Then, the well parameter is given as follows:

$$u = \frac{R_w^2 S}{4kDt} = \frac{(25.4)^2 \times 1.76 \times 10^{-2}}{4 \times 3.91 \times 10^{-2} \times 10 \times 21,600} = 3.36 \times 10^{-4}$$

With $u < 0.01$, we can use the Jacob's nonequilibrium equation. From Table 9.2, we have the following:

$$W(u) = -0.5772 - \ln u = 7.42$$

and the required total quantity of water to be pumped is calculated as below:

$$Q_{tot} = \frac{4\pi kDs}{W(u)} = \frac{4 \times \pi \times 3.91 \times 10^{-2} \times 10 \times 7.0}{7.42} = 4.64 \text{ m/min}$$

Example 9.3

The thickness of a sandy soil layer is about 40 m, below which is an impermeable rocky stratum. The groundwater level is at the ground surface. The coefficient of permeability of the sandy layer is 2.0×10^{-3} cm/s; the coefficient of storage is 0.2. Here is an excavation site of 100m × 50m; the excavation depth is 8.5m. For the convenience of excavation and restricted time of construction, the well point method is to be adopted to lower the groundwater level to 1.0m below the excavation bottom within 15 days. Design the numbers of the well points.

Solution

To lower the groundwater level by 9m within 15 days may leave the drawdown in the nonequilibrium condition. Thus, a nonequilibrium equation is to be adopted for analyses. With the drawdown within the excavation zone to be 9.5m, which is far less than the thickness of the aquifer, we can choose either the Theis' or Jacob's nonequilibrium equation. The circumference of 100m × 50m can be converted to an equivalent radius of the following:

$$R_w = \frac{a+b}{\pi} = \frac{100+50}{\pi} \approx 47.7 \text{ m}$$

The coefficient of permeability $k = 2.0 \times 10^{-3}$ cm/s $= 1.2 \times 10^{-3}$ m/min
The coefficient of transmissivity $T = kD = 1.2 \times 10^{-3} \times 40 = 0.048$ m^2/min
The pumping time $t = 15$ days $= 21,600$ min

The well parameter $u = \dfrac{R_w^2 S}{4Tt} = \dfrac{47.7^2 \times 0.2}{4 \times 0.048 \times 21,600} = 0.11$

Refer to values in Table 9.2; we have the value $W(u) \approx 1.737$.

With the pumping height limited to about 5m for well points and the required amount of dewatering about 9.0m, dewatering has to be carried out in two phases:

Assume the discharge quantity of each well point $Q_w = 0.01$ m^3/min (see Tables 9.4 and 9.5). The dewatering height of the first phase $s = 5.0$m (the groundwater level lowered from GL-0.0 to GL-5.0m) (GL: ground surface level). A possible drawdown curve is similar to Figure 9.13. The required quantity of water to be pumped will be as follows:

$$Q_1 = \frac{4\pi Ts}{W(u)} = \frac{4 \times \pi \times 0.048 \times 5}{1.737} = 1.74 \text{ m}^3/\text{min}$$

and the required number of well points will be as follows:

$$n_1 = \frac{Q_1}{Q_w} = \frac{1.74}{0.01} = 174$$

Assume the collecting pipes are arranged along the two longer sides of the excavation site, which is 200m long in total. The distance between two well points will be calculated as follows:

$$a = \frac{200}{174} = 1.15 \text{ m} \rightarrow \text{adopt } 1.0 \text{ m}$$

The dewatering height of the second phase $s = 4.5$ m (the groundwater level lowered from GL-5.0 to GL-9.0 m). A possible drawdown curve is also similar to Figure 9.13. The quantity of water to be pumped will be as follows:

$$Q_2 = \frac{4\pi Ts}{W(u)} = \frac{4 \times \pi \times 0.048 \times 4.5}{1.737} = 1.56 \text{ m}^3/\text{min}$$

$$n_2 = \frac{Q_2}{Q_w} = \frac{1.56}{0.01} = 156$$

Assume the collecting pipes are also arranged along the two longer sides of the excavation site. The distance between two well points will be as follows:

$$a = \frac{200}{156} = 1.28 \text{ m} \rightarrow \text{adopt } 1.2 \text{ m}$$

PROBLEMS

9.1   What are the objectives of dewatering in excavations?
9.2   Explicate the different dewatering methods and their application ranges, respectively.
9.3   Explain the physical meanings of the coefficients of permeability, transmissivity, and storage for confined aquifers and free aquifers, respectively.
9.4   Explicate the dewatering plan for excavations in sandy soils.
9.5   Suppose there exists a confined aquifer. Its thickness is 40 m; $k = 0.005$ cm/s; $S = 0.004$. Above the aquifer is a 9-m thick clayey layer. The piezometric level is 3 m below the ground surface. Assume a 20-cm-diameter well that fully penetrates into the confined aquifer. When the pumping rate $Q = 0.6$ m^3/min, what will be the drawdown in the observation well 10 m from the pumping well after an hour dewatering using Theis' and Jacob's methods, respectively?
9.6   Same as the previous problem. Assume the change rate in the water level in the observation well is small 10 hours later. Use the Thiem's method to compute the drawdown in the observation well.
9.7   Redo Problem 9.5 and assume $k = 0.001$ cm/s and $S = 0.001$.
9.8   Same as the previous problem. Assume the change rate in the water level in the observation well is small after a 15-hour long dewatering. Use the Thiem's method to compute the drawdown in the observation well.
9.9   Suppose soil conditions are the same as in Problem 9.5. Assume there exists an excavation, 30 m × 30 m in dimension and 10 m in excavation depth. To keep the stability of excavation, a dewatering plan is made to lower the piezometric level of the confined aquifer to 0.5 m below the excavation bottom within 5 days. Estimate the quantity of water required to be pumped and the number of wells.

9.10   Suppose soil conditions are the same as in Problem 9.7. Assume a $100\,m \times 100\,m$ excavation. The excavation depth is $12\,m$. To keep the stability of the excavation, a dewatering plan is made to lower the piezometric level of the confined aquifer to $0.5\,m$ below the excavation bottom in $10\,days$ before starting the excavation. Estimate the quantity of water required to be pumped and the number of wells.

9.11   Suppose a rock layer $40\,m$ below the ground surface underlying a highly permeable sandy layer. The groundwater level is $1.0\,m$ below the ground surface. The coefficient of permeability of sand $k = 0.05$ cm/s, and the coefficient of storage $S = 0.3$. Assume a 20-cm-diameter full penetration well going through the sand layer and reaching the rocky layer. When the pumping rate $Q = 0.7$ m^3/min, compute the drawdown in the observation well $4\,m$ away from the deep well after $2\,hours$ of dewatering.

9.12   Same as the previous problem. The groundwater level in the pumping well is $2\,m$ below the ground surface (that is, the drawdown is $1.0\,m$). Assume a small change rate of the water level in the observation well is recorded after a certain period of pumping. Compute the drawdown in the observation well using the Dupuit–Thiem method.

9.13   Redo Problem 9.11, but assume the coefficient of permeability of sand $k = 0.01$ cm/sec and the coefficient of storage $S = 0.01$.

9.14   Same as the previous problem. The groundwater level in the pumping well is $4\,m$ below the ground surface (that is, the drawdown is $3\,m$). Assuming after pumping for a certain period the water level in the observation well changes little, compute the drawdown in the observation well using the Dupuit–Thiem method.

9.15   Suppose soil conditions are the same as Problem 9.11. Assume a $40\,m \times 40\,m$ excavation base. The excavation depth is $8\,m$. To keep the stability of the excavation, a dewatering plan is made to lower the groundwater level in the sandy soil to $0.5\,m$ below the excavation bottom in $5\,days$. Assuming the radius of pumping wells is equal to $0.1\,m$, estimate the total quantity of water to be pumped and the number of wells.

9.16   Same as above. Assume the dewatering adopts the well point method. Compute the required number of wells.

9.17   Same as Problem 9.13. Assume there exists an $80\,m \times 40\,m$ excavation and the excavation depth is $6\,m$. To keep the stability of the excavation, a dewatering plan is made to lower the groundwater level to $1.0\,m$ below the excavation bottom in $10\,days$. Assuming the radius of pumping wells is equal to $0.1\,m$, estimate the total quantity of water to be pumped and the number of deep wells.

9.18   Same as the previous problem. Assume the well point method is adopted. Design the number of well points.

9.19   A pumping test is carried out in a confined aquifer. Given the pumping rate $Q = 788$ m^3/day and the observation results in an observation well $20\,m$ away from the center of the pumping well as listed in Table P9.19, use the Jacob's method to compute the coefficients of transmissivity and storage.

Table P9.19 Results of a Pumping Test for an Excavation

t (min)	s (m)	t (min)	s (m)
0	0	18	0.680
0.10	0.04	27	0.742
0.25	0.08	33	0.753
0.50	0.13	41	0.779
0.70	0.18	48	0.793
1.00	0.23	59	0.819
1.40	0.28	80	0.855
1.90	0.33	95	0.873
2.33	0.36	139	0.915
2.80	0.39	181	0.935
3.36	0.42	245	0.966
4.00	0.45	300	0.990
5.35	0.50	360	1.007
6.80	0.54	480	1.050
8.30	0.57	600	1.053
8.70	0.58	728	1.072
10.0	0.60	830	1.088
13.1	0.64		

Chapter 10

Design of retaining structural components

10.1 INTRODUCTION

With the stability analyses of an excavation (Chapter 5), the penetration depth of the retaining wall can be determined. With the stress and deformation analyses, the moment and shear of the retaining wall and the strut (or anchor) load can be obtained. Once these values are known, the detailed design of the retaining wall and struts can be carried out.

This chapter will introduce the analysis and design of the structural components used in braced excavations and anchored excavations. The involved knowledge includes soil mechanics, foundation engineering, reinforced concrete design, and steel structure design. The contents of this chapter are confined to the design of the structural components used in braced excavations or anchored excavations. Although the design methods that will be introduced in this chapter cannot cover all excavation methods (Chapter 3), the basic principles of the analyses and designs for these methods are similar, and the methods here can be applied to other excavation methods with some modifications.

10.2 DESIGN METHODS AND FACTORS OF SAFETY

The working stress method and the strength design method are the design methods for reinforced concrete. The allowable stress design method (abbreviated as the ASD method) and the ultimate strength design method are for steel structures. The working stress method for reinforced concrete is similar to the allowable stress method for steel structures in design concepts, both making the stress of the structural member under the working load smaller than the allowable stress. The allowable stress is the ultimate strength of the structural member divided by a factor of safety, which is mostly derived from the experience without rigorous theoretical foundation, and the uncertainty of loads and material strengths is implicitly included in the factor of safety.

The strength design method for reinforced concrete and the ultimate strength design method for steel structures are to transform the uncertainty or variation of external loads into load factors and the uncertainty of the strengths of materials or structural members into reduction factors following the methods of probabilities. The method is also called the load resistance factor design method, abbreviated as the LRFD method. Because the uncertainty of the loads on structures does not relate

DOI: 10.1201/9780367853853-10

to the structural materials themselves, the load factors for reinforced concrete and steel structure should be identical if the LRFD method is adopted (including both the strength design method and ultimate strength design method). Their reduction factors, on the other hand, should be different as the uncertainty of the material strength relates to the properties of the materials. This book adopts the load factors and the strength reduction factors for the LRFD method based on the codes recently announced.

Reinforced concrete and steel are frequently used in an excavation as structural components. Sometimes, wood is also used. These materials are combined for use in most cases. To design reasonably, it is necessary to adopt the LRFD method, adopting the same load factors and different strength reduction factors based on the material of the structural member. However, in most countries, the LRFD method and the ASD method are used for the design of high-rise steel structures. This chapter is going to follow the general application in practice, different design methods for different structural components.

The structural components in excavations are basically temporary structures. Some structures, for example, the diaphragm walls, are temporary structures, which also serve as permanent structures after the construction is finished. Some country building codes suggest the allowable stress for the temporary structural material should be magnified. Therefore, for diaphragm walls during construction, the allowable stress would be magnified by a factor λ without the earthquake force considered. After the construction is finished, on the other hand, with the diaphragm wall serving as the outer wall of the basement, the allowable stress would not be magnified with the earthquake force taken into consideration.

Some country building codes suggest that the load for the temporary structural member is assumed to be $\lambda = 1.25$.

10.3   RETAINING WALLS

This section introduces the design of sections and dimensions for different types of retaining walls. Table 3.2 lists the comparative values of nominal stiffness for different retaining walls for the references in the preliminary design.

To design the sections and dimensions of a retaining wall, stress analysis with the simulation actual construction sequence should be conducted. Three methods of stress analysis can be used: the assumed hinge method, the finite element method, and the beam on elastic foundation method, which have been introduced in Chapters 6–8, respectively. Because the maximum bending moment occurring at each construction stage does not take place at the same depth, the design of the wall sections should adopt the envelope of the maximum bending stresses of all the construction stages. Figure 10.1 illustrates a typical moment and shear diagrams at each stage for the stress analysis of a retaining wall.

10.3.1   Soldier piles

The commonly used types of soldier piles in excavations are the H steel, I steel, and rail piles. Concerning the dimensions and properties, the books on steel structures or the AISC Specification can be referred to. Types of the rail pile are usually expressed in terms of weight per length (kgf/m). Section 3.3.1 summarizes the strength, shortcoming, and construction of soldier piles.

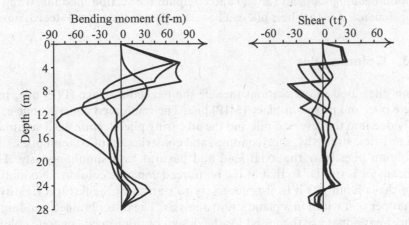

Figure 10.1 Typical bending moment and shear diagrams of a retaining wall by stress analysis.

The dimensions of soldier piles and their spacing can be determined from the results of stress analysis. Then, take the maximum bending moment (M_{max}) from the typical bending moment envelope (Figure 10.1). According to the ASD method, we can obtain the section modulus of the soldier pile as follows:

$$S = \frac{M_{max}}{\lambda \sigma_a} \tag{10.1}$$

where σ_a = the allowable stress of the steel;

λ = the short-term magnified factor of the allowable stress, which can be found from the country building codes.

The dimensions and spacing of rail piles can thus be selected based on the computed section modulus. Basically, under a certain stress, the longer the spacing, the larger the required dimension of the soldier pile and the thicker the lagging, and vice versa.

To compute the thickness of the laggings, we usually assume the lagging to be the simply supported beam on the soldier piles. The computed thickness of the lagging often comes out larger than that of the commonly used laggings in general excavations, which are around 3.0–4.0 cm thick. Considering the lateral earth pressure on the back of the wall is not necessarily uniformly acting on the laggings, sometimes it is centering on soldier piles, which are of higher rigidity, and sometimes the pressure is less than expected owing to the effect of soil arching, the 3.0 to 4.0 cm thick lagging is often adopted if the excavation is shallow.

10.3.2 Sheet piles

There are various sections of sheet piles. U-shaped and Z-shaped sheet piles are commonly used in some countries, as shown in Figure 3.17. Section 3.3.2 summarizes the strength, shortcoming, and construction of sheet piles.

The dimension of a sheet pile can be determined from the result of the stress analysis. In accordance with the envelope of bending moments (Figure 10.1), take the maximum bending moment (M_{max}) and compute the section modulus using Eq. 10.1, and the dimension of the sheet pile can be selected from the related steel design manual.

10.3.3 Column piles

Column piles used in excavations include the packed-in-place (PIP) pile, reinforced concrete pile, and mixed-in-place (MIP) pile. The reinforced concrete pile can be further divided into the reversed pile and the all casing pile. Section 3.3.3 summarizes the characteristics, strengths, shortcomings, and construction of column piles.

Column piles bear the axial load and flexural load simultaneously. Therefore, their behavior is similar to that of the reinforced concrete columns. No matter which type of the column pile it is, it is necessary to transform the flexural rigidity per pile into that per unit width in a plane strain analysis. The thus obtained bending moment and shear envelopes are then used for the design of reinforced concrete columns. For the design of the columns, one can refer to the design chart of reinforced concrete columns or the ACI code.

10.3.4 Diaphragm walls

The design of a diaphragm wall includes the wall thickness and the reinforcements, which are usually determined from the results of stress analysis, deformation analysis, and the feasibility of detailing of concrete reinforcements. Based on the experiences, the thickness of a diaphragm wall can be assumed to be 5%H_e (H_e is the excavation depth) in the preliminary design. In addition, if the diaphragm wall is treated as a one-way floor slab, in accordance with the ACI code, the minimum thickness is suggested to be 5%H_e for the braced or strutted diaphragm wall and 10% H_e for the cantilever diaphragm wall.

The design of reinforcements usually follows the strength design method (the LRFD method). The major items of the design include the vertical main reinforcement, the horizontal main reinforcement, the shear reinforcement, and the lap splice length and development length of the reinforcement. Figure 10.2 illustrates the three-dimensional view (3D), the plan, and the side view of the reinforced cage of a diaphragm wall. The design of reinforcements of a diaphragm wall is based on the moment and shear envelope obtained from the stress analysis.

The deformation of the retaining wall in the central section of the site is usually assumed to be in the plane strain condition during analysis. Therefore, the unit width ($b = 1$ m) of the diaphragm wall is usually used for the flexural stress analysis. To be consistent with common usage in the reinforced concrete design, the signs used in this section are identical to those commonly used in the design. The definitions of the designed and nominal bending moments and shears are as follows:

$$M_u = \frac{L_F M}{\lambda}, \quad M_n = \frac{M_u}{\phi} \tag{10.2}$$

$$V_u = \frac{L_F V}{\lambda}, \quad V_n = \frac{V_u}{\phi} \tag{10.3}$$

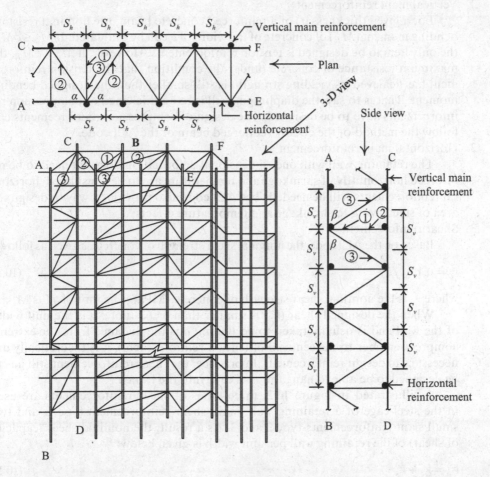

Figure 10.2 The plan view, 3D view, and side view of a steel cage of the diaphragm wall (1, 2, and 3 represent shear reinforcements).

where M_u = the bending moment for the design;

M_n = the nominal bending moment, also called the capacity of the bending moment;

V_u = the shear for the design;

V_n = the nominal shear, also called the capacity of the shear;

M = the bending moment obtained from the stress analysis;

V = the shear obtained from the stress analysis;

L_F = the load factor for soil; based on the ACI code, the L_F for the soil acting on a diaphragm wall is 1.6;

ϕ = the strength reduction factor; based on the ACI code, for bending moment ϕ = 0.9, for shear ϕ = 0.75.

As L_F = 1.6, the bending moment and shear obtained from the analysis should be multiplied by 1.6, as shown in Eqs. 10.2 and 10.3. The following excerpts the detailed reinforcement design of a diaphragm wall from the ACI code.

1. Vertical main reinforcement

   For a given thickness (t) of a reinforced concrete beam, the nominal resistant bending moment (M_R) of concrete of the width of b can be evaluated. If $M_u \leq \phi M_R$, the only item to be designed is tension reinforcements. However, if $M_u > \phi M_R$, the maximum resistance of concrete (under the condition that the tension reinforcement has achieved its yielding strength) is still smaller than the designed bending moment. That is to say, the diaphragm wall is to be thickened or compression reinforcements have to be designed. The design of compression reinforcements can follow the method of the doubly reinforced beam of the ACI code.

2. Horizontal main reinforcement

   The retaining wall with one-dimensional deformation does not need to be reinforced horizontally. If shrinkage and temperature are to be considered, horizontal reinforcements will be needed. The ACI code also provides a way to design the area of steel owing to shrinkage and temperature effects.

3. Shear reinforcement

   Based on the ACI code, the nominal shear strength of concrete is given as follows:

$$v_c = 0.17\sqrt{f_c'} \tag{10.4}$$

where v_c is the nominal shear strength of concrete (MPa). The unit of f_c' is MPa.

When the designed shear (V_u) is smaller than $\phi v_c bd$ (note: b is the unit width of the wall and is usually taken to be 100 cm; d is the distance from the extreme compression fiber to the centroid of tension reinforcement), it is theoretically unnecessary to design reinforcement. In practice, the shear reinforcement still has to be designed to be able to hang the steel cage into the trench.

As illustrated in Figure 10.2, three types of shear reinforcements are used in the steel cage of a retaining wall, main shear reinforcement-type 1, and two small slant reinforcements-type 2 and 3. As a result, the nominal shear (capacity of shear) of the retaining wall per unit width is given below:

$$V_n = V_c + V_s \tag{10.5}$$

where V_s is the nominal shear of shear reinforcements.

As the horizontal distance between any two shear reinforcements is identical, the sectional area of the shear reinforcement per unit width ($b = 100$ cm) is as follows:

$$A_v = \frac{100 A_b}{S_h} \tag{10.6}$$

where A_v = the total sectional area of all shear reinforcements on the horizontal section per unit width (cm);

A_b = the sectional area of a single shear reinforcement (cm);

S_h = the horizontal distance between shear reinforcements (cm).

The nominal shear of the type 1 (i.e., main shear reinforcement) is given as follows:

$$V_{s1} = \frac{A_v f_y d}{S_v} \tag{10.7}$$

where S_v = the vertical distance between the main shear reinforcements.

The nominal shear of the type 2 (i.e., small slant reinforcement) is given as follows:

$$V_{s2} = \frac{A_v f_y d}{S_v} \sin \alpha \qquad (10.8)$$

where α = the angle between the small slant reinforcement and the horizontal reinforcement.

The nominal shear of the type 3 (i.e., small slant reinforcement) is given as follows:

$$V_{s3} = \frac{A_v f_y d}{S_v} \sin \beta \qquad (10.9)$$

where β = the angle between the small slant reinforcement and the vertical reinforcement.

The nominal shear from all shear reinforcements per unit width is given as follows:

$$V_s = V_{s1} + V_{s2} + V_{s3} \qquad (10.10)$$

4. Lap splice length and development length

The minimum lap splice length and development length of reinforcements in a diaphragm wall can be designed based on the ACI code, or they can be determined using Eqs. 10.11 and 10.12. The coefficient 1.25 is to magnify the lap splice length and development length, considering the effects of concrete casting in bentonite, which leads to a smaller bond stress.

$$\text{Development length } L_d = 1.25(cl_d) = \frac{23.7 c A_b f_y}{\sqrt{f_c'}} (\text{m}) \qquad (10.11)$$

$$\text{Lap splice length } L_d = 1.25(cl_d) = \frac{28.8 c A_b f_y}{\sqrt{f_c'}} (\text{m}) \qquad (10.12)$$

where A_b = the sectional area of a single reinforcement (m^2); if applied to a vertical reinforcement, $c = 1.0$ and to horizontal reinforcement, $c = 1.4$. The units of f_c' and f_y are MPa.

We can then proceed to the reinforcement design on the basis of the above method. Figure 10.3 is a typical reinforcement design of a diaphragm wall. The 3D picture of a steel cage of a diaphragm wall is shown in Figure 10.2.

10.4 STRUCTURAL COMPONENTS IN BRACED EXCAVATIONS

Figure 10.4 shows the commonly used structural components of the retaining wall in a braced excavation. In addition to the retaining wall, the main components of the strutting system include horizontal struts, wales, rakers, end braces, corner braces, and

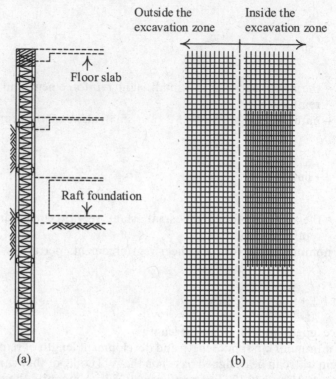

Outside the
excavation zone

Inside the
excavation zone

Floor slab

Raft foundation

(a)   (b)

Figure 10.3 The design of reinforcements in a steel cage of the diaphragm wall (a) profile (b) side view.

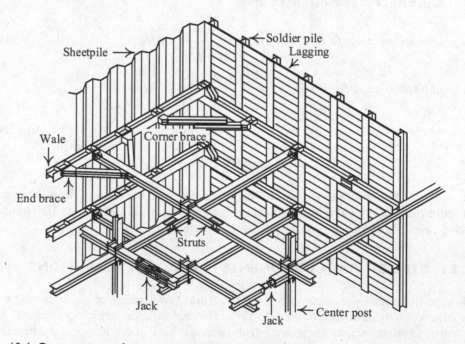

Sheetpile →

Soldier pile
Lagging

Wale

Corner brace

End brace

Struts

Jack

Jack

Center post

Figure 10.4 Components of a strutting retaining system.

center posts. Figure 10.5 shows the plan of a single strutting system. Figure 10.6a–c shows the 3D picture, the plane view, and the photo of a joint of a single strutting system with a center post, respectively. If the earth pressure on the back of the retaining wall grows too large or the strutting spacing is to be widened, the strutting system can be designed as double struts, as shown in Figure 10.7. Figure 10.8a–c shows the 3D picture, the plane view, and the photo of a join of double struts with a center post, respectively.

The load on struts can be computed using the apparent earth pressure method (Chapter 6), the finite element method (Chapter 7), or the beam on elastic foundation method (Chapter 8). In this section, the commonly used designing methods for a strut system are introduced.

10.5 STRUT SYSTEMS

10.5.1 Horizontal struts

1. Stress computation

A strut is usually subjected to the axial compressive load as well as the flexural load. The axial compressive stress can be computed as follows:

$$f_a = \frac{N}{A} \tag{10.13}$$

where A = the cross-sectional area of the strut;

$\quad N$ = axial load = $N_1 + N_2$

$\quad N_1$ = the strut load induced by an excavation, which can be computed using the beam on elastic foundation method, the finite element method, or the apparent earth pressure method;

$\quad N_2$ = the strut load induced by the temperature change = $\alpha \Delta t E A$;

$\quad \alpha$ = the coefficient of thermal expansion of struts; for the steel strut, α = $1.32 \times 10^{-5} / \,^{\circ}C$;

$\quad \Delta t$ = the temperature change of struts ($^{\circ}C$);

$\quad E$ = Young's modulus.

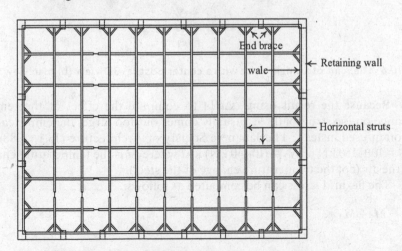

Figure 10.5 The single strutting system.

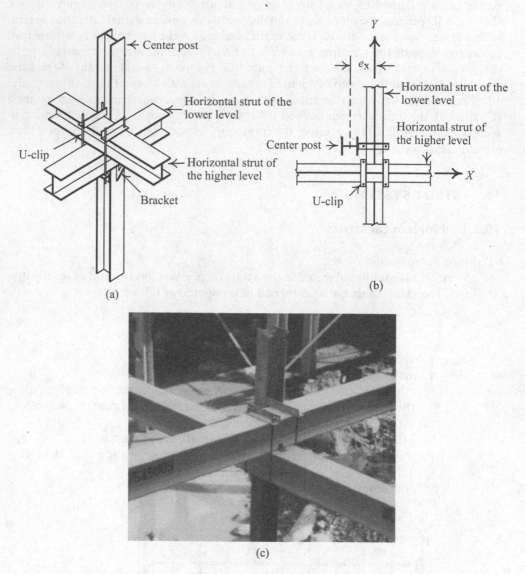

Figure 10.6 The joint of a single strut with a center post (a) 3D view (b) plan view (c) photo.

Because the result using $\alpha \Delta t E A$ to compute the effect of the temperature change on the strut load, N_2, usually comes out too large, the empirical formula is often used instead. The Japanese Society of Architecture (JSA, 1988) suggests $N_2 = 100-150\,\text{kN}$ or $N_2 = (10-40\,\text{kN}) \times \Delta t$ where Δt is the temperature change ($^\circ$C) in the air (not the temperature change of the steel).

The flexural stress can be computed as follows:

$$f_b = \frac{M_1 + M_2}{S} \tag{10.14}$$

where $M_1 =$ the bending moment produced by the strut weight and the live load; taking the center post as the simply supported hinge, then $M_1 = wL^2 / 8$;

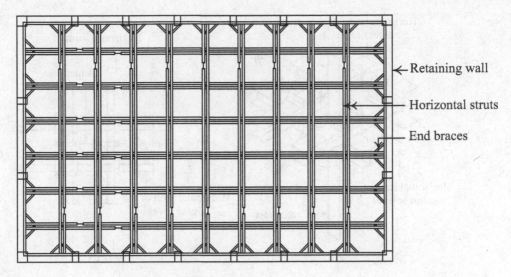

Figure 10.7 The double strutting system.

w = strut weight+ live load $\cong 5\,\text{kN/m}$;

L = the distance between two adjacent center posts;

M_2 = the bending moment caused by the uplift of the center post; because struts are constructed level by level during excavation, the influence of M_2 on the top level would be the largest whereas that on the lowest level would be the smallest;

S = the section modulus.

2. Allowable stress

Struts bear both the axial compressive force and the moment simultaneously during excavation. Thus, the allowable compressive stress has to be computed based on the beam-column theory. The allowable axial compressive stress of a strut can be selected from the tables and charts provided by the AISC Specification or computed using the following equation:

$$\frac{KL}{r_y} < C_c \qquad F_a = \frac{\left[1 - \frac{1}{2}\left(\dfrac{KL/r_y}{C_c}\right)\right]F_y}{\dfrac{5}{3} + \dfrac{3}{8}\left(\dfrac{KL/r_y}{C_c}\right) - \dfrac{1}{8}\left(\dfrac{KL/r_y}{C_c}\right)^3} \cdot \lambda \qquad (10.15)$$

$$\frac{KL}{r_y} > C_c \qquad F_a = \frac{12}{23}\frac{\pi^2 E}{\left(\dfrac{KL}{r_y}\right)^2} \cdot \lambda \qquad (10.16)$$

where KL/r_y = the effective slenderness ratio of the strut on the flexural plane where K can be taken as 1.0;

L = the unsupported length of the strut, usually the distance between the two adjacent center posts;

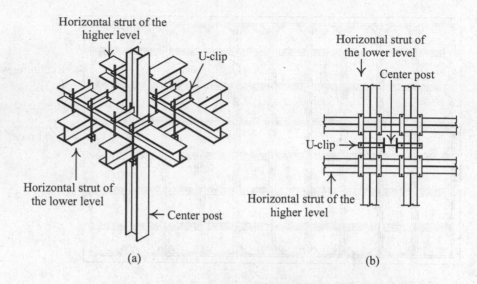

(c)

Figure 10.8 The joint of a double strut with a center post (a) 3D view (b) plan view (c) photo.

r_y = the radius of gyration of the cross section of the strut in the direction of the weak axis;

C_c = the critical slenderness ratio = $\sqrt{2\pi^2 E/F_y}$;

E = Young's modulus of struts;

F_y = the yielding stress of struts.

The allowable flexural stress (F_b) of a strut can be derived from the tables and charts provided by the AISC Specification. However, the flexural stress (f_b) of a strut, during normal excavation (that is, the uplift of the center post is not much), is not large. To simplify the design, we can assume $F_b = 0.6F_y \cdot \lambda$.

3. Examination of combined stresses

In accordance with the AISC Specification, the stress on each section of a strut should satisfy the following equation:

$$\frac{f_a}{F_a} \leq 15\% \qquad \frac{f_a}{F_a} + \frac{f_b}{F_b} \leq 1.0 \tag{10.17}$$

$$\frac{f_a}{F_a} > 15\% \qquad \frac{f_a}{F_a} + \frac{C_m f_b}{\left(1 - \dfrac{f_a}{F'_e}\right) F_b} \leq 1.0 \tag{10.18}$$

where C_m = the coefficient of modification = 0.85;

$1 / (1 - f_a / F'_e)$ = the amplification factor;

F'_e = the allowable Euler stress = $\lambda \cdot 12\pi^2 E / [23(KL / r_x^2]$;

KL / r_x = the effective slenderness ratio on the flexural plane;

r_x = the radius of gyration of the strut in the direction of the strong axis.

10.5.2   End braces and corner braces

For construction convenience or other reasons, the spacing between two adjacent struts in some positions come to be a little over the allowable distance in the arrangement of horizontal strutting. If we add one more strut, the spacing becomes too narrow. Under such a condition, an end brace or corner brace may be used to adjust the condition. The function of an end brace or corner brace is then to shorten the span of the wale without adding more struts (see the next section on the design of the wale), as shown in Figure 10.9. End braces are usually installed at 45° angles against the wale.

As shown in Figure 10.9, the axial force on the end brace is given as follows:

$$N = p \left(\frac{\ell_1 + \ell_2}{2} \right) \frac{1}{\sin \theta_1} \tag{10.19}$$

where p = the strut load per unit width estimated from the apparent earth pressure or numerical analysis;

ℓ_1, ℓ_2 = spans (see Figure 10.9);

θ_1 = the angle between the end brace and the wale (usually 45°).

Similarly, the axial force on a corner brace, as shown in Figure 10.9, is given below:

$$N_1 = p \frac{\ell_3 + \ell_4}{2} \frac{1}{\sin \theta_2} \tag{10.20}$$

$$N_2 = p \frac{\ell_5 + \ell_6}{2} \frac{1}{\sin \theta_3} \tag{10.21}$$

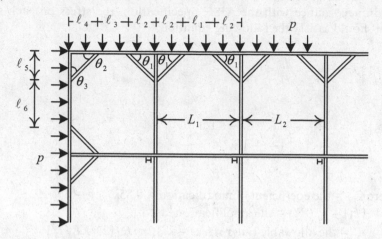

Figure 10.9 The spacing and angle between struts and end braces or corner braces.

where ℓ_3, ℓ_4, ℓ_5, ℓ_6 = the spans (see Figure 10.9);

θ_2, θ_3 = angles between the corner brace and the wale, 45° in most cases.

To be conservative, the designed load can be assumed to be the maximum value of the N_1 and N_2. The design of the end and corner braces is identical to that of the horizontal struts and would not be mentioned again.

10.5.3 Wales

The function of wales is to transfer the earth pressure on the retaining wall to the struts. For analysis, the earth pressure can therefore be assumed to act on the wale directly. The earth pressure can be obtained from the apparent earth pressure method or by transforming the strut load, computed using the finite element method or beam on elastic foundation method. As a matter of fact, the wale is usually acted on by the earth pressure as well as the axial force from the end brace or corner brace. That is to say, the wale bears the moment and the axial force simultaneously, and its design falls in the domain of the beam-column system. With ample lateral support, the analyses of secondary moment and buckling for wales can be saved.

To compute the maximum bending moment and shear of wales, the wales can be viewed as simply supported beams with struts as supporting hinges, or viewed as fixed end beams, as shown in Figure 10.10.

If a wale is viewed as a simply supported beam, we have the following:

$$M_{max} = \frac{1}{8} pL^2, \qquad Q_{max} = \frac{1}{2} pL \qquad (10.22)$$

If viewed as a fixed end beam, we have the following:

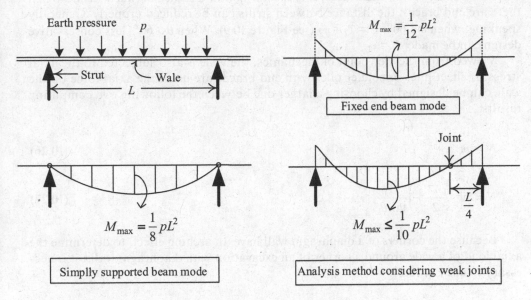

Figure 10.10 Computation of the bending moment of a wale.

$$M_{max} = \frac{1}{12} pL^2, \qquad Q_{max} = \frac{1}{2} pL \qquad (10.23)$$

where M_{max} = the maximum bending moment of the wale;
   Q_{max} = the maximum shear of the wale;
   L = the distance between struts;
   p = the earth pressure.

Because the length of a wale is limited, the wales have to be joined in the field. The strength of the joint is not fully rigid and the joint could easily become the weakest part of the structure. Thus, the joints had better be located at the places where the stress is smaller. Based on the bending moment distribution diagram of a continuous or a simply supported beam, if the joint is located at the places 1/4 of the span from the support, the maximum bending moment of the wale would be as follows:

$$M_{max} \leq \frac{1}{10} pL^2 \qquad (10.24)$$

As a result, in design practice, if the wale is assumed to be a simply supported beam, it may not work out as economical. If it is designed to be a fixed end beam, it may tend to be insecure. A more reliable method is to locate the joint at the place 1/4 of the span from the support and compute the maximum bending moment and shear as follows:

$$M_{max} = \frac{1}{10} pL^2, \qquad Q_{max} = \frac{1}{2} pL \qquad (10.25)$$

When there are no end braces, L is the horizontal distance between struts. When there are end braces, the distance between struts can be reduced properly. Generally speaking, when $\theta \leq 60°$, $L = \ell_1 + \ell_2$ (see Figure 10.9). When $\theta > 60°$, less conservative design can be made: $L = \ell_1$.

Viewed from the viewpoint of mechanics, the wale is also subjected to the axial stress. If sheet piles or soldier piles with end braces are used, the axial force on the wale can be designed by choosing a larger one between the following two computing results:

$$N = p(\ell_5 + \frac{\ell_6}{2})$$ (10.26)

$$N = p\left(\frac{\ell_1 + \ell_2}{2}\right)\frac{1}{\tan\theta_1}$$ (10.27)

Because the corners of a diaphragm wall have an arching effect, to determine the axial load of a wale around a corner of an excavation with diaphragm walls, it is necessary to use Eq. 10.27.

10.5.4 Center posts

In a braced excavation, center posts are usually set to bear the weight of struts, the materials on the struts, and other extra loading due to the movement of the retaining system. Center posts, usually H steels, are often installed by striking the piles into soils directly. Considering that the noise and vibration caused by striking may influence adjacent properties (especially when the installation is carried out in hard soils), when the embedment depths are designed deeper, or the pile driver is not powerful enough, the H steel is sometimes embedded into soils by way of preboring or inserted into a cast-in-place pile, as shown in Figure 10.11.

The design of a center post includes the design of the section and the embedment depth. Before designing the section, the load on the center post has to be determined first. The possible axial loads on each center post are as follows:

1. The weight of the horizontal strut and the live load, P_1:

$$P_1 = \sum_{i=1}^{n} w_i (L_1 + L_2)$$ (10.28)

where w_i is the strut weight of each level and its live load, n is the number of the levels, and L_1 and L_2 can refer to Figure 10.12.

2. The weight of the center post above the excavation surface, P_2.
3. The compressive force from the slant strut, P_3.

If the construction of a center post is defective or there occurs a differential settlement between the center post and the retaining wall, these conditions may cause the center post to move downward or upward by the axial force of the strut. Suppose the tilt angle of a horizontal strut is θ, as shown in Figure 10.13, the downward or upward force on the center post would be as follows:

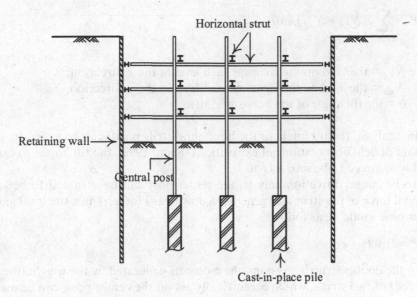

Figure 10.11 Installation of center posts onto cast in situ piles.

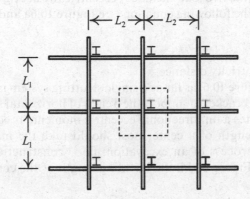

Figure 10.12 Distribution of strut weight on a center post.

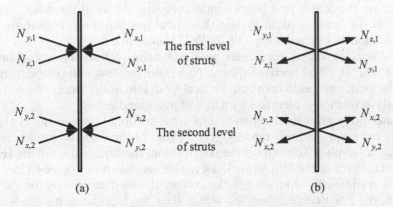

Figure 10.13 Effect of the axial force of struts on a center post (a) when the center post settles (b) when the center post heaves.

$$P_3 = \sum_{i=1}^{n} 2(N_{x,i} + N_{y,i})\sin\theta \qquad (10.29)$$

where $N_{x,i}$ = the load on the struts at each level in the x direction;
$N_{y,i}$ = the load on the struts at each level in the y direction;
θ = the tilt angle of the horizontal strut.

In analysis, the tilt angle of the horizontal strut is difficult to estimate. Based on the data of field observations of excavations (Shen, 1999), the tilt angle (θ) can be reasonably assumed to be $\sin\theta \cong 1/50$.

To be conservative in analysis, the center post can be assumed to be subject to the axial force of the strut and generate a downward force. Thus, the total load on the center post would be as follows:

$$P = P_1 + P_2 + P_3 \qquad (10.30)$$

In the double strutting system, the moments generated by the weight (the live load included) of two struts, which eccentrically act on the center post, can be assumed to be mutually offset (see Figure 10.8). In the single strutting system, the moment caused by the strut weight (the live load included) eccentrically acting on the center post can be computed using the following equation (see Figure 10.6a and b)

$$M = (P_2 + P_3)e_x \qquad (10.31)$$

where e_x = the eccentricity distance.

As shown in Figure 10.6, as far as a single strutting system is concerned, the center post, using only one bracket to support the weight of horizontal struts in both the x and y directions, generates a unidirectional eccentric moment, as computed by Eq. 10.31.

The buckling length of a center post should take the maximum unsupported length during the process of an excavation, floor construction, and strut dismantling. As shown in Figure 10.14, the buckling length of the center post L is given as follows:

$$L = \max(L_1, L_2, L_3, L_4, L_5) \qquad (10.32)$$

Because the center post bears, simultaneously, the axial force and bending moment, it is the beam-column system that should be adopted in analysis. Choose a proper section and then use Eqs. 10.28–10.32 to examine it.

As discussed above, the center post may bear vertical loads and uplift forces caused by the tilt of the horizontal strut. As a result, to design the embedment depth of the center post, one has to take both vertical loads and uplift forces into consideration. The analysis method is identical with that of pile foundations.

Center posts are usually H steel. The cross-sectional area of the H steel used to compute the point load resistance (Q_p) is the sum of the cross-sectional area of the H steel and its soil plug. To compute the skin friction, the surface area of the H steel is the total surface area of the H steel enclosed region, as shown in Figure 10.15.

This section also summarizes the commonly used equations for the design of center posts. For detailed theories, please refer to the related books, such as local building codes or books on pile foundations.

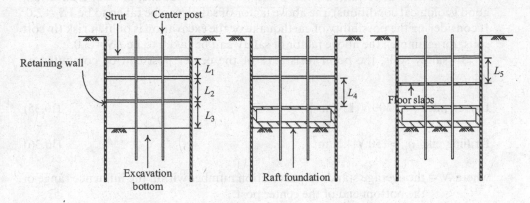

Figure 10.14 Unsupported length of center posts.

usually $L_3 > L_1, L_2$

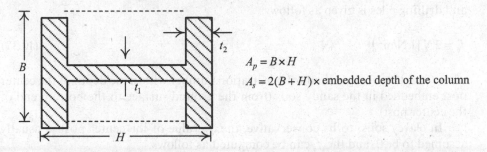

$$A_p = B \times H$$
$$A_s = 2(B + H) \times \text{embedded depth of the column}$$

Figure 10.15 The area to be adopted for the computation of the vertical bearing capacity of center posts of the H pile.

1. Vertical bearing capacity

The ultimate vertical bearing capacity can be expressed as follows:

$$Q_u = Q_p + Q_s = f_s A_s + q_p A_p \qquad (10.33)$$

where Q_u = the ultimate vertical bearing capacity of the center post;
Q_p = the point load resistance of the center post;
Q_s = the skin frictional resistance of the center post;
f_s = the unit frictional resistance of the center post;
A_s = the surface area of the center post;
q_p = the unit point resistance per unit area of the center post;
A_p = the sectional area of the center post.

The allowable vertical bearing capacity of the center post is given as follows:

$$Q_a = \frac{Q_u}{FS} = \frac{Q_s + Q_p}{FS} \qquad (10.34)$$

where Q_a is the allowable vertical bearing capacity and FS is the factor of safety.

Under general conditions (low possibility of earthquake, general excavation, good geological conditions), the above factor of safety can be taken to be $FS = 2.0$. If considering the possibility of earthquake or the excavation is of high risk (in soft soils, for example), the above factor of safety can be taken to be $FS = 3.0$.

In sandy soils, the point resistance of the center post can be computed as follows:

$$\text{Driven pile}: q_p = 400N\left(\text{kN/m}^2\right) \tag{10.35}$$

$$\text{Drilling pile}: q_p = 150N\left(\text{kN/m}^2\right) \tag{10.36}$$

where N = the average standard penetration number within the influence range of the bottom end of the center post.

Generally speaking, the N value can be taken as the average value within the range of four times of the center post diameter above the bottom end of the center post and one center post diameter below it.

In sandy soil, the frictional resistance of the center post for both the driven and drilling piles is given as follows:

$$f_s = 2N\left(\text{kN/m}^2\right) \tag{10.37}$$

where N is the average standard penetration number within the depth of the center post embedded in the sandy soils (from the ground surface to the bottom end of the center post)

In clayey soils, to be conservative, the q_p value of the center post is usually assumed to be 0, and the f_s can be computed as follows:

$$f_s = c_w = \alpha s_u \tag{10.38}$$

where c_w = adhesion between the surface of the center post and the surrounding soils;

s_u = the undrained shear strength of clay;

α = the reduction value of the undrained shear strength, which can be found from Figure 4.12.

2. Uplift resistance

The allowable uplift resistance of a center post is given as follows:

$$R_a = W_p + \frac{1}{FS} f_s A_s \tag{10.39}$$

where W_p = the weight of the center post, the influence of groundwater considered;

f_s = the frictional resistance per unit area of the center post, which can be estimated using Eqs. 10.35–10.38

10.6 STRUCTURAL COMPONENTS IN ANCHORED EXCAVATIONS

Figure 10.16 illustrates an excavation with an anchored support system. As shown in the figure, in addition to the retaining wall, the main structural component is the

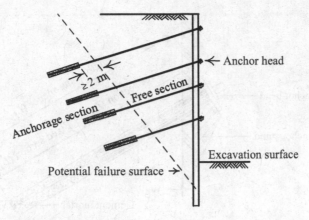

Figure 10.16 The anchored excavation.

anchor system, which consists of an anchor and a wale. The following will introduce the analysis and design of anchored excavations.

10.7   ANCHOR SYSTEMS

Anchors are categorized into permanent anchors and temporary ones. The design of an anchor is related to the soil properties, the materials of the anchor, grouting, and the construction details. The last is especially crucial in determining the quality of the anchor. Concerning the general rules of the design and construction of an anchor, please refer to the related literature and specifications referred to in this section.

Anchors in excavations are temporary. This section will only introduce the design of the anchor system of the retaining wall in excavations.

10.7.1   Components of anchors

An anchor is basically composed of the anchor head, the free section, and the fixed section, as shown in Figure 10.17. The anchor head is used to transmit the force or preload generated by the displacement of the retaining wall or wale to the free section. The free section transmits the force or preload to the fixed section and offers the elastic strain necessary for preloading. The function of the fixed section is to transmit the force from the free section to soil layers.

As shown in Figure 10.17, to make the axial center of the anchor head match that of the anchor, an anchor stand is usually designed under the bearing plate. To make sure the tendon of the free section can deform freely when preloaded, the free section is usually covered using a plastic pipe to isolate the tendon from grouts (or soils). The anchor head is made of a locking device and a bearing plate.

Anchors can be categorized into the frictional resistance type, the bearing resistance type, and the composite type based on the characteristics of the bearing force provided by the fixed section (Figure 10.18). The anchoring force of the frictional resistance type of anchors comes from the friction or adhesion between the fixed section and

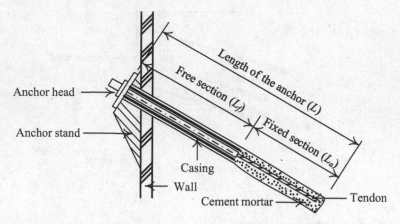

Figure 10.17 Configuration of an anchor.

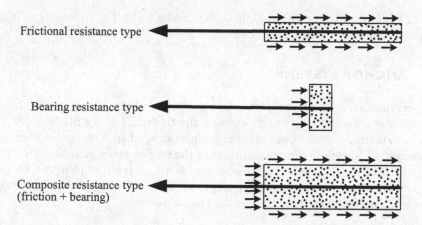

Figure 10.18 Types of anchorage body (a) frictional resistance type (b) bearing resistance type (c) composite resistance type. (Redrawn after CICHE, 1998.)

surrounding soils. The anchoring force of the bearing resistance type lies in the bearing capacity of soils surrounding the fixed section, which is usually underreamed. This type is also called the underreamed anchor. The composite type is a mixed form possessing the characteristics of both the frictional resistance and bearing resistance types.

10.7.2 Analysis of anchor load

The horizontal component of the anchor load can be computed using the apparent earth pressure method, the beam on elastic foundation method, or the finite element method. The related analytical method is identical to that for the braced excavation except that the horizontal component computed from analysis has to be transformed into the anchor load because the axial center of the anchor is not perpendicular to the retaining wall. As shown in Figure 10.19, the anchor load is given as follows:

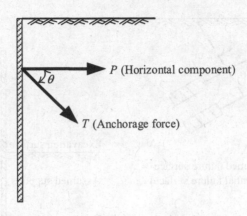

Figure 10.19 The anchorage force and its horizontal component.

$$T_w = \frac{P}{\cos\theta} \tag{10.40}$$

where T_w = the designed load of the anchor;

P = the horizontal component obtained from analysis;

θ = the angle between the axis of the anchor and the horizontal plane.

10.7.3 Arrangement of anchors

The minimum spacing between anchors has to be set so that the group anchor effect will not be produced or the installation of an anchor may not cast a bad influence on the adjacent anchors. According to Federation Internationale de la Precontrainte (FIP, 1982), the horizontal spacing between anchors in an excavation should be larger than 1.5 m or $4d_b$ where d_b is the diameter of the fixed section. The vertical spacing between anchors depends on the analytical results of the anchor load. The vertical spacing between anchors is about 2.5–4.5 m and at least 1.0–1.5 m above the floor slabs.

To maintain the overall stability of the anchor and the retaining wall, the fixed section of the anchor should be placed outside the potential failure surface at least 2 m away from it (Figure 10.20). The determination of the potential failure surface is still under study, and no final conclusion has been drawn yet. The active failure zone starting from the excavation bottom, that starting from the bottom of the retaining wall, and that starting from the assumed support, are the commonly assumed potential failure surfaces, as shown by planes **ab**, **ef**, and **cd** in Figure 10.21. The design on the basis of the active failure zone starting from the excavation bottom is not conservative, whereas the one using the basis of the active failure zone starting from the bottom of the retaining wall is more conservative. The active failure zone starting from the assumed support is in between and seems more reasonable. To determine the location of the assumed support in Figure 10.21, please refer to Section 6.10.3. The location of the fixed section can also be chosen with reference to the BSI code (British Standard Institute, 1989), as shown in Figure 10.22. The line segments DG and EF are the possible passive failures, considering the preload of the anchor.

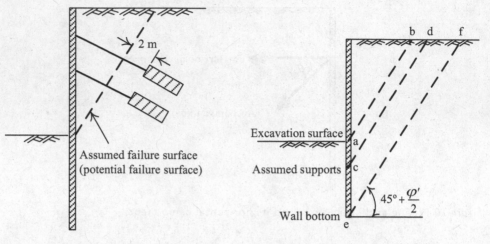

Figure 10.20 The distance between an anchorage section and the potential failure surface.

Figure 10.21 Potential failure surfaces in excavations.

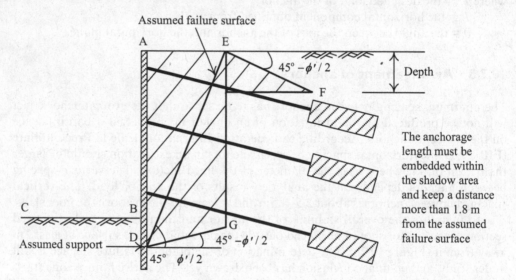

Figure 10.22 Locations of the anchorage sections in soil. (Redrawn after BSI, 1989.)

Theoretically, if an anchor is installed with the same direction of loading, it is able to develop the maximum capacity. Considering the installation quality, to clear the dregs in the drilling bore of the anchor, the installation angle should not be smaller than $10°$. The fixed section should be placed in a bearing layer (such as a sandy layer, a gravelly layer, or rocks) or soils with high strength. Generally speaking, the deeper the soils, the higher the strength. It is more suitable to place the fixed section in deeper soil. As a result, an anchor is usually installed with a certain slope. The steeper the slope, however, the larger the dragging down force on the retaining wall. When the installation angle exceeds $45°$, it becomes dangerous. That is to say, without going beyond $45°$, the installation angle should fall within the range $10° \leq \theta \leq 45°$.

10.7.4 Design of anchor stands and wales

The function of an anchor stand is to transfer the load on the wale or the retaining wall to the anchored soil layer, as illustrated by Figure 10.23. Anchor stands can be categorized into reinforced concrete anchor stands and steel anchor stands. The detailing of reinforcements of the concrete anchor stand should be carefully designed to meet safety requirements. Under the working load, the allowable compressive stress of the concrete anchor stand should be smaller than 30% of the 28th-day strength of concrete.

Generally, two parallel H steels or I steels serve as wales (Figure 10.23). Bearing only the horizontal force, wales are treated as beams. The design of wales should consider the stresses in both the weak and strong axes. Referring to Figure 10.23, the stress of a wale can be examined as follows:

1. Stress computation in the direction of the strong axis

   As shown in Figure 10.23, the wale between two anchors is as a simply supported beam structure; the maximum bending moment and shear are given as follows:

$$M_{max} = \frac{1}{8} pL_1^2, \qquad Q_{max} = \frac{1}{2} pL_1 \tag{10.41}$$

where M_{max} = the maximum bending moment in the direction of strong axis;
Q_{max} = the maximum shear in the direction of strong axis;
p = the lateral earth pressure on the wale;
L_1 = the horizontal distance between anchors.

As in the design of a wale in the braced excavation introduced in Section 10.5.3, if the joint is placed at the distance of 1/4 of the span from the support, the maximum bending moment can be computed using the following equation:

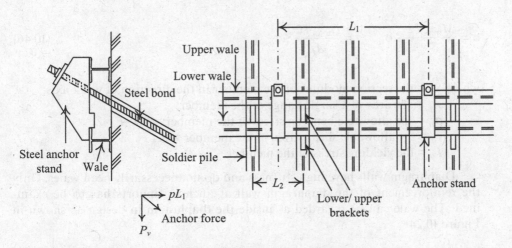

Figure 10.23 Configuration of wales in an anchored system.

$$M_{max} = \frac{1}{10} pL^2 \tag{10.42}$$

The bending stress and shear stress of the wale in the direction of the strong axis should satisfy the following criteria:

$$\sigma = \frac{M_{max}/2}{S_x} \le \sigma_a, \qquad \tau = \frac{Q_{max}/2}{A_w} \le \tau_a \tag{10.43}$$

where S_x = the section modulus of the member in the direction of the strong axis;
  A_w = the area of the web of the member;
  σ_a = the allowable bending stress of the member = $(0.6$ to $0.66)F_y \cdot \lambda$;
  τ_a = the allowable shear stress of the member = $0.4F_y \cdot \lambda$;
  F_y = the yielding stress of the member.

2. Stress computation in the direction of the weak axis

As shown in Figure 10.23b, viewing the wale between the brackets supporting the anchor as a beam, the maximum bending moment and shear of the wale are given as follows:

$$P_v = pL_1 \tan\theta \tag{10.44}$$

$$M_{max} = \frac{1}{4} P_v L_2 = \frac{1}{4} pL_1 L_2 \tan\theta, \quad Q_{max} = \frac{1}{2} P_v = \frac{1}{2} pL_1 \tan\theta \tag{10.45}$$

where M_{max} = the maximum bending moment in the direction of weak axis;
  Q_{max} = the maximum shear in the direction of weak axis;
  P_v = the vertical component of the anchorage force at the support;
  L_2 = the distance between the brackets supporting the anchor.

The bending stress and shear stress of the wale in the direction of the weak axis should satisfy the following conditions:

$$\sigma = \frac{M_{max}/2}{S_y} \le \sigma_a, \quad \tau = \frac{Q_{max}/2}{A_f} \le \tau_a \tag{10.46}$$

where S_y = the section modulus of the member in the direction of weak axis;
  A_f = the area of a single flange of the member;
  σ_a = the allowable bending stress of the member = $0.75\,F_y \cdot \lambda$;
  τ_a = the allowable shear stress of the member = $0.4\,F_y \cdot \lambda$;
  F_y = the yielding stress of the member.

Diaphragm walls have high stiffness and do not necessarily need wales. Only the reinforcement of the diaphragm wall at anchor supports has to be examined. The wale can be regarded as inside the diaphragm in design as shown in Figure 10.24.

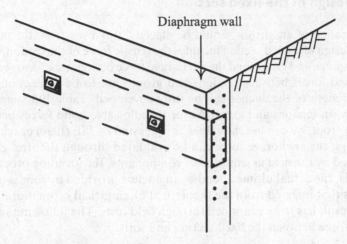

Figure 10.24 Wales located inside the diaphragm wall.

10.7.5 Design of the free section

The free section is composed of a tendon and a plastic casing. Three types of tendons are available: steel bars, steel wires, and strands. The 7-wire strand is common in many countries, usually available in sizes of 13 mm (0.5 in), 15 mm (0.6 in), and 18 mm (0.7 in). The ultimate tensile strength varies from 1,570 to 1,765 N/mm². The allowable tensile force can be computed as follows:

$$P_w = P_u/F_t \tag{10.47}$$

where F_t = the designed safety factor for the tendon. Table 10.1 lists the commonly used factors of safety;

P_u = ultimate tensile force.

The number of tendons is given as follows:

$$n = \frac{T_w}{P_w} \tag{10.48}$$

As discussed above, T_w is computed using Eq. 10.40, transforming the horizontal component, as analyzed using the apparent earth pressure method, the beam on elastic foundation method, or the finite element method.

As Figures 10.20–10.22 illustrate, the free section has to extend beyond the potential failure zone by at least 2.0 m. If the free section is too short, the stresses in soils caused by the fixed section will easily affect the retaining wall. In addition, the anchor behavior will tend to be fragile, any small displacement possibly bringing about large stress, and threatening the security of the anchor. Therefore, the free section has to be at least 4.0 m long.

10.7.6 Design of the fixed section

The fixed section of an anchor should be placed 2.0 m away from the potential failure zone. The design of a fixed section includes the tensile force of the tendon, the bond force between tendons, and grouts and the anchorage force between the fixed section and soils.

The bond forces between tendons and grout have to be large enough so that the designed strength of the anchor can be fully developed. Table 10.1 suggests the safety factors between tendons and grout. As for the allowable bond forces between tendons and cement grout, we can use the values as suggested by JSF (1990) or related literature.

Although the anchorage force can be evaluated through the strength criterion of soil. Affected by the field assemblage of components, the grouting process, and variations in soils, the actual ultimate load of an anchor provided by soils is not necessarily the same as that inferred from the theoretical or empirical estimation. Therefore, the estimated result has to be examined through field tests. The following summarizes the anchorage force between the fixed section and soils.

1. Friction type of anchor

   The ultimate anchorage force, T_u, for a friction type of anchor can be computed by the following equation:

$$T_u = \pi d_b L_a \tau_{ult} \tag{10.49}$$

   where T_u = the ultimate anchorage force;

   d_b = the diameter of the fixed section;

   L_a = the length of the fixed section;

   τ_{ult} = the average ultimate shear resistance strength per unit area (also called the frictional strength) between the fixed section and soils.

Table 10.1 Factor of safety for single anchor (CICHE, 1998)

Classification	Tensile force of tendons (F_t)	Anchoring force (F_a)	Bond force of tendons (F_b)
Temporary anchors whose working period is not longer than 6 months and which don't affect public safety when failing[2]	1.4	2.0	2.0
Temporary anchors whose working period is not longer than 2 years and which don't affect public safety, though having certain influence, when failing without alert	1.6	2.5[3]	2.5[3]
Permanent or temporary anchors which are highly risky in rusting or which affect public safety seriously due to failure[2]	2.0	3.0[4]	3.0[4]

Note:
[1] The table is from CICHE (1998), incorporating specifications made by BSI (1989) and FIP (1982)
[2] The temporary anchors in this table refer to those whose working periods are not longer than 2 years. Otherwise, they are classified as permanent anchors.
[3] With complete proving test results, the minimum factor of safety can be 2.0.
[4] If creep of soil is to be encountered, the factor of safety can be increased to 4.0.

The designed load on the anchor is given as follows:

$$T_w = \frac{T_u}{F_a} \tag{10.50}$$

where F_a = the safety factor of the designed anchorage force; the commonly used factors of safety are listed in Table 10.1.

The strength between the fixed section and the surrounding soil, τ_{ult}, changes with the types of soils where the fixed section is placed, the failure mode of the fixed section, the grouting pressure, and the installation method. Many investigators have proposed some empirical formulas to estimate τ_{ult}. Given the complexity of the computing of τ_{ult}, this section lists the commonly used methods for readers' reference, incorporating the related specifications and some investigators' research results.

a.   Anchorage in rocksv
     Littlejohn's suggestion (1970) is usually taken for the ultimate shear resistance strength between the fixed section and rocks. That is, let τ_{ult} be 0.1 q_u (for block rocks) or 0.25 q_u (for weathered rocks) where q_u is the axial compressive strength of the rock.
b.   Anchorage in sandy soils
     The τ_{ult} value of an anchor in sandy soils can be computed as follows:

$$\tau_{ult} = \sigma'_v \tan \delta \tag{10.51}$$

where σ'_v = the average effective overburden pressure above the fixed section;
       δ = the angle of friction between the fixed section and soils.
c.   Anchorage in clayey soils
     The ultimate shear resistance strength of the fixed section in clayey soils can be expressed as follows:

$$\tau_{ult} = \alpha s_u \tag{10.52}$$

where s_u = the undrained shear strength of clay;
       α = the reduction factor for undrained shear strength (refer to Figure 4.12)

As Eq. 10.49 shows, the longer the fixed section, the larger the anchorage force. However, many investigators have found that when the length of the fixed section exceeds some critical value, more lengthening of the fixed section can hardly increase its anchorage force. Thus, the length of the fixed section has to be limited. In principle, the length of the fixed section should be between 3.0 and 10.0 m. If out of this range, field tests are recommended to be carried out to examine the ultimate anchorage force. Many specifications, e.g., FIP (1982), BSI (1989), PTI (Post-Tensioning Institute, 1989), AASHTO (American Association of State Highway and Transportation Officials, 1992), and DIN (Deutsche Industrie Norm, 1988), have the requirement for the minimum and maximum lengths of the fixed section.

2. Underreamed anchors

Theoretically, the anchorage force of an underreamed anchor can also be inferred in accordance with the theory of bearing capacity of soil. However, because of the complexity of the installation of a underreamed anchor, the actual anchorage force is usually far from the inferred result. As a result, the anchorage force of an underreamed anchor is usually obtained from field tests (Section 10.8).

10.7.7 Preloading

The anchor installed usually needs to be preloaded and locked, which is called locked preloading or the locked load. The locked load is usually a little larger than the designed load of the anchor because the slipping of the wedge clips during the process of locking will cause some loss of the anchor preload. After locking, the remaining force is called the effective load. Both the tendons and fixed section are loaded after preloading the anchor. The displacement of the fixed section is not large at the initial stage of preloading (sometimes even after preloading), and thus, the preload mainly acts on the tendons, which deform accordingly and in a heavily stressed state.

10.7.8 Design of retaining walls

The stress analysis of an anchored wall can use the assumed support method, the beam on elastic foundation method, or the finite element method, which are all identical with those of a braced excavation. Because of the installation angle of the anchor, however, a downward dragging force will be produced on the retaining wall. Thus, we have to examine whether the vertical bearing capacity of the wall is larger than the total downward dragging force from all levels of anchors. The vertical bearing capacity of the wall equals the sum of point bearing capacity of the wall bottom and the frictional resistance of the wall surface. The computing is similar to that of pile foundations.

10.8 TESTS OF ANCHORS

Influenced by the process of fabrication, grouting, and geological conditions, the actual ultimate load of an anchor has to be examined through field tests. The tests are classified into many methods in terms of their different aims although their common goal is to understand the deformation behaviors and load capacities of the anchor. Anchor tests can be distinguished into the proving test, the suitability test, and the acceptance test. The major international specifications (BSI, 1989; DIN, 1988; FIP, 1982) for the three types of tests are not identical, but are all based on similar principles. This section will not discuss the test details, and readers interested on the subject can find more information from the related documents.

PROBLEMS

10.1   Figure P10.1 illustrates a 3.2 m deep excavation. The retaining wall is cantilever H piles arranged with a span of 1.5 m. The dimensions of the H piles are H300× 300×10×15 (properties, see related design manuals), and the embedment depth is 5.0 m. The laggings are 4.2 m thick. Examine the safety of the retaining wall and laggings (Note: the stress of a cantilevered retaining wall can be computed using a computer program or the methods introduced in Chapter 6).

10.2   Here is an excavation site which is 30 m wide and 50 m long. The braced excavation method is adopted. The final excavation depth is 9.5 m. Figure P10.2 illustrates the depths for the excavation stages and the locations of the struts and floor slabs. The groundwater level is rather deep. The properties of soils at the site are $c' = 0$, $\phi' = 28°$, and $\gamma = 21.6 \text{ kN/m}^3$. If soldier piles (H piles) are adopted as the retaining wall, determine suitable dimensions and distances of the soldier piles, based on the bending moment and shear diagrams obtained from the results of the finite element method or the beam on elastic foundation method, which simulate each excavation stage as well as basement construction stages, including strut dismantling and floor slab construction stages. (The dimensions of the piles and struts and spacing between the piles have to be assumed first to compute the stiffness of the retaining wall and the struts, which are then used for analysis. After analysis, suitable dimensions of the retaining wall and struts can be selected. If necessary, analyze it again).

10.3   Same as Problem 10.2, except sheet piles are adopted as the retaining wall. Select suitable sheet piles.

10.4   Same as Problem 10.2, except a 50 cm thick diaphragm wall is adopted as the retaining wall. Design the reinforcements in accordance with the bending moment and shear diagrams of the wall generated by excavation.

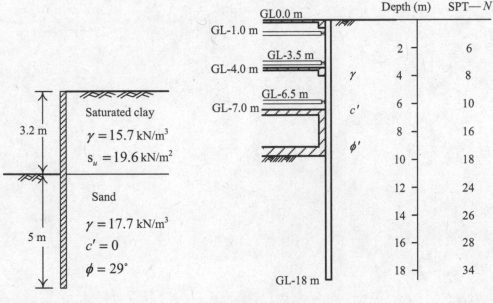

Figure P10.1   Figure P10.2

10.5   Same as Problem 10.2. Design the horizontal struts and center posts. (The apparent earth pressure method or the beam on elastic foundation method can be used to compute the strut load).

10.6   Same as Problem 10.2. If an anchor system is used in an excavation, the location of the anchor stand is the same as that of the horizontal strut in Problem 10.2 and the installation angle of the anchor being 30°, compute the working load of the anchor per unit width.

10.7   Same as Problem 10.6. Suppose the tendons of the free section are made of steel strands and the fixed section is the friction resistance type, design the number of steel strands and the lengths of the free and fixed section.

10.8   Same as above. Design the wales and examine the diaphragm wall safety generated by the downward dragging force from the anchor.

10.9   As in Example 7.2 and Figure 7.15, suppose there exists an excavation which is 40 m wide and 50 m long. The retaining strutting system is the diaphragm wall and steel struts. Analyze the strut load using the apparent earth pressure method or beam on elastic foundation method and design the struts.

10.10 Same as above. Based on the analysis results, select the suitable thickness of the diaphragm wall and design the reinforcements.

10.11 As in Problem 10.9, if the strutting system is the anchor instead, whose anchor stands are located at the same depths as the horizontal strut in Figure 7.15 and whose installation angle is 30°, compute the strut load using the apparent earth pressure method or the beam on elastic foundation method.

10.12 Same as above, suppose the tendon of the free section is steel strands and the fixed section is the friction resistance type. Design the number of steel strands and the lengths of the free section and the fixed section.

10.13 Same as above; design the wales and examine the diaphragm wall safety generated by the downward dragging force from the anchor.

Chapter 11

Excavation and protection of adjacent buildings

11.1 INTRODUCTION

Generally speaking, the protection of adjacent buildings during excavation can be made based on the following three procedures: (a) before-excavation plan, (b) monitoring and prevention during the construction, and (c) compensation after damages have been done. The most important job in a before-excavation plan is a comprehensive geological investigation, based on which analysis and design are carried out.

Before excavation, the buildings within the influence range of excavation have to be properly evaluated. For example, if there exist cracks before excavation, they have to be measured and recorded to forestall dispute. As for the determination of the settlement influence range, please refer to the empirical formulas introduced in Section 6.4.2 or solve it by way of finite element analysis. For predictable but unavoidable damages, cracks in brick or old houses induced by the construction of retaining walls or by excavation for example, risk management and construction damage insurance are important. Lastly, construction safety should be strictly maintained, and the concepts of environmental protection should be fully apprehended.

Monitoring systems should be implemented in the vicinity of excavations. The monitoring items have to be carefully chosen, and the monitoring results should be reasonably explained. The other items to be monitored are the deformation of the retaining wall and the ground surface. If excavation is carried out in sandy soils, leakage through the retaining wall should be carefully watched. Emergency measures should be taken in advance against any possible disaster or damage.

This chapter is confined to the protection of adjacent buildings by means of design and during the construction.

11.2 ALLOWABLE SETTLEMENT OF BUILDINGS

11.2.1 Allowable settlement under the building weight

Settlement will occur under the weight of a building. Too much settlement will cause the components of a building, such as beams, columns, walls, and foundations, to crack or be damaged. The amount of settlement is related to the type, area, and materials of the building; the soil properties; and the type of the foundation. The following is a synthesis of the results from previous studies (Bjerrum, 1963; Burland and Wroth,

DOI: 10.1201/9780367853853-11

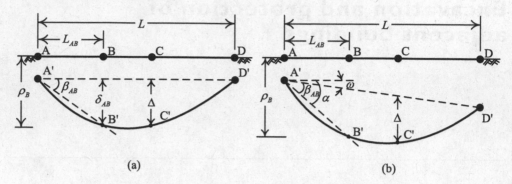

Figure 11.1 Parameters of settlement of buildings: (a) settlement without rigid rotation and (b) settlement with rigid rotation.

1974; Grant et al., 1974; Wahls, 1981; Boscardin and Cording, 1989; Skempton and McDonald, 1957; JSA, 1988).

Figure 11.1 gives the definitions of various deformation parameters. Figure 11.1a is the condition of settlement of a building without a rigid rotation. Figure 11.1b is the condition of settlement of a building with a rigid rotation. Points A, B, C, and D are the points before settlement, whereas A′, B′, C′, and D′ are those after settlement. The definitions of deformation parameters are given as follows (Wahls, 1981):

ρ_i = total settlement at point i
δ_{ij} = differential settlement between points i and j
ω = rigid rotation
β_{ij} = angular distortion between point i and j, $\beta_{ij} = \delta_{ij} / L_{ij} - \omega$
L_{ij} = distance between referent points i and j
α = tilt, $\alpha = \omega + \beta$ (β is the angular distortion)
Δ = relative deflection
Δ / L = deflection ratio

Figure 11.2 gives a more definition about angular distortion (β), rigid rotation (ω), lateral strain (ε_L), and tilt (α) for a building element. The value measured from a tiltmeter (see Section 12.4.2) in excavations is the tilt (α). When the angle distortion is equal to 0, it means that the building unit only exhibits a rigid rotation, at this time $\alpha = \omega$. A building element can be the structure between two adjacent columns (Figure 11.3). The angle distortion and tilt of the building unit ABCD (bay 1), as shown in Figure 11.3, can be obtained by measuring the horizontal and vertical displacement at each point with theodolite or total station. The calculation method is as follows:

$$\alpha = \frac{(C_\ell - B_\ell) + (D_\ell - A_\ell)}{2H} \quad \text{or} \quad \alpha = \frac{C_\ell - B_\ell}{H} \tag{11.1}$$

$$\beta = \frac{A_v - B_v}{L} - \alpha = S - \alpha \tag{11.2}$$

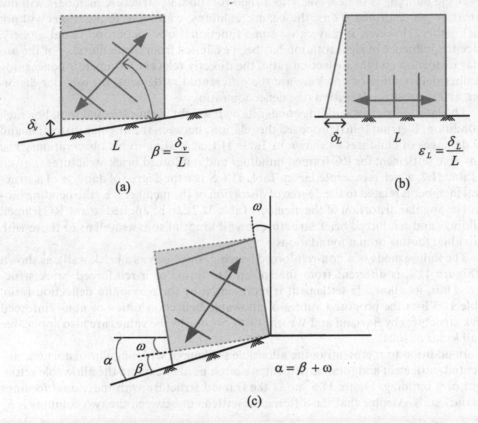

Figure 11.2 Parameters of settlement for a building element: (a) angular distortion, (b) lateral strain, and (c) tilt and rigid rotation.

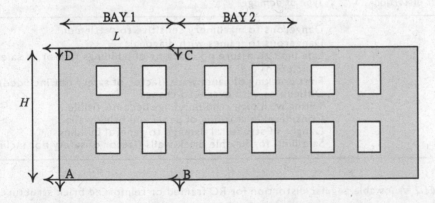

Figure 11.3 Representative of a building element.

In Eqs. 11.1–11.2, the subscript ℓ represents the lateral or horizontal displacement, and the subscript v represents the vertical displacement or settlement; L represents the length of the building unit; and H represents the height of the building unit.

If the building is subject solely to a rigid rotation, its structure members will not distort or get deformed. Thus, the beams, columns, walls, and foundations will not crack either. However, it may cause some functional obstruction or visual anxiety. Since the influence of rigid rotation has been excluded from the definitions of the angular distortion and the deflection ratio, the directly relevant parameters concerning member deformation or cracking are the differential settlement, the angular distortion, and the relative deflection (the deflection ratio).

The lateral strain of buildings under its own weight is almost equal to 0. For such a condition, Bjerrum (1963) proposed the relations between the angular distortion and the damages of buildings as shown in Table 11.1, according to his observations. The allowable settlement for RC framed buildings and reinforced brick structures is given in Table 11.2, which is excerpted from Table 11.1. Since the degree of damage of a structural member is related to the degree of distortion of the member, i.e., the bending moment or angular distortion of the member, Table 11.2 can be applied to any RC framed buildings and reinforced brick structures on all kinds of soils as well as to those with individual footing or mat foundations.

The failure mode of a non-reinforced bearing wall (such as a brick wall), as shown in Figure 11.4, is different from that of an RC frame or a reinforced brick structure. Thus, its allowable settlement is represented by the maximum deflection ratio. Table 11.3 lists the proposed values of allowable deflection ratios of non-reinforced brick structures by Burland and Wroth (1974). Similarly, the values are also applicable to all kinds of soils.

In addition to representing the allowable settlement with angular distortion, differential settlement and total settlement are often used to explore the allowable settlement of a building. Figure 11.5 shows the framed structure with individual footings in sandy soils. Assume that the differential settlement between the two columns is δ_{ij}

Table *11.1* Limiting values of angular distortion (Bjerrum, 1963)

Angular distortion	Type of damage
1/750	Dangerous to machinery sensitive to settlement
1/600	Dangerous to frames with diagonals
1/500	Safe limit to assure no cracking of buildings (factor of safety included)
1/300	First cracking of panel walls (factor of safety not included)
1/300	Difficulties with overhead cranes
1/250	Tilting with high rigid buildings become visible
1/150	Considerable cracking of panel and brick walls
1/150	Danger of structural damage to general buildings
1/150	Safe limit for flexible brick walls (factor of safety not included)

Table *11.2* Allowable angular distortion for RC framed or reinforced brick structures

Angular distortion	Behaviors of buildings
1/500	Non-structural damage (factor of safety included)
1/300	Non-structural damage, such as cracks, occur on panel walls
1/150	Structural damage

Figure 11.4 Failure patterns of unreinforced load bearing walls: (a) sagging and (b) hogging.

Table 11.3 Allowable deflection ratio for non-strengthened non-reinforced brick structures (Burland and Wroth, 1974)

Type of deformation	Allowable flexibility ratio (Δ/L)		
Sagging	1/2,500	for	L/H = 1
	1/1,250	for	L/H = 5
Hogging	1/5,000	for	L/H = 1
	1/2,500	for	L/H = 5

Note: L and H represent the length and height of a building respectively.

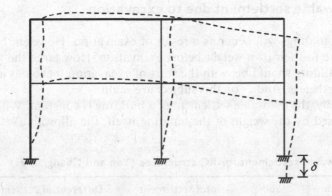

Figure 11.5 Differential settlement of a structure with individual footings.

under the effect of building weight. If non-structural damage is not generated, according to Table 11.1 or 11.2, the building would have to meet the following criterion:

$$\frac{\delta_{ij}}{L_{ij}} \leq \frac{1}{300} \tag{11.3}$$

where L_{ij} is the distance between the two columns and δ_{ij} is the differential settlement between the two columns.

In Europe and America, in most cases, the typical distance between two columns is about 20 feet (6 meters). Thus, the above equation can be rewritten as

$$\delta_{ij} \leq \left(\frac{1}{300}\right) \times 20 \text{ ft} \approx \frac{3}{4} \text{ in} \approx 2.0 \text{ cm} \tag{11.4}$$

Since differential settlement is not easily measured, by experience, the differential settlement in sandy soils is about 3/4 of the total settlement (including the required safety factor). Then,

$$\rho_i \approx \frac{4}{3}\delta_{ij} \approx 1.0 \text{ in} = 2.5 \text{ cm} \tag{11.5}$$

This is why the allowable settlement of individual footings in sand in foundation engineering books or building codes is usually set at 2.5 cm.

Settlement of buildings on clayey soils is usually more uniform. Thus, Eq. 11.5 is no longer applicable, though Eqs. 11.3 and 11.4 are still valid. Therefore, it is necessary to derive the allowable settlement of buildings on clayey soils under the effect of building weight in terms of the total settlement.

Table 11.4 summarizes the values of the allowable total settlement and the differential settlement proposed by some investigators and national professional societies (Skempton and McDonald, 1957; JSA, 1988; Terzaghi and Peck, 1967; Grant et al., 1974). The total settlements listed in Table 11.4 are basically exclusively applicable to buildings with a column distance of 6.0 m. For those with column distance longer or shorter than 6.0 m, the total settlements listed in Table 11.4 is not applicable.

11.2.2 Allowable settlement due to excavation

Settlement of buildings will occur as a result of excavation. However, buildings may have settled due to their own weight before excavation. How large the allowable settlement of a building should be, with the start of excavation, becomes a complicated problem. Nevertheless, studies on this subject are scant.

Theoretically, the allowable settlement of a building is constant. With the existing settlement caused by the weight of the building itself, the allowable settlement after

Table 11.4 Allowable settlement for RC structures (Yen and Chang, 1991)

Type of foundation	Soil	Total settlement (cm)	Differential settlement (cm)	Note
Individual footings	Sand	2.5	2.0	TP
		5.0	3.0	SM
		3.0	–	J
Individual footings	Clay	7.5	–	SM
		10.0	–	J
Mat foundation	Sand	5.0	2.0	TP
		5.0–7.5	3.0	SM
		6.0–8.0	–	J
		–	3.0	G
Mat foundation	Clay	7.5~12.5	4.5	SM
		20.0–30.0	–	J
		–	5.6	G

Note: TP- Terzaghi and Peck (1967); SM- Skempton and MacDonald (1957), corresponding to the angular distortion of 1/300; J- JSA (1988); G-Grant et al., (1974), corresponding to the angular distortion of 1/300.

excavation is started will not be large. The amounts should be smaller than those listed in Tables 11.1–11.4. Nevertheless, some investigators believe that although the inherent weight of buildings has induced some settlement, the members of buildings will gradually conform to the settlement and adjust their loading capability over a long period of time. As a result, their allowable settlement will not be too small and should not be far from the proposed values listed in Tables 11.1–11.4.

Whether a member of a building will crack should be related to the bending moment on it, that is, angular distortion or relative deflection. The allowable values listed in Tables 11.1–11.3 are expressed in angular distortion or relative deflection. Therefore, although the mechanism of damage caused by excavation-induced settlement is different from that of damage caused by building weight-induced settlement, the values in Tables 11.1–11.3 are still applicable and useful for evaluating the influence of excavation on adjacent buildings.

According to the above elucidation and the analyses of actual measured settlement of 42 buildings located next to excavation sites in Taipei (Yen and Chang, 1991), the author suggests that for RC framed buildings with individual footings on all kinds of soils, the allowable total settlement and the allowable differential settlement should be in the range of 2.5–5.0 cm and 2.0–3.0 cm, respectively. As for the mat foundation, the angular distortion and differential settlement are recommended instead of total settlement.

Note that buildings with mat foundations are mostly high-rise and most mat foundations are quite thick. With the same thickness of mat foundation, the smaller the mat area, the higher the stiffness. Under such a condition, the value measured by a tiltmeter is due to mostly rigid rotation. Thus, with no significantly observable distortion of building frames or cracks on panel walls, a building can still be observed inclining (refer to Section 11.4.4). The higher the building, the more obvious the inclination. Such inclination is not only unacceptable for the residents, and an inclining building is dubious in its earthquake resistance. In other words, the allowable inclination of a high-rise building is to be evaluated from the relative degree of inclination. For readers' reference, the allowable settlement induced by excavation proposed by the Taipei Rapid Transit System (TRTS) is shown in Table 11.5.

According to the literature (Son and Cording, 2005; Boscardin and Cording, 1989), during excavation, the soil behind the wall will not only settle, but also move laterally. The building damage is related to both settlement and lateral movement. Figure 11.6 shows the relationship between the damage of the brick building and the angular distortion and lateral strain (Son and Cording, 2005); the damage is categorized into negligible (NEGL) and very slight (VSL), slight (slight), moderate, severe, and very severe.

Table 11.5 Allowable settlement of buildings recommended by TRTS

Type of foundation	Maximum settlement (mm)	Tilt angle	Angular distortion	Deflection ratio (hogging)	Deflection ratio (sagging)
RC raft foundation	45	1/500	1/500	0.0008	0.0012
RC individual footings	50	1/500	1/500	0.0006	0.0008
Brick individual footings	25	1/500	1/2500	0.0002	0.0004
Temporary buildings	40	1/500	1/500	0.0008	0.0012

Note: Tilt angles are the measurement value on the tiltmeter (Section 12.4.2).

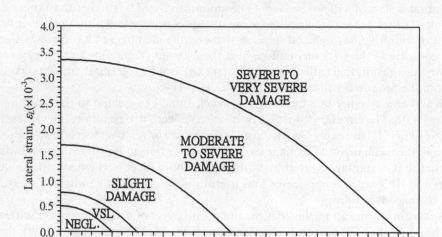

Figure 11.6 Relationship between the damage of brick buildings and the angular variable and lateral strain.

Note that the boundary line between "slight" and "moderate to severe" is consistent with the phenomenon of "first cracking of panel walls," as listed in Table 11.1.

A frame building with individual footings (Figure 11.7) has high lateral stiffness and relatively low vertical stiffness, and even exhibits a flexible stiffness behavior. Figure 11.8a shows the lateral displacements in the free field (without building case), at the footing, and at the grade beam caused by adjacent excavation obtained by the three-dimensional finite element method. The figure also shows that the adjacent excavation will produce different magnitudes of lateral soil displacement at various

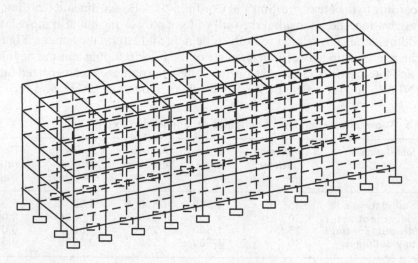

Figure 11.7 Frame structures with individual foundation.

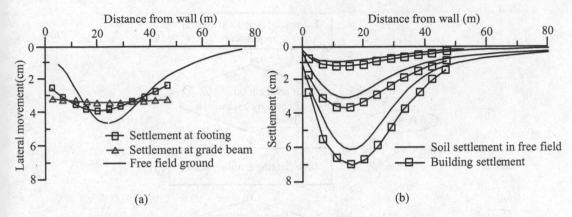

Figure 11.8 Movements of a frame structure and soil due to excavation: (a) lateral displacement and (b) settlement (Phuoc, 2014).

distances in the free field. The excavation will induce quite similar displacements at the building footing, especially at the location of the grade beam. It can be inferred that the lateral strain of the building is quite small, which can be regarded as 0. Figure 11.8b shows the comparison of surface settlement induced by excavation at the initial, intermediate, and final stages in the free field and those with building. The analysis results show that when the weight of the building is not considered, the settlement of the building is almost the same as the settlement in the free field; when the weight of the building is considered, the settlement of the building will be slightly larger than that in the free field (Phuoc, 2014).

The lateral stiffness of brick buildings is low, so the lateral displacement of the building is close to that of the soil in the free field, leading to the low lateral strain. Therefore, the lateral strain must be considered when assessing the damage caused by excavation. Most of the buildings on the raft foundation are frame structures with low lateral strain. The vertical stiffness is not only affected by the inner wall material, but also affected by the thickness of the raft foundation. If the effect of the internal brick wall is considered, the vertical stiffness of the building is equal to the sum of the stiffness of the brick building and the raft foundation.

11.3 ASSESSMENT OF BUILDING SAFETY AND DESIGN OF PROTECTION MEASURES

From the discussion in Section 11.2, it is known that the primary task of the protection of adjacent buildings in excavation projects is to predict the ground settlement caused by excavation, investigate the existing condition of the adjacent buildings, determine the allowable settlement of the adjacent buildings, and then evaluate the damage to the adjacent buildings. Finally, measures should be designed to protect adjacent buildings by controlling the movement induced by excavation. The detailed program is shown in Figure 11.9.

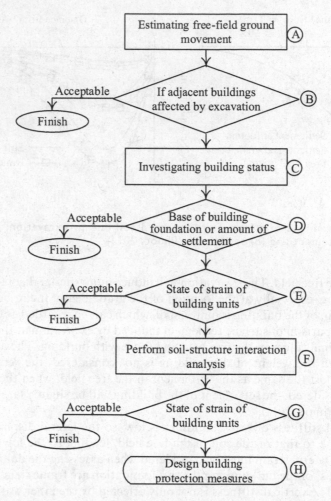

Figure 11.9 Procedure for assessment of the damage of an adjacent structure induced by excavation.

A. Prediction of ground settlement (free field) caused by excavation: The prediction method can be empirical method or finite element method.

B. Analyze and judge the extent to which adjacent buildings are affected by excavation: if adjacent buildings are outside the influence range of the excavation, or the angle distortion of the soil at the location of adjacent buildings is less than 1/500, or the maximum ground settlement is less than 10 mm (Son and Cording, 2005), then the safety of adjacent buildings can be ensured; otherwise, the next stage of assessment should be carried out.

C. Conduct a survey of existing adjacent building conditions, including the distance to the excavation site, type of foundation of adjacent buildings, depth of foundations, type of buildings and the materials (including partition wall materials).

D. Determine if the adjacent building is within the zone of excavation-induced settlement where the depth of building foundations is also above the excavation bottom. If it is, then proceed to the next stage of assessment.

E. Calculate the angular distortion and lateral strain of the adjacent building unit according to the computed settlement curve in the free field, that is, estimate the state of strain of the adjacent building unit, and then evaluate the damage of the building according to Figure 11.6. If the damage is serious, proceed to the next stage of assessment.

F. Include the buildings in the finite element mesh with consideration of the stiffness of adjacent buildings, and then perform the analysis, the so-called soil-structure interaction analysis. The angular distortion and lateral strain of the building units should be smaller than the state in the free field. The soil-structure interaction analysis can also be conducted by the semi-empirical method, and readers can refer to related literatures.

G. Calculate the angular distortion and lateral strain of the adjacent building units, that is, estimate the strain state of the adjacent building units, and then evaluate the damage of the building according to Figure 11.6.

H. Design building protection measures according to the assessment in the previous stage if the adjacent buildings expect to be damaged seriously.

11.4 ADJUSTMENT OF CONSTRUCTION SEQUENCE

11.4.1 Reduce the unsupported length of the retaining wall

Figure 11.10a and b are identical in the final excavation depth, the number of excavation stages, the number of strut levels, and the location of struts. The only difference between them is the distance between the location of strut and the excavation surface. In Figure 11.10a, each level of struts is installed 0.5 m above the excavation surface. When the first stage of excavation is completed, the unsupported length of the wall is 2.0 m. When the first level of struts is installed at the depth of GL-1.5 m and excavation

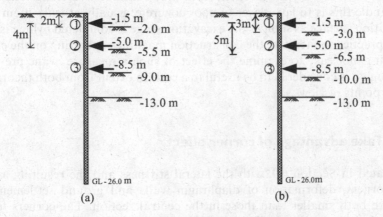

Figure 11.10 Distance between the struts and the excavation: (a) 0.5 m and (b) 1.5 m.

proceeds to GL-5.5m, the unsupported length of the wall is 4.0m. When the second level of struts is installed and excavation proceeds to the third stage, the unsupported length of the wall is also 4.0m. With the completion of the third level of struts, the unsupported length is 4.5m. However, in Figure 11.10b, each level of struts is installed 1.5m above the excavation surface. Evaluated similarly, the unsupported lengths of the wall at the second and the third stages are both 5.0m. The unsupported length after completion of the final stage of excavation is 4.5m.

As elucidated above, except for the last stage of excavation where the unsupported lengths in the two cases are identical, the unsupported lengths illustrated in Figure 11.10a are smaller than those in Figure 11.10b at other stages of excavation, with the same earth pressure on the back of the retaining wall. Actually, the total deformation of the retaining wall or ground settlement is the accumulated deformation at every stage of excavation. Thus, the deformation in Figure 11.10a should be smaller than that in Figure 11.10b. That is to say, the location of struts should be as close to the excavation surface as possible. Generally, considering the convenience of strut installation, the distance between struts and the excavation surface is about 0.5m.

11.4.2 Decrease the influence of creep

The definition of creep is given as the condition where deformation increases with time, provided the stress remains constant. Creep usually occurs in clayey soils. The softer the soils, the more obvious the characteristics of creep. Creep is related to time and stress. It increases with the increase in the stress level and time. Since soils near the retaining wall and the excavation surface may be close to failure condition, the stress level must be high. As a result, soils with prominent characteristics of creep are susceptible to a large settlement. As for the influence of creep of soils on excavation, please refer to Section 6.6.

To prevent creep from occurring, especially when excavating in soft clayey soils, struts have to be installed as soon as each stage of excavation is completed. Strut installation usually takes a few days in braced excavations. If the top-down construction is adopted, it takes a few weeks to construct floor slabs. During the process of strut installation or floor slab construction, creep may continue worsening. The expedient way to handle this is to lay a layer of poor concrete, usually at least 10cm thick, on the excavation surface as soon as the excavation stage is completed (which is also necessary in practice to facilitate the construction machines operating on the excavation surface). It is not easy to examine the effect of such a measure in the prevention of creep though it is considered to be useful to a certain extent from both theoretical and empirical points of views.

11.4.3 Take advantage of corner effect

As elucidated in Section 6.9, with the lateral stiffness and the resulting arch effect around corners, deformation of diaphragm walls and ground settlement around corners are both smaller than those in the central section. The corners for soldier piles, sheet piles, and bored piles, which are not continuous in the horizontal direction and have much less lateral stiffness, accordingly, do not have much difference in deformation or ground settlement from the central section. Thus, if the building is located at a corner or on the shorter side (see Figure 6.32), diaphragm walls can be adopted to take advantage of corner effects for the protection of buildings.

11.4.4 Take advantage of the characteristics of ground settlement

If ground settlement can be accurately predicted or observed, the characteristics of ground settlement can also be utilized for building protection. The TNEC case history will be used for illustration. For the geological conditions, excavation process and monitoring results of TNEC, please refer to Appendix B in this book.

Figure 3.35a diagrams the positions of the excavation site and Building A, which is a 12-floor building with one level of basement. The building is about 40m high with a basement of 6m depth. It adopts mat foundations, which covers an area of about 16m × 17m. As shown in the figure, the building is right next to the west side of the excavation site, separated from the retaining wall by only 1.5m. During trench excavation and diaphragm wall construction, there appeared several cracks on the ground surface in the vicinity of the building. Therefore, the contractor installed two tiltmeters, B-1 and B-2, on the building.

When the construction of the diaphragm wall was finished, the measurements of B-1 and B-2 showed that Building A had inclined east by 1/838. To reduce settlement, the contractor improved the soils below the mat foundation with jet grouting of 35cm in diameter in hope of reducing movement. Figure 11.11a and b illustrate the plan and profile of jet grouting, respectively. The jet grouting started from corner **a** on the northeast

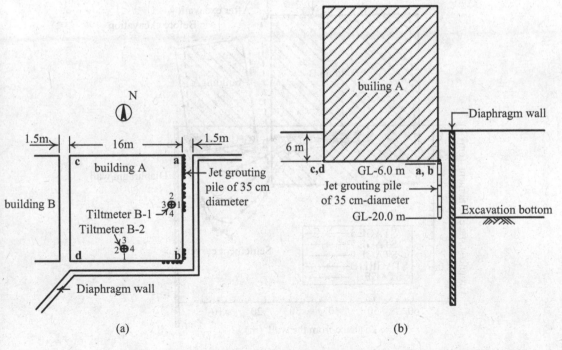

Figure 11.11 Arrangement of grouting at the building in the vicinity of the TNEC excavation: (a) plan and (b) profile.

side of the building and was implemented at every other pile. The depths ranged from GL-6 to GL-20 m. Tiltmeters were strictly monitored during the process of jet grouting.

Unfortunately, the plan of jet grouting failed since the inclination of the building worsened. According to the measurements from B-1 and B-2, the building continued to incline toward the excavation site (eastward) during jet grouting, as shown in Figure 11.12. One of the results of B-1 showed that the maximum angle of inclination achieved 1/491 toward the excavation zone, while Building A did not show any cracking on the panel walls.

Though the angle of inclination 1/491 for Building A was not a large amount, nor did the panel walls in the building begin cracking, it could be observed to incline visually. This is because Building A covered a relatively small area compared to its height and was very close next to Building B, which was little affected by excavation and therefore was implicitly seen as a vertical benchmark.

The reason for the inclination of the building was either insufficient control of the implementation of jet grouting or the grouting-induced disturbance, which made the soil strength weaken, according to the preliminary judgment. Right during the process of grouting when the grout materials (cement grouts) had not congealed, the inclination of the building continued worsening with the execution of jet grouting. Though the building could be observed to incline visually, the interior was kept intact. The reason might be that the mat foundations, which covered a relatively small area, had great rigidity. As a result, the inclination was basically a rigid rotation.

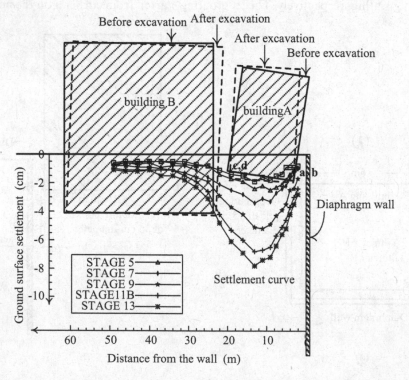

Figure 11.12 Tilting conditions of the adjacent buildings in the TNEC excavation.

With the continuation of excavation, the angle of inclination of Building A, however, began declining. When the excavation was finished, the readings on B-1 and B-2 were separately 1/1,375 and 1/2,864. The phenomenon of the declining angle of inclination of Building A can be explained by the shape of the settlement profile. As shown in Figure 11.11, though no settlement was observed on a section perpendicular to the line segment $\overline{ab}$ in Figure 11.11a, it can be assumed to be similar to the settlement profile of the main observation section (note: main observation section can refer to Figure 3.34a). Plot the settlement profile of the main observation section (Figure 6.10) and the relative positions of Buildings A and B on Figure 11.12. As shown in the figure, the settlements at corners **a** and **b** of the building were smaller than those at corners **c** and **d** during excavation. Therefore, as the excavation went deeper, the buildings that have been inclined toward the excavation area were gradually pulled back, and the inclination angle was gradually reduced.

From the above discussion on TNEC, it follows that an accurate prediction of settlement during the process of excavation will help in building protection. The characteristics of ground surface settlement can be predicted by the way of the simplified method (Section 6.8) or the finite element method. They can also be derived from measurements.

11.5 STRENGTHENING THE STRUT-RETAINING SYSTEM

Theoretically, the increase of the stiffness of a retaining wall can reduce the wall deformation and the ground settlement. The methods of increasing the system stiffness of the retaining-strut system include increasing the stiffness or thickness of the retaining wall, decreasing the horizontal spacing of struts, increasing the stiffness of each strut, and decreasing the vertical spacing of struts, etc. As shown in Figure 6.6, the increase in thickness of a retaining wall does not help much in decreasing deformation as long as the wall thickness is sufficient to maintain its design requirement; for example, the wall thickness is increased from $t_w = 0.9$–1.3 m.

When the axial stiffness of the strut is not large enough, increasing the stiffness of struts (such as by cutting the horizontal spacing or increasing the stiffness of each strut) can effectively reduce the deformation of a retaining wall. However, when the axial stiffness of struts is already rather large, more increase in stiffness will not reduce the deformation accordingly, the reason for which is as shown in Section 6.3.6.

Decreasing the vertical spacing of struts is an effective way of reducing the deformation of a retaining wall. The reason is that cutting the vertical spacing implies the increase in number of strut levels. That is to say, the rigidity of the retaining system is increased. Deformation will decrease as a result. From another perspective, since the deformation of a retaining wall is the accumulated result of the excavation stages, the cutting of the vertical spacing will decrease the unsupported length of the retaining wall at each excavation stage (see Section 11.4.1). Therefore, deformation will also decrease.

Some engineers wrongly believe that increasing the penetration depth of a retaining wall can also help decrease the deformation. Actually, as explicated in Section 6.3.4 and by Figure 6.5, when the penetration depth has gone deeper than the depth required to maintain the stability of the retaining wall, more increase in penetration depth won't help prevent deformation. Thus, increasing the penetration depth cannot contribute to building protection.

11.6 SOIL IMPROVEMENT

Soil improvement is the most commonly used measure to control the ground movement in excavations. This section will introduce the soil improvement method and its analysis and design in excavations.

11.6.1 Soil improvement methods

The chemical grouting method, the deep mixing method, and the jet grouting method are the three main methods of soil improvement in excavations. In the following sections will be introduced their methods, analysis, and design.

11.6.1.1 Chemical grouting method

The chemical grouting method is to inject the grouting material into soil strata by low pressure. With the condensation of the grouting material, the soils will be strengthened and the permeability and settlement be decreased. The grouting pressure for the chemical grouting method is usually lower than 2 MPa, and the method is, therefore, also called the low-pressure grouting method. Figure 11.13 diagrams the implementation of the chemical grouting method (the single packer method in the example).

Various materials can be used for grouting, such as water glass (sodium silicate solution), bentonite, bentonite mixed with cement, etc., depending on the purpose of grouting and the properties of rock and soil. When chemical grouting is used in sandy soil with a pressure lower than 1 MPa, the grout will permeate into the voids of the sand and achieve a good effect of soil improvement without destroying the soil structure. Basically, chemical grouting is not applicable to clayey soil because of its low permeability.

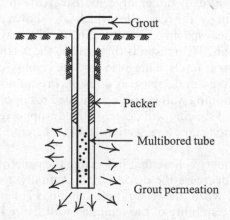

Figure 11.13 Schematic diagram of chemical grouting (single packer method).

11.6.1.2   Jet grouting method

The jet grouting method implants a grouting pipe, with a jet, into the soil to a certain depth with the help of a boring machine and emits high-pressure grout or water (about 20 MPa), pounding and cutting soils at the same time. When the pulsing current, at high pressure and speed, exceeds the strength of soils, the soil particles will be separated from the soil body. Some fine soil particles will flow with water or grout out of the ground, while other soil particles will mix with the grout, under the influence of pounding, centrifugal force, and gravity. The mixed soil particles will be rearranged into a grout-soil mixture. When the grout is congealed, it forms a solid body. Figure 11.14 diagrams the sequence of jet grouting.

Some commonly used methods of jet grouting are the single-tube method, the double-tube method, and the triple-tube method, as shown in Figure 11.15. The single-tube method is to implant the boring rod with a special jet fixed on the side of

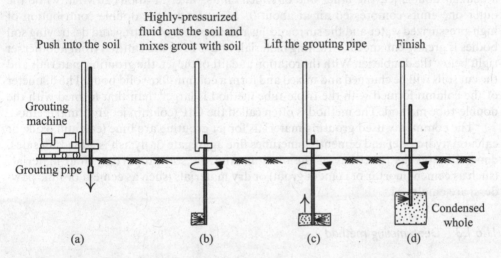

Figure 11.14 Procedure of the jet grouting method.

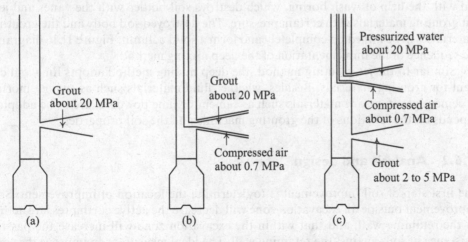

Figure 11.15 Types of the jet grouting method: (a) single-tube method, (b) double-tube method, and (c) triple-tube method.

the rod bottom with the help of the boring machine and emit pressurized grout (about 20 MPa) into soil to cut and destroy soil bodies. With the rotation and lift of the rod, the grout and the cut soils will be churned and mixed. After a certain period, the whole of the grout and soil will congeal and form a column.

As shown in Figure 11.15b, the configuration of the double tube is a coaxial double jet. The inner jet emits grouting materials at 20 MPa, while the outer one pumps compressed air at 0.7 MPa. Under the double effect of jetting grout and the surrounding air pressure, the capability to destroy and cut soil is highly enhanced. With the rotational jetting and lift of the double jet, grouting materials and cut soils are churned and mixed and form a column-like solid body together. The diameter of the column formed with the double-tube method is larger than that formed with the single-tube method. The method is often called JSG (Jumbo Special Grouting) method, or JSP (Jumbo Special Pile) method.

The triple tube is a jet that emits water, air, and grouting materials simultaneously. As shown in Figure 11.15c, on the side of the bottom of the grouting pipe is attached a coaxial double jet, the inner one of which emits water at about 20 MPa, while the outer one emits compressed air at about 0.7 MPa. Under the double contribution of high-pressurized water and the surrounding air, the power of cutting and destroying soil bodies is greatly enhanced. Grouting materials of 2–5 MPa are emitted from another jet right below the double jet. With the rotation and lift of the jet, the grouting materials and the cut soils will be churned and mixed and form a column-like solid body. The diameter of the column formed with the triple-tube method is larger than that formed with the double-tube method. The method is often called the CJP (column jet grouting) method.

The commonly used grouting materials for jet grouting are lime (calcium oxide or calcium hydroxide) and cement. Sometimes fine aggregate or fly ash is added. Besides, depending on the functions of the machine and the soil properties, fluid materials (such as cement mortar or cement grout) or dry materials (such as cement or lime powders) are employed.

11.6.1.3 Deep mixing method

The deep mixing method, DMM for short, installs mixing vanes connected to a hollow rod with the help of wash boring, which destroys soil bodies with the vanes and jets out grouting materials with certain pressure. The destroyed soil body and the grouting materials will be mixed up completely and form a solid column. Figure 11.16 diagrams the sequence of the implementation of the deep mixing method.

Similar to the jet grouting method, the deep mixing method adopts lime and cement for grouting materials. Besides, whether fluid materials such as cement mortar or cement grout, or dry materials such as cement or lime powders, are to be adopted depends on the functions of the grouting machine and the soil properties.

11.6.2 Analysis and design

The first step of soil improvement is to determine the location of improvement. Soil improvement outside an excavation zone will decrease the active earth pressure acting on the retaining wall, and that within the excavation zone will increase the passive resistance of soils against the retaining wall. The ideal measure is to improve the soils

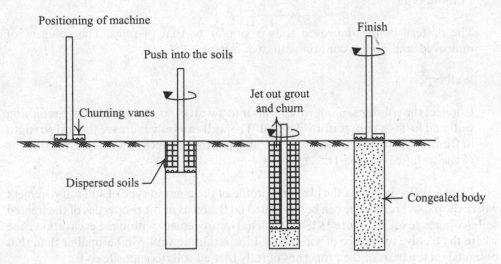

Positioning of machine

Push into the soils

Jet out grout and churn

Finish

Churning vanes

Dispersed soils

Congealed body

Figure 11.16 Procedure of the deep mixing method.

both inside and outside the excavation zone. Nevertheless, the cost may be too high. According to finite element parametric study results (Ou and Wu, 1990), the effects within the excavation zone are better than those outside, given the same conditions. Judged from mechanisms, the soil improvement inside the excavation zone directly resists the movement of the retaining wall.

Some commonly used arrangements, including the block type, the column type, and the wall type, are shown in Figure 11.17. The arrangements are elucidated as follows.

1. Block type

Within a specific area, improve the soils fully. Replace the soil bodies within the area completely or have them completely combined with chemicals into treated soils.

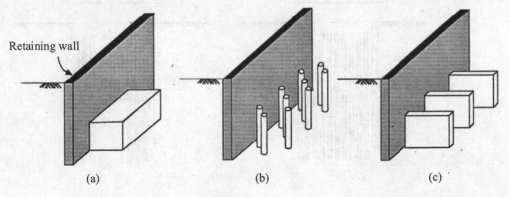

Retaining wall

(a) (b) (c)

Figure 11.17 Typical arrangement of soil improvement in excavations: (a) block type, (b) column type, and (c) wall type.

2. Column type

The pattern of the improved soils is similar to that of piles. The columns of improved soils do not connect with each other.

3. Wall type

Connect the columns of improved soils into a wall shape, which joins the retaining wall and forms a counterfort-like wall. The wall can only increase the soil strength in front of the retaining wall. It is not able to raise the moment-resistance stiffness of the wall (see Section 11.8.1).

Figure 11.18a diagrams the plan and profile of the column type of soil improvement where the passive resistance can be computed on the basis of the properties of the treated soils (composite soils). Figure 11.18b is a partial improvement within the excavation zone where the passive resistance of soils against the retaining wall will be smaller than that obtained on the basis of the properties of fully treated soils (composite soils).

To ensure soil improvement capable of building protection, improvement should be analyzed in terms of the strength of the treated soil, its diameter, spacing, depth, location, and range. The soils within the area can be viewed as a composite material in analysis, and its material properties should be evaluated reasonably. If the Mohr-Coulomb model is used for analysis, four parameters are required, namely, Young's modulus (E), Poisson's ratio (μ), and strength parameters (c, ϕ). According to research (Ou et al., 1996), the Young's modulus and Poisson's ratio of composite

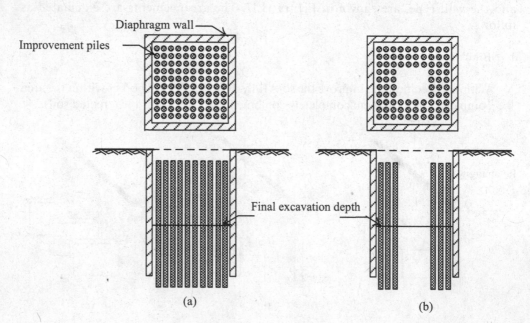

Figure 11.18 Soil improvement within excavation zone: (a) improvement over the whole zone and (b) partial improvement.

soil can be calculated using the weighted average of the area of treated soil and untreated soil, as follows:

$$P_{eq} = P_t I_r + P_s (1 - I_r) \tag{11.6}$$

where P_{eq} = equivalent Young's modulus or Poisson's ratio of the composite material
P_t and P_s = Young's modulus or Poisson's ratio of the treated and untreated soil, respectively
I_r = improvement ratio, that is, the area of the treated soil divided by the total area

When the composite soil body bears the vertical load (see Figure 11.19a), according to the principle of force equilibrium, the equivalent undrained shear strength ($s_{u,eq}$) of the composite soil can be written as

$$s_{u,eq} = s_{u,t} I_r + s_{u,s} (1 - I_r) \tag{11.7}$$

where $s_{u,t}$ and $s_{u,s}$ are the undrained shear strength (compressive) of the treated soil and untreated soil, respectively.

The above equation is applicable to an area with loading condition similar to Figure 11.19a, such as the area outside the excavation zone. The compressive strengths of the treated and untreated soils are used for such a condition. However, the area inside the excavation zone is mainly subjected to the vertical unloading or extension (see Figure 11.19b). The tensile shear strength of the treated and untreated soils should be used instead of compressive strength. However, in engineering practice, compressive strength tests are often carried out rather than extension tests. Therefore, some engineers prefer to adopt the following formula to estimate the strength of the composite material for the area inside the excavation.

$$s_{u,eq} = \alpha s_{u,t} I_r + s_{u,s} (1 - I_r) \tag{11.8}$$

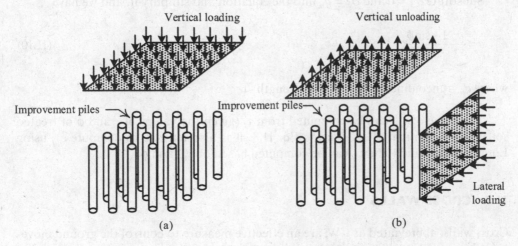

Figure 11.19 Loading on improved soil area: (a) vertical loading and (b) vertical unloading or lateral loading.

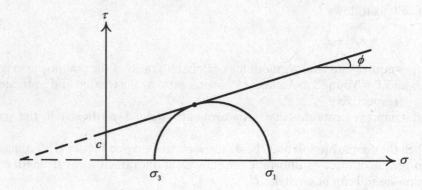

Figure 11.20 Mohr's circle and the envelope for the failure condition of the treated soil.

where $s_{u,t}$ is the undrained shear strength of the treated soil obtained from the compression test and $s_{u,s}$ is the undrained shear strength of the untreated soil; α is the modification factor, around 0.3–0.35, which is about equal to the ratio of tensile strength to compressive strength of the treated soil.

Besides, according to Clough and Tan's (1980) study, if the untreated soils are granular soils (such as sand, gravel), the friction angle of the treated soil will be close to that of the untreated soil. Assuming the friction angle for the treated soil is ϕ, according to the Mohr's circle representing the failure of the treated soil in Figure 11.20, we have

$$\frac{\sigma_1 - \sigma_3}{2} = \left(c \cot\phi + \frac{\sigma_1 + \sigma_3}{2} \right) \sin\phi \qquad (11.9)$$

Substitute $\sigma_3 = 0$ and $\sigma_1 = q_u$ into the equation and simplify it, and we have

$$c = \frac{q_u}{2} \frac{1 - \sin\phi}{\cos\phi} \qquad (11.10)$$

where q_u = unconfined compression strength

The value of τ_{eq} can be computed from τ_t (given the values of c and ϕ of treated soils) and τ_s (untreated soils) using Eq. 11.8. It is also feasible to compute c_{eq} using Eq. 11.7 or 11.8, and then τ_{eq} can be computed based on c_{eq} and ϕ.

11.7 CROSS WALLS

Cross walls, abbreviated as CW, are an effective measure to control the ground movement in excavations. This section will introduce the mechanism, analysis, and design and construction.

11.7.1 Mechanism

The arrangement of cross walls is schematically shown in Figure 11.21a and b. One constructs a wall connecting the opposing two retaining walls before excavation. It can be constructed using soil improvement techniques (such as jet grouting or deep mixing). To obtain a better construction quality or compressive strength, concrete wall can be constructed using a diaphragm wall construction technology (the unconfined compression strengths of treated soils are usually between 1 and 2 MPa, whereas that of concrete wall can achieve 28 MPa with a minimum of 10 MPa). The mechanism of the cross wall for reducing wall deformation is quite different from that of soil improvement. The designing principle of soil improvement is to enhance the passive resistance of the soils in front of the wall. The cross wall should be viewed it as a lateral support or strut, which exists before excavation and bears a great deal of compression strength (especially concrete walls). Theoretically, the locations where cross walls have been placed are less susceptible to deformation because they are restrained from moving. The effects of the cross wall on reducing lateral deformation of a retaining wall are quite obvious.

Figure 11.21a is the plan of an excavation case located in Taipei. The excavation was about 76 m long, 25 m wide, and 13.5 m deep. The retaining wall was a 70-cm-thick, 30-m-deep diaphragm wall. To protect buildings in the vicinity, two cross walls were constructed. They were constructed in the same way as the diaphragm wall except

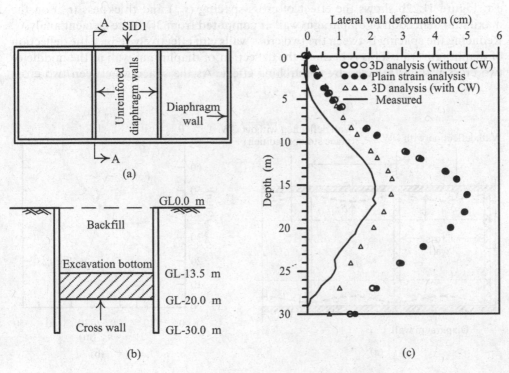

Figure 11.21 A case study for excavation with cross walls: (a) plan, (b) profile, and (c) comparison of measured and computed wall deformations.

not reinforced. The thickness of the cross walls was the same as the diaphragm wall. It was constructed from GL-13.5 m to GL-20 m. Soils were backfilled above the cross walls (GL 0.0 to GL-13.5 m), as shown in Figure11.21b. Note the rigid joint as shown in Figure 3.24a is adopted between the diaphragm walls in this project.

Figure 11.21c compares the measured values with the computed values using the three-dimensional finite element method for the wall deformation in the central section of the longer side of the excavation site (SIDI in Figure 11.21a) at the last excavation stage. To understand the effects of the cross walls on reducing the deformation of the diaphragm wall, the conditions without cross walls were also analyzed using the three-dimensional and the plane strain finite element methods. The analysis results indicated that the plane strain analysis yields almost identical results with the three-dimensional analysis for the condition without cross walls. The results from three-dimensional analyses were very close to those from field measurements. The cross walls were effective in reducing the deformation of the diaphragm wall. The three-dimensional finite element method is useful in simulating wall deformation with cross walls, too.

11.7.2 Performance, analysis, and design

Figure 11.22a shows the maximum deflection of diaphragm wall for an excavation with cross walls as compared with those without cross walls. The deflection of diaphragm wall is smallest at the cross wall section, while that between the cross walls is the largest. Figure 11.22b shows the effect of cross spacing (s_{cw}) and thickness (t_{cw}) on the maximum deflection of diaphragm wall as computed from 3D finite element analysis. Reducing the spacing between the two cross walls can effectively reduce the deflection of diaphragm wall. This is because the deflection of diaphragm wall at the middle of two cross walls is affected by the arching effect. As the spacing between two cross

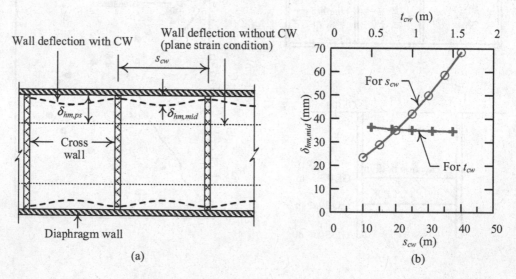

Figure 11.22 Relationship between the maximum deformation of the retaining wall and various CW parameters.

walls is reduced, the arching effect that affects the deflection of diaphragm wall at each section between two cross walls can reduce the wall deflection significantly. On the other hand, it is found that reduction of the thickness of cross walls has a little effect on the deflection of diaphragm wall. As long as the compressive force acting on the cross walls does not exceed the compressive strength of the concrete of cross walls, the thickness of cross walls does not affect the deformation of diaphragm wall.

Figure 11.23 shows the deflection of diaphragm wall at the middle section of two cross walls for four different scenarios of cross wall depths computed from 3D finite element analysis. Compared with no cross wall case, installation of cross walls obviously can reduce wall deflections. Of those scenarios, an excavation with full cross wall depth (the case of $H_{ch}/H_e = 1$ and $H_{ce}/H_p = 1$) yields the best reduction in wall deflections of diaphragm wall. Cross walls installed only above the excavation bottom (the case of $H_{ch}/H_e = 1$ and $H_{ce}/H_p = 0$) or only embedded below the excavation bottom (the case of $H_{ch}/H_e = 0$ and $H_{ce}/H_p = 1.0$) all exhibit a different deflection pattern and a larger deflection of diaphragm wall as compared with that of full cross wall depth. The wall deflections for the case of $H_{ch}/H_e = 0$ and $H_{ce}/H_p = 0.3$ are similar to those for the case of $H_{ch}/H_e = 0$ and $H_{ce}/H_p = 1.0$. To further reduce wall deflections, it is necessary to install a certain cross wall above the excavation bottom. Therefore, a scenario of $H_{ch}/H_e = 0.3$–0.5 and $H_{ce}/H_p = 0.3$–0.5 may be a good choice when both economy and safety are considered.

To ensure that cross walls can effectively reduce deformation of the diaphragm wall, the design of cross walls has to consider compression strength, depth, and spacing of two cross walls. Since the behavior of the diaphragm wall with cross walls is three dimensional, to design cross walls, one has to resort to the three-dimensional finite element method or successful case histories. In addition, the author and his

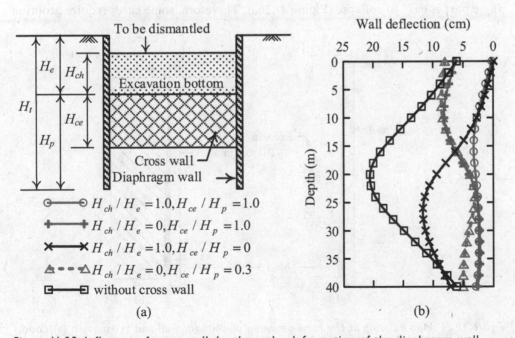

Figure 11.23 Influence of cross wall depth on the deformation of the diaphragm wall.

working team have studied the simplified method and two-dimensional finite element analysis method for the design of cross walls. Interested readers are advised to refer to the related literatures (Hsieh and Ou, 2016; Hsieh et al., 2012).

11.7.3 Construction

The cross wall must be closely connected with the main diaphragm wall in order to exert its effect. If the main diaphragm wall is constructed first, and then cross walls are constructed with the diaphragm wall construction technology such as excavators, there will exist mud between the main diaphragm wall and the cross wall (Figure 11.24). As excavation causes the diaphragm wall to move forward, the mud will be compressed before the cross wall is stressed, so the deflection of the diaphragm wall may be greater than expected. In order to make the cross wall and the main diaphragm wall tightly connected, there are two types of joints between the cross wall and the main diaphragm wall: end plate joint and T-shaped joint, as shown in Figures 11.25 and 11.26.

The end plate joint is to attach an end plate (steel plate) connected to the side of the reinforcement cage of the main diaphragm wall (Figure 11.25a). As the main diaphragm wall is completed, then excavate the trench of the cross wall, and then use a steel brush or other methods to remove the mud on the end plate, and then cast the Tremie concrete to complete the joint between the cross wall and the main diaphragm wall.

The T-joint is to excavate the trench of the main diaphragm wall and the cross wall at the same time, and then place the T-shaped steel cage in the trench, and then cast the Tremie concrete at the same time. The stability of the T-joint is poor, and the right-angle part is easy to collapse (Figure 11.26a). Therefore, some cases require grouting

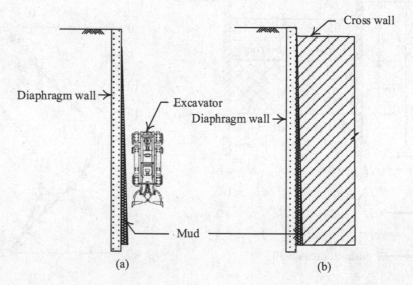

(a)   (b)

Figure 11.24 Mud existing at the joint between diaphragm wall and cross wall: (a) condition of trench excavation and (b) formation of mud.

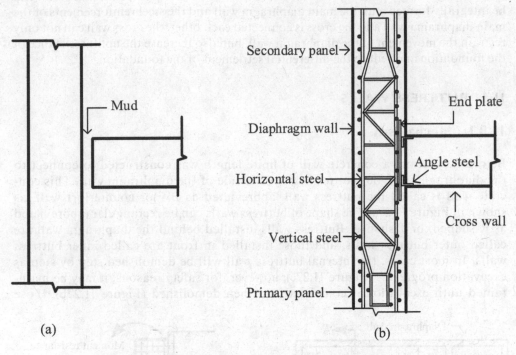

(a)

(b)

Figure 11.25 End plate joint: (a) formation of mud in joint and (b) reinforcement diagram of end plate joint.

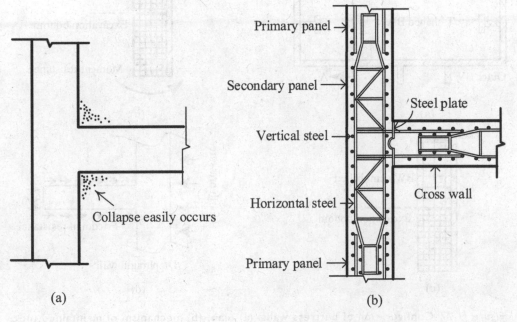

(a)

(b)

Figure 11.26 T-joint: (a) trench excavation of diaphragm wall and cross wall and (b) T-reinforcement cage.

near the right-angle part to avoid the collapse of the T-joint. Since the cross wall can be integrally formed with the main diaphragm wall and the steel reinforcements of the main diaphragm wall and the cross is connected each other, the cross wall can not only restrain the movement of the diaphragm wall, but also increase the uplift resistance of the foundation and reduce the differential settlement of the foundation.

11.8 BUTTRESS WALLS

11.8.1 Mechanism

Before excavation, a concrete wall of finite length was constructed to connect to the diaphragm wall, but not to the opposite side of the diaphragm wall. This concrete wall is called the buttress wall (abbreviated as BW)or counterfort wall, as shown in Figure 11.27a. The shape of buttress walls can be rectangular (abbreviated as R-shaped) or T-shaped. Buttress walls installed behind the diaphragm wall are called outer buttress walls, and those installed in front are called inner buttress walls. In most cases, the internal buttress wall will be demolished step by step as excavation progresses (Figure 11.27c); however, for safety reasons, it may be maintained until excavation is completed and then demolished (Figure 11.27b); those

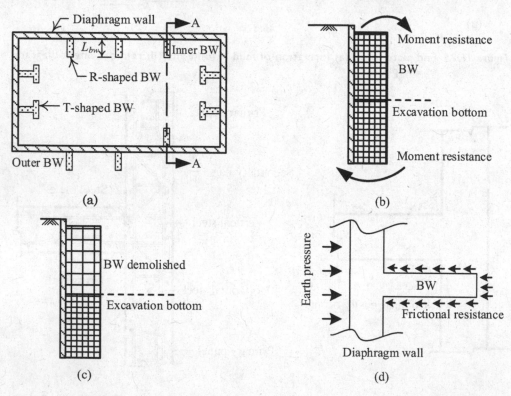

Figure 11.27 Configuration of buttress walls: (a) plan, (b) mechanism of maintained (not demolished) buttress walls, (c) buttress wall demolish, and (d) mechanism of demolished buttress walls.

that are not demolished can be used as permanent structures or partition walls. The outer buttress wall is located outside the excavation area, so it will not hinder construction activities.

If inner buttress walls above the excavation surface are demolished step by step as excavation progresses (Figure 11.27c), the main mechanism of buttress walls to restrain the lateral movement of the diaphragm wall is the friction between the buttress walls and the surrounding soil (Figure 11.27d). The friction should be mobilized by the relative displacement of the buttress walls and the surrounding soil. As excavation progresses, both the buttress walls and the soil in front of the diaphragm wall are pushed forward by the diaphragm wall and move forward. The two produce almost the same amount of displacement, resulting in little friction between the buttress walls and the surrounding soil. However, as the soil farther away from the wall, the compression or movement of the soil decreases while the buttress wall still maintains the same amount of the movement. The relative movement between the soil and the buttress wall gradually increases, and the friction force is gradually mobilized. Therefore, only the distal part of the buttress wall has the effect of restraining the movement of the diaphragm wall, while the effect of that near the diaphragm wall is quite poor.

The friction force can only be generated when the buttress and the soil have a relative displacement. If the deformation of the diaphragm wall is originally small, friction will be little generated, so the buttress wall has a low effect in restraining the deformation of the diaphragm wall. On the other hand, if the length of the buttress is not enough so that the friction force on the buttress is fully mobilized before the final excavation depth is reached, the remaining friction force on the buttress wall is too little to restrain the diaphragm wall from movement.

When outer buttress walls or inner buttress walls are cast integrally with the diaphragm wall and are not demolished with excavation of soil (Figure 11.27b), the stiffness from the composite section formed by the buttress wall and the diaphragm wall is a main source to restrain the diaphragm wall from the movement.

As the inner buttress wall is combined with the cross wall, it is called the U-shaped wall, abbreviated as UW (Figure 11.28). The UW takes advantage of strength of both buttress walls and cross walls, it can reduce the wall movement significantly. Interested readers are advised to refer to the literature (Lim et al., 2020).

Figure 11.29a is the plan for the construction of buttress walls of an excavation. The excavation was about 44m long, 42m wide, and 8.6m deep. The diaphragm wall was 0.6m thick with a penetration depth of 21m. Excavation was carried out at four stages. From the ground surface to GL-20.0m, there was soft clay. From GL-20.0 to GL-26.0m, there was silty sand. Below GL-26.0m, there was gravelly soil. To protect buildings in the vicinity of the excavation site, the T-shaped buttress walls, as shown in Figure 11.29a, were constructed along the 42m side.

Figure 11.29b illustrates the measured deformations of the diaphragm wall with (as shown as SI4 in Figure 11.29a) and without (as shown as SI6 in Figure 11.29a) buttress walls in the central section at the last excavation stage. The deformation of the retaining wall at SI4 was much smaller than that at SI6. Use of buttress walls was certainly effective in reducing wall deformation in this case.

In addition to the case history in Figure 11.29, more case studies regarding the effectiveness of buttress walls in excavations can be found in the literature (e.g., Hsieh and Ou, 2018; Lim et al., 2016; Hsieh et al., 2016).

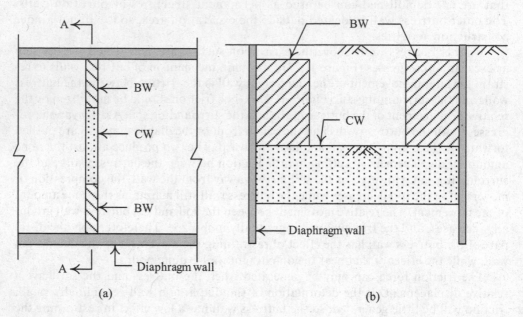

Figure 11.28 Configuration of U-shaped wall (UW): (a) plan and (b) profile.

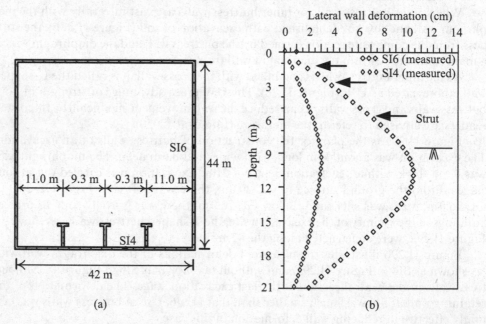

Figure 11.29 A case study of an excavation with buttress wall: (a) plan and (b) wall deformation measured at SI4 and SI6.

11.8.2 Performance, analysis, and design

Buttress walls have more complicated performance in reducing the deflection of the diaphragm wall than cross walls. The performance of the buttress wall is related to many factors, such as its length, thickness, spacing, demolished or maintained, shape, and penetrating into hard soil or not. This section will summarize the research results of the influence of the thickness, length, and spacing of buttress wall on the deformation of diaphragm walls.

Figure 11.30a shows the influence of the length of the buttress wall on the deformation of the diaphragm wall when the buttress wall is demolished step by step along with excavation (the spacing between the buttress wall is assumed 20 m), where $L_{bw} = 0$ implies that no buttress wall is installed. The figure shows that the deformation of the diaphragm wall decreases with the increase in the buttress wall length. When the buttress length wall (L_{bw}) is less than 2.5 m (inclusive), the deformation of the diaphragm wall is almost the same as that without the buttress; even when $L_{bw} = 5$ m, the deformation of the diaphragm wall is only slightly reduced. The main reason can refer to Figure 11.27d and literature (Hsieh and Ou, 2018).

Figure 11.30b shows the influence of diaphragm wall deformation on various buttress wall lengths when the buttress wall is not demolished with excavation. The figure shows that the deformation of the diaphragm wall decreases as the buttress wall

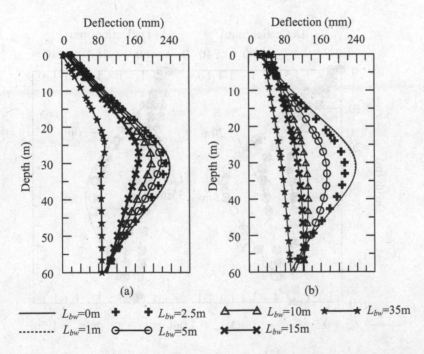

Figure 11.30 Influence of various buttress wall lengths on diaphragm wall deformation: (a) buttress wall demolished with excavation and (b) buttress wall maintained during excavation.

length increases. When $L_{bw} = 1\,m$, the deformation of the diaphragm wall is almost the same as that of the diaphragm wall without the buttress wall. When $L_{bw} = 5\,m$, the deformation of the diaphragm wall is smaller than that of demolished step by step with excavation (refer to Figure 11.30a), and the deformation of the diaphragm wall has been reduced to a considerable extent. When it is greater than 15 m, the diaphragm wall presents a linear and rigid deformation, and the entire continuous wall is pushed.

Figure 11.31a shows the influence of the thickness of the buttress on the deformation of the continuous wall when the buttress is demolished along with excavation. The figure shows that if the thickness of the buttress is increased, the deformation of the diaphragm wall is not affected in any way. This is because the main mechanism of the buttress wall in restraining the deformation of the diaphragm wall comes from the friction between the buttress wall and the surrounding soil, increasing the buttress wall thickness without increasing the friction. Figure 11.31b shows the results when the buttress wall is not demolished with excavation. When the L_{bw} is only 5 m, increasing the thickness of the buttress reduces the deformation of the retaining wall, but the impact is limited. This is because the increased stiffness of the composite section formed by the buttress wall and the diaphragm wall will also restrain the deformation of the diaphragm wall. When the L_{bw} is 30 m, increasing the thickness of the buttress will also reduce the deformation of the continuous wall, but the reduction is smaller (Hsieh et al., 2016).

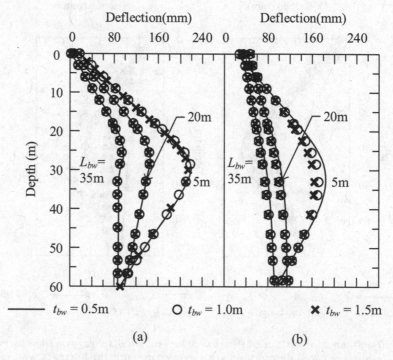

Figure 11.31 Influence of various buttress wall thicknesses on diaphragm wall deformation: (a) buttress wall demolished with excavation and (b) buttress wall maintained during excavation.

As mentioned in this section, deep excavations with buttress wall, the deformation of the diaphragm wall and soil is a three-dimensional behavior. In addition, the mechanism of buttress wall affecting deformation is complex. At this moment, there is no reliable simplified two-dimensional analysis method. On the other hand, the three-dimensional finite element method can fully consider the bending stiffness of diaphragm walls and buttress wall, the degree of mobilized friction between the buttress walls and the soil on both sides, and the bearing capacity of the soil at the buttress wall ends. Therefore, three-dimensional finite element analysis can determine the reasonable depth, thickness, length, and spacing of buttress walls.

11.8.3 Construction

Similar to cross walls, buttress walls can be constructed using diaphragm wall construction technology (Section 11.7.3). As buttress walls are demolished step by step with excavation of soil, the joint between buttress walls and diaphragm walls as shown in Figure 11.25 can be adopted. However, if buttress walls maintained (not demolished) step by step with excavation of soil, it is recommended to adopt the joint as shown in Figure 11.26 because it is more rigid to generate a better moment resistance.

11.9 MICRO PILES

Micro piles are also called soil nails. Since they were first applied in Europe to strengthen or underpin existing buildings, they have been used for more than 40 years. They have also been applied to building protection in many countries. Some successful case histories have been documented in the literature (Woo, 1992), and many more unsuccessful cases were not reported.

Because the mechanisms of micro piles for building protection are indirect and systematic studies are also lacking, though successfully applied in some excavations, most of them are designed on an empirical basis and without theoretical support.

In practice, the diameter of a micro pile varies from 10 cm to 30 cm. The reinforcements can be steel bars, steel rails, H steels, or even steel cages. The construction process of micro piles is as follows. First, bore to the designed depth with casings or by other drilling measures and then place reinforcements into the bores. Inject cement mortar into bores under a certain pressure. Pull out the casing gradually, and simultaneously, add more mortar. The micro piles are usually arranged in a single row or multiple rows. The distance between piles is 3–5 times the pile diameter, depending on the soil strength. No matter if the arrangement is the single row or the multiple row type, they should be intermingled by 5°–30° as shown in Figure 11.32.

The design principles for micro piles use are two. The first is they have to pass the potential failure surface so that the shear strength of micro piles and the pull-out resistance can restrain the failure of soils and reduce the possibility of ground settlement accordingly, as shown in Figure 11.32. The potential failure surface, however, is usually rather large and the micro piles passing it are limited in number. The shear strength and pull-out resistance of the micro piles are not too large. Thus, whether the design is useful remains to be evaluated.

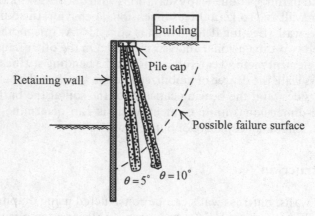

Figure 11.32 Mechanism of micro piles in protecting buildings.

The other principle is to design many small-diameter micro piles enveloping the retaining wall to reinforce the soils through the process of steel placing and grouting. Especially in sandy soils, grouts may permeate into soils extensively and reinforce a larger area of soils. They may even form a quasi-gravity retaining wall. Figure 11.33a diagrams the micro piles in two rows, which perform as gravity retaining wall-like structure, as shown in Figure 11.33b. The method may be capable of stabilizing the retaining wall and reducing the active earth pressure on the retaining wall or increasing the resistance to the failure surface. Thus, ground settlement would be reduced.

Nevertheless, neither of the principles has been proved useful through systematic studies. Theoretically, micro piles can be applied to sandy and clayey soils alike.

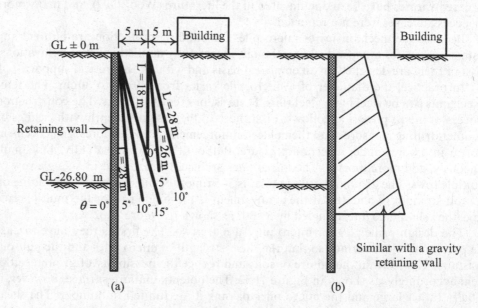

Figure 11.33 Mechanism of micro piles in protecting buildings: (a) arrangement of micro piles and (b) serving as a gravity-retaining wall.

Actually, with the larger strength of sand and the extensive permeation of grouts into voids, grouting would be effective to certain extent in sandy soils. In clayey soils, on the contrary, the low soil strength and small voids in soils, which would prevent the grout from permeation, render the effects of micro piles dubious.

11.10 UNDERPINNING

Underpinning is to strengthen existing foundations of a building, to improve the soils, or to replace original foundations. The applications of underpinning are quite extensive, including building protection in excavations, the prevention of natural settlement of heavy buildings, the strengthening of the foundations of buildings which have been unsuitably designed or constructed, and the underpinning of the bottoms of buildings through which a new tunnel has just been constructed, etc.

The methods of underpinning commonly applied in building protection in excavation are as follows:

1. Improve the soils beneath the foundations of existing buildings

As shown in Figure 11.34, to prevent over-settlement of the adjacent buildings near excavations, the soils beneath the foundation are treated before excavation. Once excavation is started, the soils behind the retaining wall come to the active condition and the maximum deformation usually occurs near the excavation surface. The potential failure surface outside the excavation zone normally develops from below the excavation surface. Thus, to underpin existing buildings, the depth of the soil improvement should extend from below the foundations to outside the failure surface.

When carrying out soil improvement, the over-pressurized grouting often heaves the building. It also disturbs the soil structure below the building foundations and reduces the soil strength. If the method is not implemented with prudence, it may not only turn out to be a failure in strengthening the foundation soils but may also worsen inclining conditions and damages buildings. Thus, a detailed plan referring to successful case histories is required before implementation.

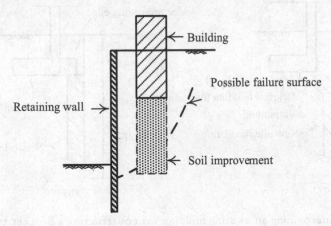

Figure 11.34 Underpinning an existing building by soil improvement.

2. Add an extra foundation to existing buildings

For fear of the insufficient bearing capacity of the foundations of existing buildings, an extra foundation could be constructed near the original foundation before excavation. Such an underpinning measure is usually implemented before excavation. The implementation, without weakening the original foundations, does not need to be accompanied by the measure of load transfer, which will be introduced in the next section. Figure 11.35 shows some possible underpinning methods. Figure 11.35a shows a method of constructing new piles beside the retaining wall to support the foundations of existing buildings. Figure 11.35b shows a method of constructing an additional part of the foundation of an existing building to join a newly constructed pile. Figure 11.36 shows the underpinning an existing government building during the construction of the Singapore mass transit system (Huang, 1992). In Figure 11.36, the government building is a brick building with wood piles of 4.5 m deep and 50 to 100 mm in diameter. The mass transit system passes near the government building. The construction followed the open-cut method with an excavation depth of 27 m. The retaining system consisted of steel sheet piles and soldier piles. The outside column of the government building was only 3 m away from the soldier piles. To protect the building during the construction of the mass transit system, an underpinning measure was used. The method is as follows: four micro piles were constructed next to the outside columns.

The micro piles were constructed through the pile caps to 26–28 m below the ground surface, which was 5 m below the 45° failure surface. The resulting void in the ground was filled fully with cement mortar. Field monitoring data showed that the building was still subject to a significant settlement after excavation, which might be due to the construction problem.

3. Construct new foundations under existing buildings

Figure 11.37 diagrams the underpinning method in which an additional foundation is constructed beneath the building's original one. With the new foundation constructed, the original one may weaken or even become useless. The method is as follows:

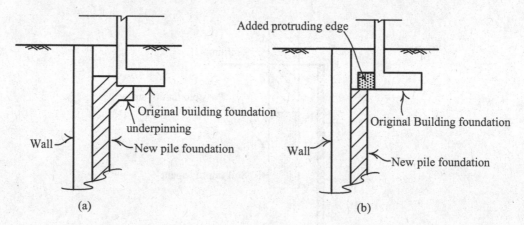

Figure 11.35 Underpinning an existing building: (a) constructing a bracket and foundation piles and (b) constructing an additional foundation and new piles.

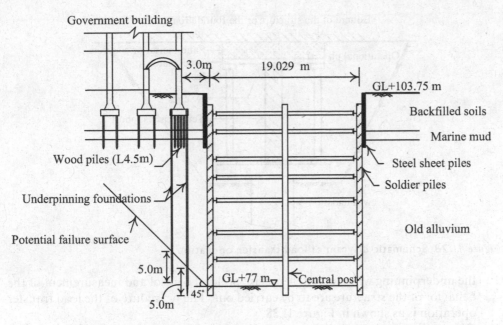

Figure 11.36 Underpinning foundations of a government building near an excavation of Singapore's mass transit system.

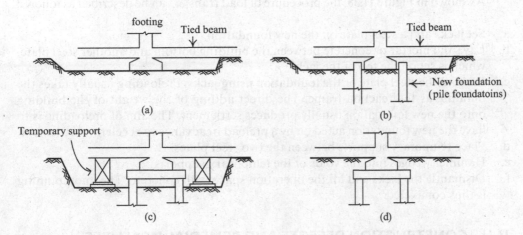

Figure 11.37 Underpinning an existing building by constructing new foundations: (a) excavation, (b) building new foundations, (c) installing bearing beams and temporary supports, and (d) load transferred to the new foundations and backfilling.

a. Excavate an operation space beside the footing and below the foundation.
b. Construct new foundations (piles).
c. Install temporary supports.
d. Load transfer operation—transfer the weight of the building to the new foundation or the strengthened one. This step is the crucial point in determining whether

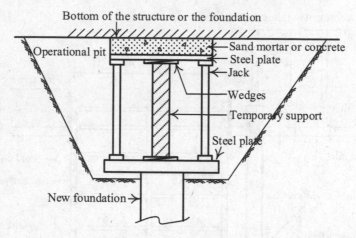

Figure 11.38 Schematic diagram of load transfer operation.

the underpinning will succeed. A precise preload control and measurement of the behavior of the structure are to be carried out. The procedure of the load transfer operation is as shown in Figure 11.38.

e. Dismantle the temporary supports. Proceed to grouting and backfilling.

As shown in Figure 11.38, the procedure of load transfer can be described as follows:

a. Set jacks and a steel plate on the new foundation.
b. Lay sand mortar or concrete between the building bottom and another steel plate, which is set on the top of the jack
c. Preload the steel plate or the foundation using jacks. Preloading usually takes the building as the reaction frame. The direct adding of the weight of the building onto the new foundation usually produces settlement. The aim of preloading is to have the new foundation acted on by a preload in advance to accelerate settlement.
d. Place temporary supports between the two steel plates.
e. Hammer wedges into the voids of the temporary supports.
f. Dismantle the jacks and fill the operation space with concrete. The underpinning is thus completed.

11.11   CONSTRUCTION DEFECTS AND REMEDIAL MEASURES

If possible construction defects in excavations are known in advance, a remedial measure for such defects can be prepared. Some commonly found construction defects are described as follows:

11.11.1   Leakage through the retaining wall

In excavations in sandy soils with high groundwater levels, if the retaining wall is broken or not satisfactorily watertight, one will encounter the problems of water and/or sand leaking through the wall, which can be extremely dangerous. If the leaking amount becomes large enough, it will enlarge the holes in the wall rapidly and a certain part of the excavation site will collapse as a result.

Even when leaking is not serious, it is still dangerous to a certain extent. The reason is that water is often intermingled with fine soils (such as silt). Though the fine soils flowing to the excavation zone are not much at first, they will be accumulated day by day without causing any alarm (for example, they may be swept away once they leak into the excavation zone). As a result, the soils outside the excavation zone will badly lack fine soils and the soil structure will grow unstable. Once the ground is subject to any extra load or vibration, the ground surface will collapse, as shown in Figure 11.39. Because the flowing directions of groundwater are complicatedly intertwined, from where the fine soils have been taken away can hardly be known. As a result, the settlement may occur near or far and may be local or overall. The outcome is possibly disastrous.

Settlement and the losing of fine soils caused by leakages in the retaining wall will only occur in soils with high permeability (such as sandy soils). Clayey soils have low permeability, and there exists cohesion between particles. Therefore, even when the retaining wall leaks, no possibility of settlement or losing fine soils has to be considered.

The causes and remedial measures due to leakage are generally as follows:

1. The soldier piles or laggings are not satisfactorily watertight. Excavation below the groundwater level is susceptible to water leaking even though grouting is implemented in advance as a rule. That is because the quality of grouting is not easily guaranteed. Though sheet piles are much more watertight, if they are used too many times or are unsuitably placed, they may still leak at the joints. Soldier piles and sheet piles are to be avoided when excavating in sandy soils with high groundwater tables.

2. If the quality of the concrete used in a diaphragm wall is not well controlled, water leakage is likely to happen, especially at the joints between panels. Besides, during the process of casting, if the process of lifting Tremie pipes is flawed for example, the position of the flaw on the plan and its elevation should be recorded immediately so that grouting can be implemented at the place where the concrete

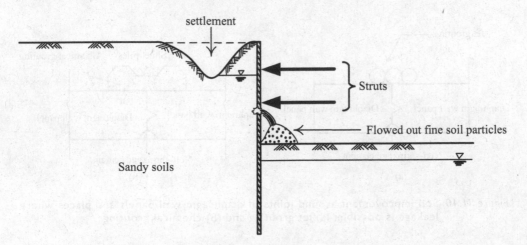

Figure 11.39 Ground settlement induced by leakage in the retaining wall.

casting is flawed before excavation. The joints between the panels of the diaphragm wall may be similarly treated sometimes in case leaking from the panel joints is suspected.

3. When two retaining walls with different rigidities are joined, different deformations will be produced as a result of earth pressure. The condition is especially susceptible to cracking and has to be remedied by grouting at joints.

4. If there exist pipes under the retaining wall, the construction will be confined by insufficient space. As a result, the construction quality of the bottom of the wall is not easily controlled, which may produce flaws that lead to leakage. These places should be treated by grouting before excavation.

5. When installing a ground anchor in granular soils with high groundwater levels, cutting the diaphragm from inside the wall with a drilling machine would easily bring about leakage through the drilled place. Leakage may occur directly through the diaphragm wall, through the gap between the bore and the casing of the anchor, or from the void of the cable bundle. Special techniques of implementation have to be adopted to prevent leaking or grouting should be implemented at the place where the anchor is installed.

As discussed above, if engineers find a possible leak, grouting can be carried out in advance before excavation. Jet grouting piles can be used, which should be installed closely and in contact with the outside of the wall to achieve water tightness. Chemical grouting (or low pressure grouting) can also be adopted to achieve water tightness. When chemical grouting is carried out, in order to confine the flow of the grout, sheet piles can be installed at the outside of the joint or leaking hole, and then grouting is performed between the sheet piles and the wall, as shown in Figure 11.40.

If a small, slow leak occurs during excavation, we can use quick-gelling agent such as a mixture of sodium silicate and cement to stop the water first, and then remove brittle inclusions. Then, use non-shrinkage cement and other filling materials to fill the crevices. If leaking amount is great, leaking is serious, and safety is concerned, the first thing is to stop the leaking. The general emergency treatment method is to first stuff straws or geotextile into the leaking hole, and surround the leaking hole with sand bags to reduce the outflow of soil and/or water and to reduce the rate of water

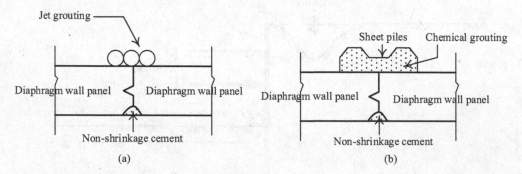

Figure 11.40 Soil improvement around joints of diaphragm wall panels and places where leakage is possible: (a) jet grouting and (b) chemical grouting.

leakage. In addition, chemical grouting should be applied to the outside of the wall. The gelling time of the grouting material must also be controlled to block the water inlet and achieve the purpose of stopping the water. When grouting is applied to the outside of the wall, improper operation of grouting pressure will have an adverse effect on the retaining support system, for example, distortion of the strut and/or damage to the wall. Thus, the grouting operation has to be carefully performed. Installation of suitable instruments to monitor the process closely should be carried out. After the emergency stop of the water leaking hole, remove the poor concrete and fill the hole with non-shrinking cement.

Generally speaking, jet grouting is unsuitable for the treatment of water leaking because its high pressure grouting often disturbs the soils and brings about the failure of soil structures, even damages to the strut and wall. Instead, chemical grouting is recommended. The choice of grout should be limited to those with high strength and the capability of controlling gelling time.

11.11.2 Dewatering during excavation

Dewatering is usually necessary to keep the excavation surface dry for excavations in sandy or gravel soils. It is basically not required for excavations in clayey soils except for lowering the groundwater pressure of the permeable soil below the clayey soil to prevent upheaving failure. Pumping outside the excavation zone is likely to lower the groundwater level around the excavation site. If the site consists of clayey soils, it may raise the effective stress of the clay and bring about consolidation settlement of soils outside the excavation zone. The settlement may be uniform or nonuniform, depending on the geological conditions. Pumping, if not carried out carefully, could carry fine soils out of the ground, which will cause settlement, too, as shown in Figure 11.41a.

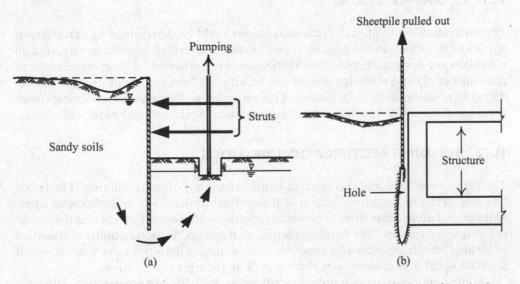

Figure 11.41 Ground settlement induced by (a) pumping and (b) pulling out of used piles.

11.11.3   Construction of the retaining wall

Constructing a diaphragm wall, under normal conditions, will produce settlement inevitably because of the unloading caused by excavation of trenches. For settlement caused by the construction of diaphragm walls, please refer to Section 6.2. On the other hand, if the level or concentration of bentonite in the trench is insufficient, it is likely to bring about movement or even collapse of the trench, which further induces ground settlement, during the construction of a diaphragm wall.

When the PIP pile (see Section 3.3.3) is adopted as the retaining wall, the loosening of soils caused by boring or the defective backfill of sandy grouts, which both produce voids in soils, will cause the movement of the surrounding soils and ground settlement accordingly.

Driving a sheet pile into sandy soils with high groundwater levels by vibration usually causes settlement because of the effects of compaction. If excavation is carried out in clayey soils, pile driving will cause heaves of the nearby soils. What's more, the ground settlement will occur due to the consolidation effect.

11.11.4   Pulling out the used piles

If sheet piles or soldier piles are used as the retaining walls in an excavation, after the construction of the basement is finished, to save cost, the sheet piles or soldier piles are often pulled out for repeated use. Pulling out the piles, however, will take out soils around the piles and generate voids in the soil, as shown in Figure 11.41b. The generated voids are not easily filled by way of grouting or with sand. As a result, settlement is produced. To prevent this, leave the piles close to buildings in place.

11.11.5   Over-excavation

The excavation depth at each excavation stage should be determined by carrying out stress and deformation analyses before excavation. Nevertheless, some contractors do not follow the designed depth of excavation and over-excavate for their own benefit or convenience. Over-excavation should not be allowed because it will produce unpredicted large amounts of deformation of the retaining wall or ground settlement under good conditions. Under worse conditions, it may cause the collapse of the excavation.

11.12   BUILDING RECTIFICATION METHODS

The principle of building rectification is similar to that of underpinning. The latter, however, refers to the strengthening of the original foundations or addition of a new foundation before excavation to prevent excavation-induced settlement and the resulting building damages. The building rectification method is used to uplift over-settled or leaning buildings, caused by excavation, construction defects, or the liquefaction of soils below the foundations after earthquakes, to their proper location.

Some commonly used building rectification methods are compaction grouting, chemical grouting, and underpinning, etc. This section will explicate these methods in the following.

11.12.1 Compaction grouting

The compaction grouting method injects pressurized cement mortar with high consistency, low slump, and low plasticity into soils (Figure 11.42). The cement mortar does not easily flow into the voids of soils and will therefore form a grouting bulb, which will in turn make an interface with the soils around. Squeezing the cement mortar into soils continuously, the grouting bulb will expand and press the soils around to compact them. Besides, the grouting bulb expands in radiation though faster in lateral direction (because the lateral stress of soils is smaller than the vertical stress). When the soils are compacted to a certain degree, nevertheless, the expansion in the lateral direction will stop. Instead, the grouting bulbs begin to expand upward and generate an upheaving force, which will make a cone-shaped shear zone above. When the upheaving force exceeds the weight of overburden or the building, the phenomenon of ground surface heave or building upheaval will occur as shown in Figure 11.42.

The materials used in the compaction grouting method have to be designed to achieve high consistency, low slump, and low plasticity. Otherwise, they may flow into the voids of soils. On the other hand, they have to have a certain fluidity to be well transported. Sand mortar and cement mortar are two widely used grouting materials. Fly ash and bentonite are widely adopted additives.

Theoretically, the compaction grouting method is limited in its application to buildings under a certain weight. No related studies in the literature indicated that the maximum building weight in the compaction grouting method is capable of uplifting. Graf (1992) once lifted up a 4.5 m × 4.5 m square footing by 5 cm using the compaction grouting method. The footing was 1.2 m deep and bore a dead load of 500 tons and a live load of six driveways. The grouting bulb was 5.75 m deep below the ground surface. Wong et al. (1996) uplifted one of the footings of a four-floor building by 24 mm, which was about 1.0 m below the ground surface. Grouting was implemented from the bottom-up, starting at the depth of 8 m below the ground surface up to 3 m below the ground surface. When the grouting depth grew higher than 4.5 m, the foundation could be observed uplifted apparently.

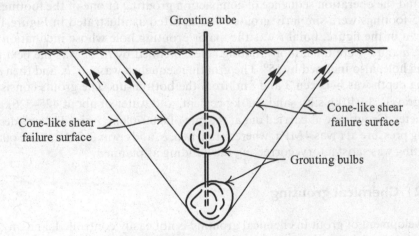

Figure 11.42 Compaction grouting method.

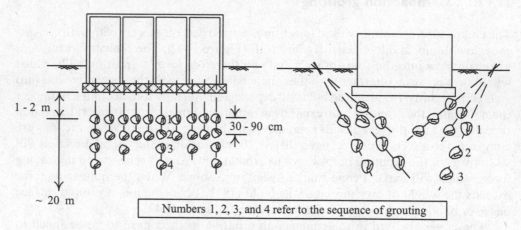

Numbers 1, 2, 3, and 4 refer to the sequence of grouting

Figure 11.43 Procedures and grouting positions of the compaction grouting method. (Redraw from Nonveiller (1989).)

A detailed plan of the grouting pressure, grouting rate, grouting depth, arrangement of the grouting points, and the implementation method is required before proceeding to the building rectification operation. Besides, the building should be equipped with complete monitoring instruments for immediate adjustment of the operation.

To uplift a building, the ways of grouting are two: either bore through the floor of the building and grout vertically or grout slantingly from the building side, as shown in Figure 11.43. The former has better effects since it is more direct. Nevertheless, it will affect the use of the building. The latter is less effective but has the benefit of not disturbing the use of the building.

Here is a four-floor building with 16 columns in Taipei, each of which is supported by an individual footing embedded 1.0 m below the ground surface. Influenced by the construction of CP263 of the Taipei rapid transit system (TRTS), the footings of the buildings settled by 2.0–4.4 cm. To rectify the building, compaction grouting was implemented at all of the footings. Figure 11.44a diagrams the arrangement of the grouting points and the operation sequence of compaction grouting at one of the footings. The other 15 footings were similarly grouted and uplifted as illustrated in Figure 11.44b. As shown in the figure, point i was the major grouting hole whose inclination angle was $12°$. a and b were the minor grouting holes, inclined by $15°$. c was the next minor grouting hole, also inclined by $15°$. The grouting sequence was i, a, b, and then c. The grouting depth was between 3 and 8 m, from the bottom-up. The grouts consisted of 320 kg gravels, 1,040 kg silty sand, 160 kg cement, and water of about 425–526 kg each cubic meter. Grouting was carried out at the rate of 0.00708 m^3/min. The pre-designed grouting pressure at i was 4 MPa, whereas that at a, b, and c was 3.5 MPa. The outcome of grouting was satisfactory, correcting the building as planned.

11.12.2 Chemical grouting

The development of grout in chemical grouting is not easily controlled, and any small incaution might lead to the unpredictable flowing of grouts, which would cause damage to structures and pipes and reduce the effects of grouting. A detailed plan of the

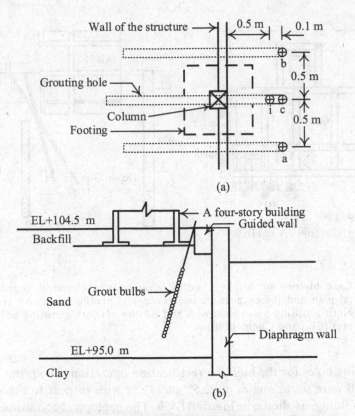

Figure 11.44 Case history of building rectification by the compaction grouting method: (a) plan and (b) profile (Wong et al., 1996).

grouting pressure, the arrangement of injecting points, the amount of injection, and the measures for preventing fugacious flowing of grouts should be made before implementing chemical grouting to rectify a building. Generally speaking, the grouting pressure, the arrangement of injection points, and the amount of injection are all determined on the basis of empirical experience and adjusted according to the monitoring data of the heave of the building. Referring to past case histories is recommended. For the prevention of grout flows going where they are not wanted, there are two methods: installing sealing piles (such as sheet piles) within the grouting range and reducing setting time of grouts.

The cost of chemical grouting to correct an inclining building is relatively low. Nevertheless, the operation of carrying out a chemical grouting for building rectification is highly technical and ingenious, any incaution easily leading to damage to structures and pipes.

Take a case history from the construction of the TRTS for illustration. The inclining building had 12 floors of superstructure and one floor of basement with a mat foundation located 4.9 m below the ground surface (Chu and Chou, 1998). The building was influenced by the construction of the TRTS so that it settled 100 mm at the southeastern corner. The building inclined about 1/100 eastward and 1/200 southward. Figure 11.45a diagrams the plan of the building and the arrangement

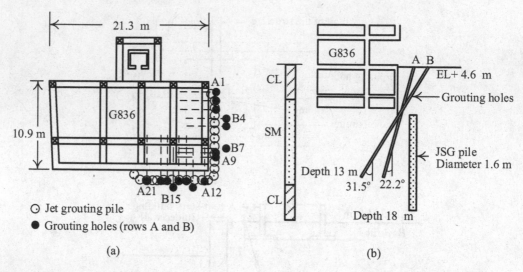

Figure 11.45 Case history of building rectification by the chemical grouting method: (a) plan and (b) excavation and geological profile (note: A9 stands for the ninth grouting point in row A and B7 the seventh grouting point in row B, etc) (Chu and Chou, 1998).

of the grouting holes for the building rectification operation. The grouting holes on row A and B were set at angles of 22.5° and 31.5° with respect to the vertical line below the building, as shown in Figure 11.45b. The geological conditions are shown in Figure 11.45b. To prevent the grouts from fleeing, the contractor installed a row of continuous jet grouting piles for sealing (called the JSG method) and limited the setting time to 60 seconds. The grouts consisted of cement of 250kg, sodium silicate solution of 0.25 m³, and water of 0.671 m³ per cubic meters. The congealing time was between 34 and 60 seconds.

To set the grouting pressure uniformly on the base of the foundation, 3–4 grouting machines were used simultaneously, each set to pump between 0.02 and 0.04 m³/min of grouting. First row A was grouted and then row B after row A was totally grouted. The grouting depth of each row was between GL-13.0 and GL-9.0 m, starting from the bottom and lifting the drilling rod gradually up. With the clayey soil below the depth of GL-13.8 m (see Figure 11.45b) and the sandy soil above, the grouting operation took the sandy soil below the depth of GL-13.0 m as the bearing stratum where a larger amount of pressurized grouts was injected to constitute a solid ground in preparation for uplifting the building. With the lifting of the drilling rod, the grouting rate and pressure were lowered gradually. According to the above principles and experience, the control values of the grouting pressure and flow were determined.

After 11 days' building rectification, the building was uplifted. The contour map of lifting heights is shown in Figure 11.46. As it shows, the building was satisfactorily corrected and the lifted height was distributed lineally, which shows that the influence of the rectification operation on the structure of the building was slight.

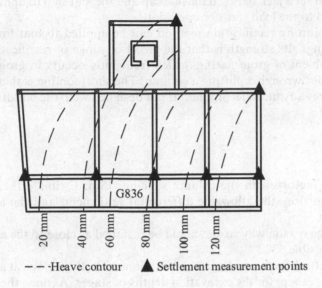

− − -Heave contour   ▲ Settlement measurement points

Figure 11.46 Distribution of the heave of the building (Chu and Chou, 1998).

11.12.3   Underpinning

The principle of the underpinning for rectifying buildings is similar to that of the underpinning method introduced in Section 11.10. The only difference is that the latter aims at prevention before excavation, transferring the load of the building to a new foundation which is not to be influenced by excavation. Therefore, jacks are not necessarily required to uplift the building. The former is to solve an existing problem, correcting a building problem due to excavation. Since the building has already settled, jacks are required to uplift the building. The procedure of the underpinning method is shown in Figure 11.47. First excavate a sufficient working space, where underpinning piles are to be constructed, around the mat foundation or the individual footing under

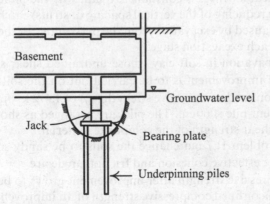

Basement

Groundwater level

Jack

Bearing plate

Underpinning piles

Figure 11.47 Rectification of buildings by the underpinning method.

the column, then set a jack between the pile cap and the mat slab to uplift the building, and last proceed to the load transfer operation.

The underpinning rectification method can be applied to mat foundations and individual footings. Its strength is that the target or range of rectification is specific, without the problem of grout fleeing, which frequently occurs in grouting methods, or worrying about wrongly uplifting a column. The shortcoming is that it costs much more and requires advanced techniques. If not designed well, the building might settle more.

PROBLEMS

11.1 Here is a factory with the column spacing, 15 m. Estimate the allowable angular distortion, the allowable differential settlement, and the allowable total settlement.

11.2 Explain the reason why struts should be installed as close to the excavation bottom as possible.

11.3 Figure 11.10 shows two identical excavations in both geological and excavation conditions except for the excavation depths of stages. Assume the soil at the site is the normally consolidated clay and the groundwater level is at the ground surface. The basic properties of the clay are as follows: $\gamma_{sat} = 18.6$ kN/m^3, $c' = 0$, $\phi' = 30°$, and $s_u / \sigma'_v = 0.3$. The thickness of the diaphragm wall is 60 cm. The stiffness of the struts is 30,000 kN/m^2/m (per unit width). The preload is 45.0 kN/m (unit width). Analyze the excavations as shown in Figure 11.10a and b through the computer program and compare the deformation characteristics of the two cases to find out which one is to produce more deformation of the retaining wall.

11.4 Explain why the deformation of the retaining wall or ground settlement with the top-down method is larger than that with the bottom-up method.

11.5 The increase of the strut stiffness can reduce the wall deformation or ground settlement caused by excavation. Explain why, however, the increase of the strut stiffness cannot help reduce the wall deformation or ground settlement once the strut stiffness has reached a certain level.

11.6 Assume the strut stiffness is constant. Explain why the increase of the levels of struts (i.e., the reducing of the vertical spacing of struts) can help reduce the wall deformation caused by excavation from the perspective of the deformation characteristics at each excavation stage.

11.7 Here is an excavation in soft clay whose undrained shear strength is $s_u = 98.1$ kN/m^2. A soil improvement is to be carried out on the soil. The improvement pile has a unconfined compressive strength (q_u) of 1,472 kN/m^2. The diameter of the improvement pile is 60 cm. The piles are arranged as shown in Figure P11.7. Estimate the shear strength of the composite material.

11.8 Same as in Problem 11.7 but change the soils to be sandy soils. Before the improvement, the effective cohesion and friction angle are $c' = 0$ and $\phi' = 30°$. The uniaxial compressive strength after improvement grows to be 5,000 kN/m^2.

11.9 Assume the unconfined compressive strength of an improvement pile to be 1,472 kN/m^2 and the diameter of the pile is 80 cm. The piles are arranged as shown in Figure P11.9. Rework Problem 11.7.

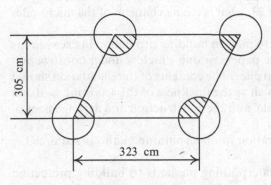

305 cm

323 cm

Figure P11.7

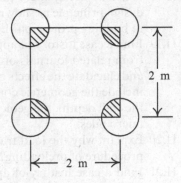

2 m

2 m

Figure P11.9

11.10 Find a case history of soil improvement in an excavation from related journals or conference papers and elucidate the effects of the improvement. The contents of the elucidation should include the geometric conditions (such as the thickness and depth of the retaining wall), the geological profile, the soil improvement method, the improvement range, and the effects.

11.11 Find a case history of constructing buttress walls in an excavation from related journals or conference papers or one which is under construction and elucidate the effects of buttress walls. The contents of the elucidation should include the geometric conditions (such as the thickness of the retaining wall and the wall depth), the geological profile, the sizes of the buttress walls, and the effects.

11.12 Find a case history of constructing cross walls in an excavation from related journals or conference papers or one which is under construction and elucidate the effects of the cross walls. The content of the elucidation should include the geometric conditions (such as the thickness of the retaining wall and the wall depth), the geological profile, the sizes of the cross walls, and the effects.

11.13 A 50 m × 50m excavation in normally consolidated clay where the ground water table is at the surface. The final depth of the excavation is 20 m and adopts a 1.0-m-thick, 40-m-deep diaphragm wall as the retaining wall. It is known that a 30-year-old building is close to the excavation. (1) Estimate preliminarily the maximum settlement of the building using the empirical formulas or charts in Chapter 6. (2) Schematically design a protective measure using soil improvement against the damage of the building during excavation, for example, the improvement area, depth, ratio, etc.

11.14 Same as Problem 11.13. Schematically design cross walls, for example, cross wall spacing, depth, etc.

11.15 Same as Problem 11.13. Schematically design buttress walls, for example, treatment (demolished or not), spacing, etc.

11.16 Similar to Problem 11.13 but excavation in sand. What is possible construction defect during excavation? Propose a possible remedial measure.

11.17 When applying micro piles to building protection in excavations, assuming the design principle as shown Figure 11.32, what is the mechanism of the micro piles in building protection? Elucidate it.

11.18 When applying micro piles to building protection in excavations, assuming the design principle as shown Figure 11.33, what is the mechanism of the micro piles in building protection? Elucidate it.

11.19 Find a case history of applying micro piles to building protection in excavations from related journals or conference papers or one which is under construction and elucidate the effects of the micro piles. The contents of the elucidation should include the geometric conditions (such as the thickness of the retaining wall and the wall depth), the geological profile, and the construction and arrangement of micro piles.

11.20 Explain why the load transfer operation in underpinning methods requires the procedure of preloading?

11.21 Find a case history of applying underpinning methods to building protection in excavations from related journals or conference papers or one which is under construction and elucidate the effects of underpinning. The contents of the elucidation should include the geometric conditions (such as the thickness of the retaining wall and the wall depth), the geological profile, the construction of underpinning, and the effects.

11.22 When excavating in sandy soils, what types of construction defect may happen and cause building damage in the vicinity of the excavation?

11.23 When excavating in soft clay, what types of construction defect may happen and cause building damage in the vicinity of the excavation?

11.24 Find a case history of building damage caused by defective construction in excavation from related journals or conference papers or one which is under construction and elucidate the causes of damage. Elucidate the geometric conditions (such as the thickness of the retaining wall and the wall depth), the geological profile, and investigate the causes of building damage

11.25 Find a case history of building rectification from related journals or conference papers or one which is under construction, and elucidate the effects of the method. Include the causes of building settlement, the process and effects of rectification, etc.

Chapter 12

Monitoring systems

12.1 INTRODUCTION

In recent years, with a high development of society, excavation depths grow deeper and deeper. What's more, these excavations are usually located in densely populated areas and more and more problems of adjacent buildings damage are encountered. Though the technologies of analysis, design, and construction have advanced significantly, they are not necessarily capable of coping with changes during excavation. Thus, a well-arranged monitoring system will be helpful not only to monitor the excavation condition but to ensure the safety of excavations.

Generally, the objects of an excavation monitoring program can be summarized in the following:

1. To ensure the safety of excavation. To ensure the safety of excavation is the first thing for an excavation monitoring program. Before a failure happens, there will be some signals, such as an extraordinary increase of the wall or soil deformation or stress. A monitoring system can issue an immediate warning to help engineers adopt effective measures to forestall a failure when these signals appear.
2. To ensure the safety of the surroundings. Excavations are usually located in urban areas. It is almost impossible to avoid influencing the surrounding buildings, underground pipes, and public utilities. The designer has to consider all these factors and establish a monitoring system to ensure the safety of adjacent buildings or properties during excavation.
3. To confirm the design conditions. Since the existing analysis theories are not satisfactorily mature and the geological investigations cannot fully represent the *in situ* conditions and the complicated construction environment, the analysis results do not necessarily meet the actual conditions. Back analyses based on monitoring results can help modify and correct the original design, reduce the cost, shorten the excavation period, and change the basis of design. Further, they can serve as a reference for similar designs in the future and help enhance the excavation techniques.
4. To follow long-term behavior. An important construction project finished, the monitoring system can be retained for the long-term follow-up, studying whether the long-term behavior of the case conforms to the original hypotheses.

DOI: 10.1201/9780367853853-12

5. To supply factual materials for legal judgment. Information obtained from monitoring systems, along with the construction records, can serve as an evidence when a calamity or property damage occurs. Understanding the true causes of the events can avoid unnecessary disputes and help the restoration and compensation work. The excavation can be least delayed.

Basically, the commonly adopted geotechnical instruments for monitoring programs can also be applied to excavation projects. Considering the particularities of excavations, such as the monitoring time period, the rapid change of stress and strain of structures and soils, serious impact on environment, and the economic factors, some special instruments and their installation methods have been developed or have proved suitable for excavations. The purpose of this chapter is to introduce the basic measurement items and principles of an excavation monitoring program. For more detailed instrument types or principles, please refer to the literature (Hanna, 1985; Dunnicliff, 1988; ICE, 1989).

12.2 ELEMENTS OF A MONITORING SYSTEM

Monitoring field performance is basically the measurement of physical quantities of some objects, such as the deflection, stress, and strain, etc. The common monitoring items in excavations are (1) movement of the structure or soil, (2) stress or strain of the structure or soil, and (3) water pressure and level. The monitoring objects of movement include the lateral deformation of the retaining structure and soils, the tilt of the building, the settlement of the ground surface and the building, the heave of the excavation bottom, and the uplift of the central post, etc. The measurements of stress and strain include those of the strut load, the stress of the retaining structure, and the earth pressure on the wall. The measurement of water pressure refers to the pore water pressures in soil and on the retaining wall.

In principle, the types of monitoring instruments in an excavation are not special. As long as they meet the budget and the criterion of accuracy, any instruments or methods are feasible. Most of the instruments can be divided into electronic and non-electronic types. The electronic type is more sensitive and easy to read. They can also be composed into an automatic or semi-automatic monitoring system. Nevertheless, the precision is easily influenced by installation process and the surroundings. The durability of electronic instruments needs examination because an excavation usually needs to be monitored over a long period. Thus, when installing an electronic instrument in the field for long-term monitoring, durability needs special attention.

Generally speaking, the instruments for plane surveying, such as tapes, levels, and theodolites, can be used for the measurement of movement. The measurement of the tilt angle can also resort to those for plane surveying (same as above) or electronic instruments. Electronic instrument sensors include strain gauges and force balance accelerometers, etc. The most common instruments to measure stress or strain are the strain gauge or the electronic transducer that takes the strain gauge as the measurement unit. Since the strain gauge is widely used directly or indirectly in monitoring systems, Section 12.3 will introduce the commonly used types and basic principles of strain gauges. This chapter will omit general instruments for plane surveying. Please refer to related books on plane surveying if interested in the subject.

The types of monitoring systems can be categorized into manual, automatic, and semi-automatic systems. A manual monitoring system is one that requires human reading of the measured values, no matter if it is electronic or mechanical. An automatic system is one that is equipped with electronic sensors, which are connected to a computer by cable or wireless connection for every monitoring instrument on the excavation site. When excavating, these values are transferred to the computer periodically and then computed and shown on computer monitors. Thus, continuous values of measurement can be obtained. The method is expensive, though effective, and saving the trouble of human reading of measured values.

An automatic monitoring system can supply continuous measured values but is usually expensive. Manual monitoring systems don't cost much but are incapable of supplying continuous measured values. As a result, a semi-automatic monitoring system is adopted in some excavations instead. The so-called semi-automatic system refers to one that employs electronic sensors for crucial parts, but counts on human reading for the others. The values from the electronic sensors are collected periodically by cable or wireless to a data logger near the excavation site. Engineers then collect the data from the data logger. The method can save cost and obtain continuous values as well.

Since the manual monitoring system is still most widely used in excavations, the monitoring instruments and methods that will be introduced in this chapter are basically those that count on human reading.

12.3 PRINCIPLES OF STRAIN GAUGES

The strain gauge is often used as a measurement unit in a monitoring system. This section will introduce the commonly used strain gauges for reference.

There are many ways to measure strains. They can be obtained directly by a strain gauge or computed from other physical quantities. The types of strain gauges are many. The commonly used in excavations are the resistance strain gauge and the vibrating wire strain gauge. This section will introduce the principles of the two types of strain gauges.

12.3.1 Wire-resistant type of strain gauges

Figure 12.1 shows the basic configurations of three commonly used resistance strain gauges. Figure 12.1a shows a bonded wire resistance strain gauge, which consists of a base (insulating plate) and a fine resistance wire (with a diameter between 0.02 and 0.025 mm) attached to it. The resistance wire, 2–60 mm long, is usually made of copper and nickel alloy or nickel and chromium alloy with its resistance set at 120 Ω. Since the strain gauge adopts a metal wire for the resistor, it is called the bonded wire resistance strain gauge. If the resistor is a thin piece of foil (made of the same material as the bonded wire strain gauge), which is directly attached to the base and photo-etched to set the resistance value (also 120 Ω in most cases), the strain gauge is called the foil resistance strain gauge, as shown in Figure 12.1b. Another type adopts a fine metal wire circling two sets of insulating rods, which are fixed onto the object of measurement. This type of strain gauge is called the unbounded wire resistance strain gauge, as shown in Figure 12.1c.

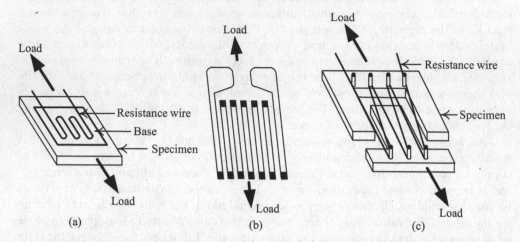

Figure 12.1 Commonly used resistance strain gauges: (a) bonded wire resistance strain gauge, (b) foil resistance strain gauge, and (c) unbonded wire resistance strain gauge.

Though different in constitution, the principles are similar. The resistor will produce resistance when electrified, the values of which vary with the materials of resistance wire. When the wire grows finer and longer, resistance will be increased. Otherwise, it will be decreased. When the object is lengthened due to loading, the wire becomes finer and longer, and resistance will be increased. When the object is compressed by a load, the resistance reduces.

Assume the length, section area, diameter, and resistance of the metal wire (the resistance wire) before it is strained are ℓ, A, d, and R. The resistance, R, can be expressed as follows:

$$R = \rho \frac{\ell}{A} \tag{12.1}$$

where ρ is the coefficient of resistance of the metal wire

Differentiating the above equation on both sides, we have

$$\frac{\Delta R}{R} = \frac{\Delta \rho}{\rho} + \frac{\Delta \ell}{\ell} - \frac{\Delta A}{A} = \frac{\Delta \rho}{\rho} + \frac{\Delta \ell}{\ell} + 2\mu \frac{\Delta \ell}{\ell} \tag{12.2}$$

where μ is Poisson's ratio of the metal wire.

Let the strain of the wire be ε. Then $\varepsilon = \Delta \ell / \ell$. The above equation can be simplified as

$$\frac{\Delta R / R}{\varepsilon} = (1 + 2\mu) + \frac{\Delta \rho / \rho}{\varepsilon} \tag{12.3}$$

Assume that the change rate of the coefficient of resistance of the wire has a direct proportion to that of its volume. Then

$$\frac{\Delta\rho}{\rho} = m\frac{\Delta V}{V} \tag{12.4}$$

where m is a constant, which can be obtained from tests on the metal material.

On the other hand,

$$\frac{\Delta V}{V} = (1-2\mu)\frac{\Delta\ell}{\ell} \tag{12.5}$$

Substituting the above two equations into Eq. 12.3, we have

$$\frac{\Delta R/R}{\varepsilon} = (1+2\mu) + m(1-2\mu) = (1+m) + 2\mu(1-m) \tag{12.6}$$

Let $K = (1+m) + 2\mu(1-m)$. Then, the above equation can be simplified as

$$\frac{\Delta R/R}{\varepsilon} = K \tag{12.7}$$

where K can be viewed as the constant of the instrument, which can be obtained from the tests on the material properties of the wire.

As illustrated above, the change rate of the resistance of the wire has a proportional relation with the strain. If the change in amount of resistance is measured, the strain can be derived.

Since the resistance type of strain gauge takes advantage of the change in resistance to measure the strain, any pollution such as dust, the change of temperature or humidity, or the length of the wire will all influence the measurement of strain. If water invades the interior of the strain gauge, the measured value of strain will become invalid because water is an electric conductor. Therefore, the strain gauge should be waterproofed to obtain good accuracy. The strength of the resistance type is that it is sensitive and can be miniaturized to serve as a small instrument-measuring unit.

12.3.2 Vibrating wire type of strain gauges

The basic configuration of the vibrating wire type of strain gauge is illustrated in Figure 12.2. The vibrating wire is fixed to an object. The length of the wire changes with the deformation of the object influenced by external factors (such as temperature or force), which causes the change of the natural frequency of the wire. The strain of the object is thus obtained through the measurement of the natural frequency of the wire.

Assume the mass of the wire per unit length is m, the wire length is ℓ_w, and the tensile or compressive force acting on the wire is F. We can compute the natural frequency of the wire as follows:

$$f = \frac{1}{2\ell_w}\sqrt{\frac{F}{m}} \tag{12.8}$$

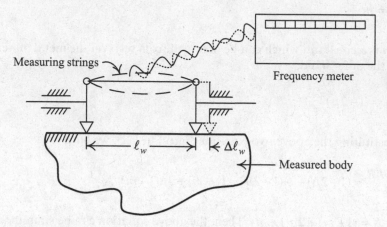

Figure 12.2 Configuration of a vibrating strain gauge.

According to mechanics of material, the tensile or compressive force can be expressed as follows:

$$F = \sigma_w a = \varepsilon_w E_w a \tag{12.9}$$

where

σ_w = tensile stress or compressive stress of the wire
a = section area of the wire
E_w = Young's modulus of the wire
ε_w = strain of the wire
Substituting Eq. 12.9 into Eq.12.8, we have

$$f = \frac{1}{2\ell_w} \sqrt{\frac{\varepsilon_w E_w a}{m}} \tag{12.10}$$

From the above equation, we can further derive the relation between the strain of the wire and its frequency as follows:

$$\varepsilon_w = \frac{f^2}{K} \tag{12.11}$$

where $K = E_w a / (4\ell_w^2 m)$ = constant of the instrument, which can be directly determined according to the material properties of the wire.

If the object and the strain gauge are well welded, the deformation of the object and that of the wire will be identical. That is to say,

$$\varepsilon_w \ell_w = \varepsilon_s L_s \tag{12.12}$$

where

ε_s = strain of the object
L_s = length of the strain gauge.

As a result,

$$\varepsilon_s = \frac{\varepsilon_w \ell_w}{L_s} \tag{12.13}$$

As elucidated above, as long as the natural vibrating frequency of the wire can be measured, the stress and strain of the wire and the object can also be derived. To measure the strain, connect the strain gauge with the sensor where the direct current produces a magnetic field that will vibrate the wire. The natural frequency of the strain gauge is measured by the sensor simultaneously. After conversion by the constant of the instrument, the stress and strain of the wire and the object can be derived.

Since the vibrating type of strain gauge takes advantage of the vibrating frequency of the wire to obtain the strain of an object, the frequency is not influenced by resistance, and the cable (the signal cable) can be extended as long as needed for field measurement. The vibrating type of strain gauge, however, is not as sensitive as the resistance type. If applied for long-term monitoring, humidity or water may invade the gauge and erode the wire. The problem of creep is also to be considered for long-term use. These factors will change the frequency of the strain gauge and influence the measured values. Applying the vibrating type to such conditions, one has to take these problems into consideration.

12.4 MEASUREMENT OF MOVEMENT AND TILT

Excavation will cause the structures and soils within the influence range to move. The directions of movement may be horizontal or vertical or both. According to the characteristics of the object to be monitored, the monitored items of movement, such as horizontal movement, vertical movement, or both or tilt of the object, are thus able to be determined. Generally, the lateral deformation of the retaining structure, the lateral deformation of soils, the tilt of the structure, the ground settlement, and the settlement of the building are the basic items of measurement for an excavation. This section will introduce the principles and the details of measurement.

12.4.1 Lateral deformation of retaining walls and soils

The lateral deformation of a retaining wall is one of the important monitoring items in excavations. The lateral deformation of a retaining wall is closely related to the ground settlement (or the settlement of the buildings). The magnitude and shape of lateral deformation of the retaining wall can be used for the judgment of the safety of the retaining wall or the buildings in the vicinity.

To explore the characteristics of an excavation or for some special objectives, it is sometimes required to measure the lateral deformation of soils. The strain at a specific point in the soil can be computed if extensometers, used to measure the vertical movement at a point in the soil, are installed near an inclinometer casing (Ou et al., 2000). The results can be used to understand the tendency to the movement of soil. This data is important for excavation studies.

The inclinometer casing and the inclinometer are commonly used devices for the measurement of the lateral deformation of retaining walls or soils. An inclinometer casing has four tracks, which form two perpendicular axes, along each of which two sets of wheels of an inclinometer can be inserted. Figure 12.3 shows a photo of some inclinometer casings and their cross sections. Some commonly used materials for inclinometer casings are ABS pipes, aluminum pipes, and PVC pipes.

Figure 12.4 shows the photo and the basic configuration of an inclinometer. As shown in the figure, the inclinometer is a four-foot (two pair of wheels) instrument containing a tilt measuring sensor (also called an electronic pendulum). The top of the inclinometer is connected to a cable, which is, in turn, connected to a readout on the ground surface. According to the type of measuring unit of the sensor, the inclinometer can be divided into resistance type, vibrating type, and force balance accelerometer type. The basic principle of the resistance and vibrating types is elucidated in Section 12.3. The force balance accelerometer type places a pendulum amid the magnetic field of coils of a location detector. When a tilt is produced, the coil magnetic field of the location detector can then detect the displacement of the pendulum with respect to the vertical line. The signal is then converted into a voltage by the detector. The voltage passing through the coil will generate a force to have the pendulum recover its original place. The tilt angle can be then obtained by converting the measured value of the voltage.

When installing an inclinometer casing, one pair of tracks of the inclinometer casing should parallel the direction of deformation of the retaining wall; in other words, it should be perpendicular to the retaining wall. The pair of tracks perpendicular to

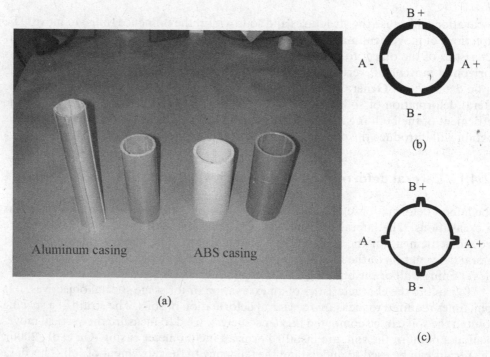

(a)

(b)

(c)

Figure 12.3 Inclinometer casings: (a) photo, (b) tracks made inside the casing, and (c) external tracks protruding from the casing.

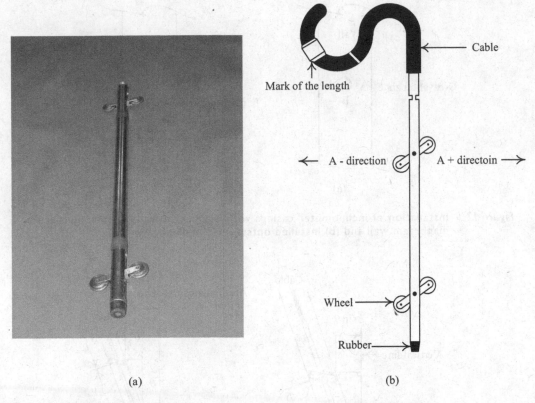

Mark of the length

Cable

← A - direction

A + directoin →

Wheel

Rubber

(a)   (b)

Figure 12.4 An inclinometer.

the retaining wall is usually called the AA axis. The other pair parallel to the retaining wall is called the BB axis, as shown in Figure 12.5. When measuring the lateral deformation of the wall, the wheels of the inclinometer should be inserted into the AA track. When measuring the other direction of deformation, they have to be inserted into the BB track. To eliminate systematic errors, after the first measurement is finished, the inclinometer should be reversed to re-measure the deformation. That is to say, the high foot (wheel) of the inclinometer should be inserted into A+ and A- directions, respectively, to take the measurement. Then, take the average value of these two measurements.

As shown in Figure 12.6, assuming the distance between the two points (or two wheels) is L and the tilt angle measured by an inclinometer is θ, the relative horizontal distance between the two points (or two wheels) is $L\sin\theta$. If the tilt angles taken from three serial measurements by the inclinometer are θ_1, θ_2, and θ_3, the relative horizontal distance between points (or wheels) A and B is $\sum L\sin\theta = L\sin\theta_1 + L\sin\theta_2 + L\sin\theta_3$.

As discussed above, the value taken from an inclinometer is the relative horizontal displacement between two points. To obtain the real displacement curve, an adjustment of the displacement of the top end of the inclinometer casing must be made or the bottom end of the inclinometer has to be placed at a real fixed point, such as a point in rocks or cobble-gravelly soils.

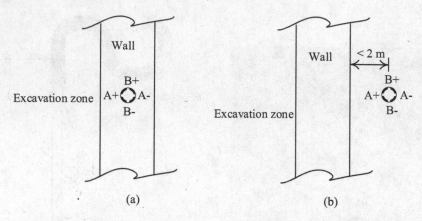

Figure 12.5 Installation of inclinometer casings with diaphragm walls (a) installed in the diaphragm wall and (b) installed outside of the diaphragm wall.

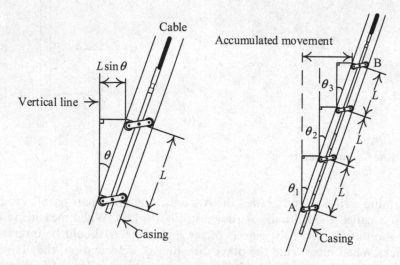

Figure 12.6 Principle of the measuring of lateral movement by an inclinometer.

The displacement at the top of the casing can be measured using a tape, level, or theodolite against a datum line, which should be set before excavation, as shown by d_1 in Figure 12.7a. The reason is that the displacement obtained from the measurement of an inclinometer is the relative displacement with respect to the bottom end of the inclinometer casing. Suppose the displacement of the casing top measured using an inclinometer is d_2. The amount of d_2 is the displacement of the casing top relative to the bottom. If we move the curve laterally by $(d_1 - d_2)$, we can obtain the real displacement curve. The datum line should not be affected by excavation. Generally speaking, the longer the two ends of the datum line extend, the less likely it is to be affected.

Figure 12.8 shows that the bottom of the inclinometer casing is placed in hard soil, which can be treated as a real fixed point. Thus, the displacement curve $A'B$ is the real displacement curve.

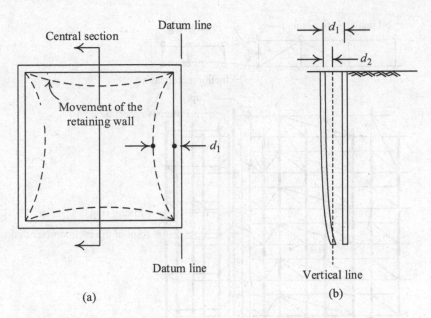

Figure 12.7 Modification of movement of the top end of casings: (a) plan and (b) profile.

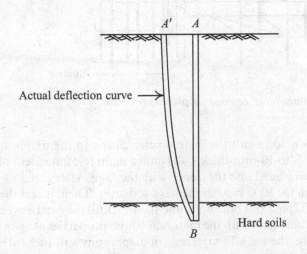

Figure 12.8 Deflection curve when the inclinometer casing is placed in hard soils.

In case of diaphragm walls, the inclinometer casing can be fixed on a main reinforcement of the steel cage of the diaphragm wall, which is then placed in the trench and cast with concrete using Tremie pipes. This finishes the installation of the inclinometer casing, as shown in Figure 12.9. The inclinometer thus installed will not be embedded deeper than the bottom of the wall. As discussed earlier, if the wall bottom is not embedded in hard soil, the bottom may move during excavation. Thus, the adjustment of the casing top should be made to obtain the real displacement curve.

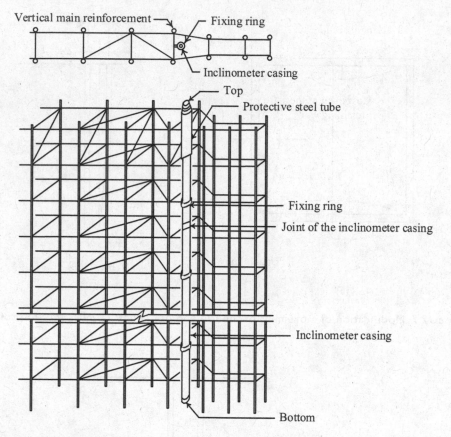

Figure 12.9 Inclinometer casing in a diaphragm wall.

Another way to install the inclinometer casing in the diaphragm wall is to fix a PVC pipe with 5- to-10-mm-thick walls onto a main reinforcement of the steel cage; the PVC pipe is then placed into the trench with the cage. After concrete casting, use a drill to bore through the PVC pipe to the hard soil layer. Then, insert the inclinometer casing into the PVC pipe and bore hole, and then backfill the gap between the inclinometer casing and the bore hole with the materials whose properties are similar to the material there. Therefore, the backfill materials in diaphragm wall or hard soil can be cement grouts, whereas those in soils can be a mixture of bentonite and cement (Figure 12.10a).

When casting the trench, the placement of the PVC pipe is completed. The PVC pipe is then used as a drill guide, allowing a drill to reach the hard soil. The inclinometer casing is then inserted into the PVC pipe or the bore. Between the inclinometer casing and the bore is filled by suitable filling materials whose properties are similar to that of the material there. The filling materials in diaphragm wall or hard soil can be cement grouts, whereas those in soils should be a mixture of bentonite and cement (Figure 12.10a). If the hard soil is too deep, the bottom end of the inclinometer casing should at least extend beyond the bottom of the diaphragm wall by more than the excavation width, i.e., the total depth of the casing should be larger than $H_e + B$, where H_e = excavation depth, B = excavation width.

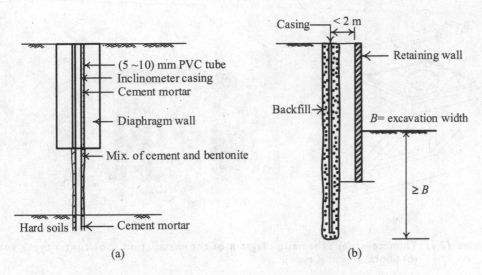

Figure 12.10 Installation of inclinometer casings: (a) in diaphragm wall and (b) in soil.

Since the movement of the soils near the retaining wall is quite close to that of the wall, the inclinometer casing can also be placed in the soils within 2 m of the outer side of the wall to measure the lateral displacement of the wall (Ou et al., 1998, 2000), as shown in Figure 12.5b. Figure 12.10b diagrams the embedment of the inclinometer casing outside the diaphragm wall. The installation is especially effective for sheet piles and soldier piles because the inclinometer casing is not easily fixed directly on sheet piles or soldier piles.

The inclinometer casing should be installed on the section where the lateral displacement is largest. The largest displacement usually occurs in the central section of the excavation (Figure 12.7a). The initial value should be taken before excavation as a baseline, which is to be deducted from the values taken during excavation.

12.4.2 Tilt of buildings

Buildings will tilt because of ground settlement and thereby be damaged. The tilt of a building can be estimated by the relative settlement between two reference points with plane surveying such as the level, or theodolite, divided by the horizontal distance between these two reference points. If the two reference points are on the two adjacent columns or foundation footings, the measured value is the angular distortion (see Section 11.2). Besides, the tilt of a building can also be monitored using a tiltmeter, and the value produced by which is called the tilt angle.

A commonly used device to measure the tilt of a building is one consisting of a datum plate, a tiltmeter, and an indicator. The datum plate is usually made of ceramics, which are less affected by temperature. The plate is fixed to the object that is going to be measured. The tiltmeter is mainly used to measure the degree of tilt of the datum plate, which represents the tilt of the structure the plate is fixed to. The tiltmeter is similar to the inclinometer (Section 12.4.1) in measurement principle. Similar to the inclinometer, the tiltmeter contains a sensor, which can be categorized into resistance type, vibrating type, and force balance accelerometer type, etc.

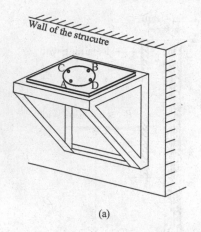

(a)   (b)

Figure 12.11 Tiltmeters: (a) schematic diagram of the installation (horizontal type) and (b) photo (vertical type).

According to the type of tiltmeter, the datum plate can be divided into horizontal and vertical types. Figure 12.11a diagrams the basic design of a horizontal-type datum plate. Figure 12.11b shows a photo of a tiltmeter with a vertical datum plate. As shown in the figure, on the plate are four protruding objects that form two axes perpendicular to each other. One of the axes is perpendicular to the wall of the building, whereas the other is parallel to it, the AB and CD axes in Figure 12.11a, respectively.

The tilt angle in the AB direction is taken first. Then, turn the sensor by 180° and measure the tilt angle in the AB direction again. Average the two measured values to eliminate the instrumental systematic error. The averaged value is thus the tilt angle in the AB direction. Use the same method to measure the tilt angle in the direction parallel to the wall (CD axis).

As discussed in Section 11.2, the tilt of a building contains the rigid rotation and the angular distortion. The value taken by a tiltmeter is the tilt of the structure at the place where the datum plate is fixed. To make the measured value that represents the safety of the building, the most suitable location of the datum plate has to be determined according to the structural behavior of the building and the convenience of measurement. The roof, column, and walls are the most commonly selected locations.

Generally speaking, the buildings within the influence range of excavation should be equipped with tiltmeters. For the determination of the influence range of excavation, refer to Section 6.4.2.

With the datum plate placed, take the initial measurement to obtain the initial value to be deducted from the value taken after excavation is started. The difference is the tilt angle of the building at the excavation stage.

12.4.3 Ground settlement and building settlement

The simplest method to measure ground settlement is marking the ground surface with steel nails (called settlement nails), to make a foresight to a fixed point and backsight to the nails using a level to obtain the ground settlement at the points where the nails are

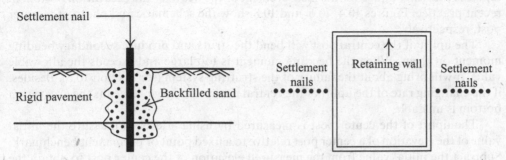

Figure 12.12 Schematic diagram of the installation of a settlement nail.

Figure 12.13 Layout of settlement marks in order to obtain the representative settlement profile.

driven in. The fixed point is the datum point outside the influence range of excavation. A certain position of a building with the pile foundation can be assumed to be a fixed point either. If there are not any buildings with pile foundations nearby, a marking object in the distance or outside the influence range of settlement has to be adopted (for the determination of the influence range of settlement, see Section 6.4.2). If necessary, a permanent benchmark has to be set to serve as a fixed point.

The surrounding ground of an excavation may be soil, asphalt pavement, or concrete pavement. The asphalt pavement and concrete pavement have relatively high rigidities. That is, even though the soil below them has settled, the pavements do not necessarily show signs of settlement. Thus, the settlement nails have to be driven through the pavements so that the settlement of the nails can represent the real settlement of the soil. Figure 12.12 illustrates a possible way of setting the settlement nails.

In addition to the vicinity of the buildings, settlement nails should also be set in the central area of the excavation zone and along a direction perpendicular to the retaining wall (Figure 12.13) to obtain the representative settlement profile (Figure 6.10b for example). That is because the behavior of the retaining wall and the soil near the central section is similar to plane strain conditions, and the ground settlement and lateral deformation of the wall there are larger than those in any other sections. If the initial settlement profile of the central area of the excavation is known, it will be easier to predict the ground settlements at the last excavation stage and in other sections by performing back analysis (see Section 7.8).

The measurement of the settlement of buildings is the same as that of ground settlement except that the settlement nails are to be set on the buildings themselves, on the wall, or on the columns for example.

12.4.4 Heave of excavation bottoms or center posts

Excavation will cause the excavation bottom to heave. Too much heaving is usually dangerous for the strutting system and causes settlement of the soils in the vicinity. Under worse conditions, it may lead to the failure of the excavation. The magnitude of heave or the rate of heaving of the excavation bottom can be measured using a heave gauge for reference for the judgment of the excavation safety. However, the heave gauge is often damaged during excavation. As a result, the uplift of the central post often replaces the heave gauge for the measurement of the heave of the excavation bottom in recent practice. Figures 10.4, 10.6, and 10.8 show the schematic and photo of a center post, respectively.

The uplift of the central post will bend the struts and produce secondary bending moment on them. When the bending moment is too large and exceeds the allowable value, it will bring about the failure of the strutting system (see Section 10.5). Besides, if the changing rate of the uplift of the central post is too large, it means the excavation bottom is unstable.

The uplift of the center post is measured by using a level to measure the initial value of the elevation of a center post relative to a fixed point or permanent benchmark. Subtract the initial value from the measured elevation of the center post to obtain the uplift of the center post at the excavation stage. Usually, there are many central posts in an excavation zone. The one with the largest uplift or the two adjacent ones with the largest differences of uplift can be selected for measurement.

12.5 MEASUREMENT OF STRESS AND FORCE

12.5.1 Strut load

The loading of the struts has to be monitored constantly during excavation lest it may exceed the allowable value and endanger the safety of the excavation. Some commonly used devices to measure the strut load are the strain gauge, the load cell, and the hydraulic jack gauge.

The strain gauge is a widely used device to measure the strut load. Placed outdoors and exposed to sun, rain, and dust, the vibrating type is more suitable and favored. If the exterior of the resistance type is well waterproofed, it can also be used to measure the strut load. Though the resistance type is more accurate, the installation cost is higher. Concerning the resistance and vibrating types of strain gauges, please refer to Section 12.3.

To install a strain gauge, weld the strain gauge onto the web of a strut using a point welding instrument and have the track of the sensor fitted with the strain gauge; install the sensor onto the strut with two iron plates and protect the sensor from water with water repellent glue around it, as shown in Figure 12.14. Figure 12.15 shows the photo of a strain gauge fixed onto a strut. Since the loading on a strut is not necessarily even, to measure the strut load correctly, on both sides of the web of a strut are installed strain gauges, and the averaged value of the two gauges is taken for the result.

The load cell can be divided into the mechanical type, the hydraulic type, and the electronic type but mechanical load cells are seldom used to measure the strut load. A hydraulic load cell consists of a liquid-filled chamber connected to a pressure measurement unit. By recording the change in pressure of the fluid on the pressure pad, the load can be determined.

The electronic type can be further divided into the resistance load cell and the vibrating one. Figure 12.16 diagrams the basic configuration of an electric load cell. Figure 12.17 shows a photo of a load cell. As shown in the figure, the main part of the load cell is a circular box made of steel or aluminum alloy where about four strain gauges are fixed inside or attached to the exterior. The accuracy of the electric load cell is the same as the strain gauge but it costs more. The load cell is usually applied when the strain gauge is not easily installed, on a wood or steel pipe of struts for example.

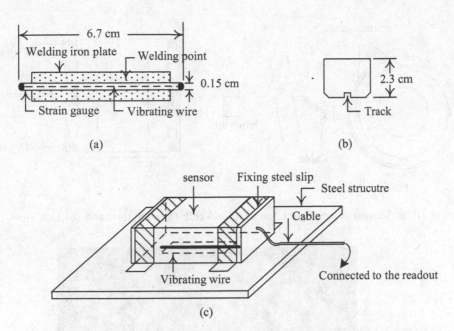

(a)

(b)

(c)

Figure 12.14 Schematic diagram of the installation of a vibrating wire strain gauge: (a) a vibrating wire strain gauge, (b) section of the sensor, and (c) fixed sensor.

Figure 12.15 Photo of a strain gauge installed on a strut.

After struts have been installed in an excavation, hydraulic jacks are often necessary to preload the struts so that they can be tightly connected with the wales and the lateral deformation of the retaining wall can be decreased. After preloading, the pressure gauge of the hydraulic jack can also be used for the measurement of the strut load. Figure 12.18 shows a photo of a hydraulic jack between struts. Nevertheless, misalignment of struts, off-center strut loading, non-parallel bearing plates, and transverse

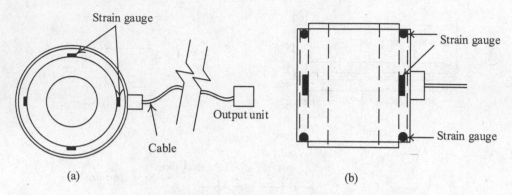

Figure 12.16 Schematic configuration of a load cell: (a) plan view and (b) side view.

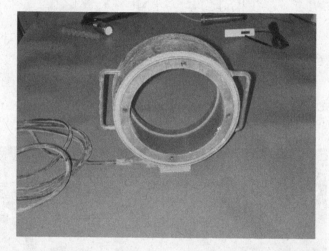

Figure 12.17 Photo of a load cell.

Figure 12.18 Photo of a hydraulic jack between struts.

relative movement of bearing plates all cause friction between the valve and cell cylinder and render the measurement of the strut load less accurate. Temperature changes and pressure gauge inaccuracy may cause an additional error. Using pressure gauges of hydraulic jacks to measure the strut load is not reliable as a result.

The strain gauge or load cell should be installed at the position where the strut load is about the largest. Similar to the measurement of the lateral deformation of the wall, the central area of an excavation site is usually the place where the largest strut load occurs. Thus, the central area of an excavation site should be equipped with the strain gauges or load cells (see Figure 12.7a).

12.5.2   Stress of the retaining wall

The stress of the retaining wall sometimes has to be measured constantly during the process of excavation lest the stress may exceed the allowable value and endanger the excavation safety.

The stresses of soldier or sheet piles can be measured by fixing commonly used strain gauges on them. As for column piles or diaphragm walls, the rebar stress meter has to be installed onto the steel inside the retaining wall to measure the load. The rebar stress meter contains a strain gauge, which is usually the resistance type. The exterior of the strain gauge has to be enveloped in many layers of water-repellent membranes, because, when the steel or the retaining wall is acted on by a force and bends, repellent membranes may break, causing the strain gauge to contact water. Since water is also an electric conductor, water will change the resistance of the strain gauge and influence the measurement.

Figure 12.19 diagrams the installation of a rebar stress meter. As shown in the figure, cut a main reinforcement at the designed depth and connect a rebar stress meter onto the main reinforcement by way of welding or a coupler. Last, pull the wire out of the steel cage to finish the installation of the rebar stress meter. Figure 12.20 shows the photo of a rebar stress meter installed on a steel cage.

The rebar stress meter should be installed at the place where the stress of the wall is the largest. As shown in Figure 12.7a, the retaining wall and soils at the center of an excavation site have the largest lateral deformation. The load on the retaining wall here should also be larger than other sections. The largest lateral deformation of the retaining wall, in the same section, is usually found near the excavation bottom (see Section 6.3). Thus, the largest stress of the reinforcement in the section should occur within several meters above or below the excavation bottom, which is also the place for the installation of the rebar stress meter.

12.5.3   Earth pressure on the retaining wall

Before excavation, the earth pressure on the retaining wall is the at-rest earth pressure. After excavation, the earth pressure on the front and back of the wall changes toward the passive and active earth pressures, respectively. Measuring the earth pressure on the wall is helpful to understand the safety of the excavation, bearing behavior of the wall and the surrounding soils.

Generally speaking, with the shallow depths of soldier piles and sheet piles, the earth pressures on them will not be large. Also, as it is difficult to install measuring

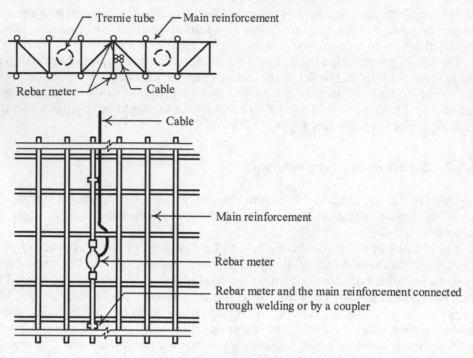

Figure 12.19 Installation of a rebar stress meter.

Figure 12.20 Photo of a rebar stress meter on a steel cage.

devices on them, the earth pressures on soldier piles and sheet piles are not measured in engineering practice. This section will only introduce the earth pressure cell used on diaphragm walls.

The commonly used earth pressure cells are the direct earth pressure cell and the indirect earth pressure cell. The direct earth pressure cell uses strain gauges (of the resistance or vibrating type) in a cell to measure the displacement of the bearing plate, which is then converted into the earth pressure. The indirect earth pressure cell is a cell filled with oil or mercury, which will transmit the earth pressure on the bearing plate. The earth pressure is then measured by a sensor on the exterior of the instrument. Figure 12.21 diagrams the basic configurations of the two types of earth pressure cells.

Figure 12.22 diagrams the installation of the earth pressure cell on a diaphragm wall. Figure 12.23 shows a photo of an earth pressure cell installed on a steel cage. The installation of the earth pressure cell is elucidated as follows:

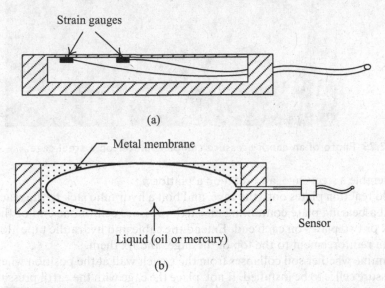

Figure 12.21 Configurations of the earth pressure cell: (a) direct type and (b) indirect type.

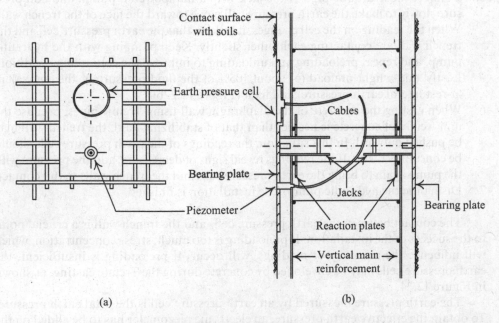

Figure 12.22 Installation of the earth pressure cell/piezometer: (a) earth pressure cell/pi-
ezometer on a steel cage and (b) earth pressure cell/piezometer in a trench.

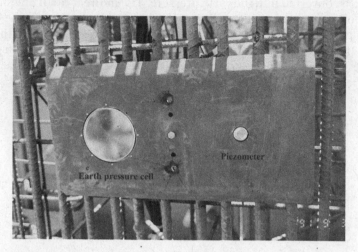

Figure 12.23 Photo of an earth pressure cell/piezometer on a steel cage.

1. Assemble a steel cage, and set it on a platform.
2. Weld reaction plates onto the cage, and bolt a hydraulic jack to the reaction plates.
3. Add a bearing plate containing an earth pressure cell to one end of the hydraulic jack or two plates on each end. Extend the cable and hydraulic pipe along with the main reinforcement to the top of the cage, and fix them.
4. Examine whether soil collapses from the trench wall at the position where the earth pressure cell is to be installed. If not, place the cage with the earth pressure cells into the trench and measure the values of the earth pressure in the stabilizing fluid.
5. Connect the hydraulic pipe of the jack with a manpowered pump, and add pressure slowly to make the earth pressure cell move toward the face of the trench wall. When the reading on the cell changes, it follows that the earth pressure cell and the trench wall are contacting each other slightly. Keep pumping with the hydraulic pump, and repeat preloading and unloading to improve the contacting condition. Lastly, add a light preload (of about 105% of the fluid pressure or the equivalent at-rest lateral earth pressure) and fix it (Figure 12.22b).
6. When casting the concrete of the diaphragm wall using Tremie pipes, because the unit weight of concrete is heavier than that of stabilizing fluid, the trench wall may be pushed outward. In this situation, the readings of the earth pressure cells should be constantly taken. If the readings reveal signs of decreasing, add the pressure with the pump again to bring the earth pressure cells and the trench walls to full contact.
7. Dismantle the hydraulic pump. The installation is finished.

The contact between the earth pressure cells and the trench wall is a crucial point to the success of the installation. If preloading is too much, stress concentration, which will influence the accuracy of readings, will occur. If preloading is insufficient, the earth pressure cells may be enveloped by concrete during the Tremie casting, as shown in Figure 12.24.

The earth pressure measured by an earth pressure cell is the total earth pressure. To obtain the effective earth pressure, an electronic piezometer has to be added on the bearing plate next to the earth pressure cell, as shown in Figure 12.22a.

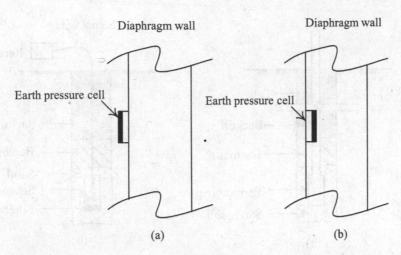

Figure 12.24 Failure of the installation of an earth pressure cell: (a) stress concentration caused by the protruding of the earth pressure cell and (b) earth pressure cell enveloped by concrete.

According to the purpose, the earth pressure cell can be set on the active side, the passive side, or both. Similar to the inclinometer, rebar stress meter, strain gauge, and settlement points on the ground surface, the central section of the excavation is the one where the earth pressure cell should be installed, considering the largest earth pressure on the section.

The maximum value of the earth pressure on the front of the retaining wall is the passive earth pressure, whereas the minimum on the back of the wall is the active earth pressure. Earth pressures on the wall usually fall between the two limiting values. An excavation design usually has taken these values into consideration. Thus, though measuring the earth pressures on the retaining wall is helpful to understand the bearing behavior of the wall and the surrounding soils, the results are not easily applied to the judgment of the excavation safety.

12.6 MEASUREMENT OF WATER PRESSURE AND GROUNDWATER LEVEL

12.6.1 Water pressure

Some commonly used piezometers in excavations are the open standpipe piezometer and the electronic piezometer. The piezometer can be installed in the soils within the excavation site, and/or outside the excavation site, and/or on the diaphragm wall. Those installed on a diaphragm wall are usually electronic piezometers. The installation of piezometers in the diaphragm wall is explicated in Section 12.5.3. In the present section, only piezometers installed in soils are to be introduced.

Figure 12.25a shows the schematic configuration of an open standpipe piezometer, also called the Casagrande piezometer. As shown in the figure, the main part of the open standpipe piezometer is a permeable or perforated pipe. The top of the

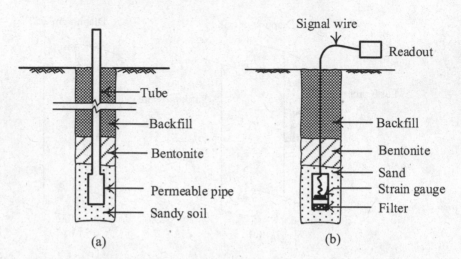

Figure 12.25 Schematic configuration of piezometers: (a) open standpipe piezometer and (b) electronic piezometer.

permeable/perforated pipe is connected to a standpipe. When groundwater enters into the permeable/perforated pipe, it will go up along the standpipe. When it goes up to a height which balances the water pressure at the level where the permeable/perforated pipe is placed, the height of the water in the standpipe is then the water pressure at the center of the permeable/perforated pipe.

The height in the vertical pipe can be measured using a water-level indicator, which is a coaxial cable of negative and positive poles in which the groundwater can be seen as the electrolyte. Put the indicator into the vertical pipe, contacting the water, a low-current circuit would then be formed and a signal be generated to reveal the height of water in the vertical pipe. The open standpipe piezometer is simple in its principle and reliable in its result, with easy installation. It is required, however, to wait till the groundwater has fully flowed into the vertical pipe before the height is measured. As a result, measuring the water pressure in clayey soils usually takes a long time in waiting.

The electronic piezometer is similar in its basic design to the earth pressure cell (see Figure 12.21a) except it excludes the influence of earth pressure. Therefore, a porous stone is placed to the front of the bearing plate/strain gauge which is acted on by groundwater flowing through the porous stone. Thus, the water pressure is measured. The electronic piezometer can be divided into the resistance and vibrating types. Figure 12.25b shows the schematic configuration of an electronic piezometer. Figure 12.26 shows the photo of electronic piezometers.

The strengths and shortcomings of the open standpipe piezometer and the electronic piezometer are listed in Table 12.1.

The installation of a piezometer in soils is elucidated as follows:

1. Bore a hole using a drill machine to 50 cm below the designed depth.
2. Place the main part of the piezometer (open standpipe type or the electronic type) in the bored hole and have the center of the piezometer located at the designed depth.

Figure 12.26 Photo of electronic piezometers.

Table 12.1 Strengths and Shortcomings of Various Types of Piezometers

Type	Strengths	Shortcomings
Open standpipe piezometer	1. reliable 2. can measure the permeability of soils 3. can take groundwater sample 4. applicable to sandy layers 5. durable	1. takes long time to read and is not suitable to be applied in clayey soils. 2. easily damaged during construction process. 3. the shaft often obstructs construction. 4. the filter stones in the permeable pipe are easily clogged.
Electronic piezometer	1. does not take long time to read; applicable to sandy and clayey soils. 2. easy to read. 3. does not obstruct construction. 4. not easily damaged or destroyed. 5. though usually not durable, some specially designed types of electronic piezometers are also good in durability. 6. fits automatic systems.	1. For long-term usage, durability and accuracy of readings have to be considered. 2. is more expensive

Fill the bored hole with sandy soil. The sandy soil is used to help groundwater permeate into the main part of the piezometer and be measured.

3. Seal the top (or both ends) of the sandy soil with bentonite to avoid groundwater from other elevations reaching the permeable pipe and influencing the measurement results. Above the bentonite is again filled with sandy soils to provide sufficient overburden weight to stabilize bentonite and sealing.

In the same borehole can be placed more than one piezometer, all of them requiring sealing to avoid water pressure from other elevations, as shown in Figure 12.27.

Depending on the monitoring aim, the piezometer can be embedded in clayey or sandy soils. Using one in clayey soils can derive the effective stress by way of measuring excess pore water pressure induced by soil deformation during excavation. Neither the excess pore water pressure nor the effective stress is directly related to excavation safety. They are, nevertheless, helpful to understand the deformation behaviors and stability characteristics. The piezometer aimed as above elucidated should be embedded in the central section of the excavation site.

The piezometer embedded in sandy or permeable soils can monitor the variation of water pressure, which is helpful for the diagnosis of excavation safety. If the water pressure changes abruptly while the stress on the reinforcement of the retaining wall is abnormal, it must get extraordinary attention and an effective remedial measure should be adopted to safeguard the excavation safety. Piezometers aimed as above should be embedded around the excavation site. They can also be placed within the excavation zone to help judge the safety factor regarding upheaval failure or sand boiling (see Sections 5.9 and 5.10).

12.6.2 Groundwater level

The instrument for measuring groundwater level is the water observation well where a perforated standpipe enveloped with two layers of nylon net is placed. The installation procedure of the water observation well is elucidated in the following (see Figure 12.28):

1. Bore to the designed depth using a drill machine.
2. Fill the bottom of the borehole with sand.

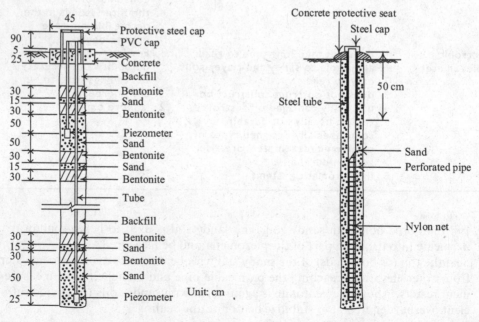

Figure 12.27 Two piezometers installed in the same borehole.

Figure 12.28 Installation of a water observation well.

3. Place the standpipe into the hole, and fill the space with sand.
4. Take some protective measures on the top of the standpipe. Thus finish the installation of the standpipe.

After groundwater flows into the pipe and the water level inside reaches the stable state, the water level in the pipe is then the groundwater level. Similar to the open standpipe piezometer, the water level in the pipe can be measured using a water-level indicator.

The aim of the water observation well is to monitor the change of the groundwater before or during excavation and supply information for the determination of the excavation method, the design of water pumping, and the management of construction period.

12.7   OTHER MEASUREMENT OBJECTS

The monitoring instruments or devices introduced above are the commonly used ones for excavations. In addition to them, to observe other characteristics of an excavation or for other special objectives, some other types of monitoring instruments can also be employed in excavations. For example, the extensometer is used for the measurement of settlement of the soils below the ground surface. The crackmeter can be used for the measurement of the cracks on the walls or columns of the buildings. The concrete stressmeter can be used to measure the stress of concrete. Because these instruments or devices are not frequently used in general excavations, the book is not going to introduce them at length. Interested readers can refer to the related literature for details.

12.8   PLAN OF MONITORING SYSTEMS

A complete plan of a monitoring system should include the following items: (1) determination of monitoring parameters, (2) determination of locations of monitoring instruments or devices, (3) analysis of the prediction values of the parameters, (4) choice of the specifications of monitoring instruments, (5) determination of installation specifications of the instruments, (6) setting of the alert and action values, and (7) determination of the measurement frequency. According to the sequence of execution, they are elucidated as follows:

1. Determination of monitoring parameters

   Excavation-related materials are soils, retaining walls, and struts. Excavation will produce physical quantities such as stress and strain on, and displacement of, these materials, which can be measured using instruments. Considering the cost, it is impossible to measure every physical quantity within the excavation influence range. Thus, suitable parameters have to be determined according to the excavation scale, geological conditions, and the situations of adjacent buildings or properties.

2. Determination of locations of monitoring instruments or devices

   The measurement parameters determined what follows is the selection of the locations and embedment depth of the instruments. The proper locations and embedment depths of the instruments should be able to unveil the excavation behaviors and the physical quantities representing the critical conditions of the excavation.

3. Analysis of the prediction values of the parameters

According to the soil properties, the type of the retaining wall, and the excavation conditions, predict the monitoring parameters for each of the excavation stages. The methods of prediction, including the simplified method, the beam on elastic foundation method, and the finite element method, are elucidated in Chapters 6–8. The prediction results are not only helpful in the determination of the specifications of instruments but also useful in the setting of the alert and action values.

4. Choice of the specifications of monitoring instruments

With the monitoring parameters for excavation stages, the specifications of the instruments can be then determined.

5. Determination of the installation specifications of the instruments

Whether the monitoring parameters are accurate affects the judgment of the excavation safety. Though the precisions of the instruments are important, whether they have been correctly installed also influences the measurement results. As a result, appropriate installation specifications have to be mapped out according to the characteristics of the instruments and the geological conditions.

6. Setting of the alert and action values

The alert value is an important indicator for engineers to determine whether the excavation condition is abnormal. The action value is the indicator for engineers to take emergency actions or measures. Before excavation, the alert and action values for each excavation stage should be determined in advance. For the meanings and determination methods, refer to Section 12.9.

7. Determination of measurement frequency

The measurement frequency refers to the times of the measurement taken within a certain period. Basically, the measurement frequency has to be increased during the process of excavation. During the waiting time, in which no excavation activities are in progress, it may be reduced properly. Table 12.2 shows the measurement frequencies of the CH218 contract of the Taipei rapid transit system (TRTS). Generally speaking, the measurement items and frequencies of TRTS are quite rigorous because they are excavations of large scale and any property damage may cause serious loss. For general excavations, Table 12.2 can be used as a reference with some proper modification on the basis of excavation scales.

12.9 APPLICATION OF MONITORING SYSTEMS

From Sections 12.4 to 12.7 are introduced the commonly used monitoring instruments and items in excavations. Among them, the groundwater level, earth pressure, and the water pressure outside the excavation zone are indexes for the diagnosis of abnormal behavior in excavations. The results from those monitoring items are not directly used for the judgment of excavation safety.

The lateral deformation of the retaining structure and soils, the tilt and settlement of the building, the ground settlement, and the heave of excavation bottom, the uplift

Table 12.2 Measuring Frequency of the Monitoring System in the CH218 of Taipei Mass Transit System (Chang, 1991)

Instrument	Minimum monitoring frequency
1. bench mark	Check the permanent bench mark every three months
2. settlement point	
a. on the building near the excavation	Once a week
b. on the buildings next to excavations of tunnels	Once a month before a shield passes; during grouting operation, monitor constantly, then once a week till settlement comes to stability
c. on the pavement	Once every two weeks
d. on the pipes of public facilities	Once every two weeks
3. inclinometers in soils and diaphragm walls	Once a week. Take extra readings before preloading, installing and dismantling of struts.
4. settlement gauges in soil	Once a month before a shield passes.
5. strain gauges	Once a day after installing and preloading struts for ten days; after that, once a week. Constantly monitor during casting of diaphragm walls.
6. observation wells/ piezometers	Once a week; Once a day during dewatering
7. earth pressure cells/electronic piezometers on diaphragm walls	Once a week
8. rebar stress meters	Once a week; take extra readings around the preloading of struts and after the dismantling of them.
9. vibrating wire strain gauges	Constantly monitor during the grouting operation

of the central post, the strut load, the stress of the retaining structure, and the water pressure within the excavation zone are directly related to the safety of both excavations and adjacent buildings. Each monitoring item can be set with an alert and action values separately. These measured values and alert and action values all contribute to the understanding of existing conditions of excavations and serve as bases to judge whether and when emergency measures are to be taken.

The alert value represents that the excavation safety has been reduced to the minimum required level of the design. When the measured value reaches this physical quantity, it indicates that the excavation may be in some abnormal conditions. Engineers must study the cause of the abnormality and be alert, but excavation still can continue, and the monitoring frequency should be increased. Emergency measures should enter the stage of preparation. The action value represents that the excavation safety has reached the limit state that excavations or adjacent buildings can tolerate. When the measured value reaches this physical quantity, it indicates that excavations or adjacent buildings may be in a dangerous state. Generally speaking, excavation must be suspended and emergency measures must be initiated. Construction can be resumed only after the excavation conditions improve or the bad conditions are eliminated.

The setting of alert and action values is based on empirical experiences so far, and no generally accepted standard formula is available. In engineering practice, the "design value" or "allowable value" or "analysis value" is usually taken as the alert value, that is, the alert value is an expected value to designers. Approximately 1.25 times the allowable value is used as the action value. The design value or allowable value or analysis value is a physical quantity acceptable to engineers and contains a safety factor. For example, if the allowable settlement of a building with the individual foundation is 2.5 cm (see Section 11.2), the alert value for the building settlement is then 2.5 cm and the action value 3.0 cm. Table 12.3 illustrates the alert and action values of CH218 of the Taipei rapid transit system and the corresponding emergency measures. Note that in Table 12.3, the allowable value and 80% times allowable value were taken as the active value and alert value, respectively, which are different from the above statements.

In fact, there are many objects of measurement related to the safety of an excavation. That some of measured values grow beyond the action values does not necessarily lead to the conclusion that the excavation is really under unsafe conditions. To our certainty, however, the more measured values grow beyond the action values, the closer the excavation is to failure. Engineers have to make their own judgment of the present conditions of safety on the basis of the relations between the measured values and the alert/action values and the characteristics of the excavation.

The above-discussed alert and action values do not vary with the excavation depth. As a result, some illogical conditions may occur. For example, for a 20-m-deep excavation, when excavation reaches the depth of 12 m, the settlement of an adjacent building is measured to be 3.0 cm, which is equal to the action value. According to the criterion, excavation has to be suspended immediately and emergency

Table 12.3 Control Values for the Monitoring System of the CH218 of Taipei Mass Transit system (Chang, 1991)

Area	Instrument	Alert value	Action value
Excavations	1. inclinometers (inside the wall)	60 mm	85 mm or 1/350
	2. inclinometers (in soil)	30 mm	40 mm
	3. observation wells/piezometers (outside of the wall)	−2 m	−3 m
	4. observation wells/piezometers (within the excavation zone)	Within 1m below excavation surface and F.S. against sand boiling lower than 1.25	Dewatering fails
	5. strain gauges	90% of the designed load	125% of the designed load
	6. rebar stress meter	250 MPa	350 MPa
Adjacent buildings	7. settlement points (on buildings)	22 mm	25 mm
	8. differential settlements of buildings	1/600	1/500
	9. tiltmeters (buildings)	1/800	1/500
	10. vibrating wire strain gauges	Any strain monitored during grouting	1.0 mini-strain

measures have to be taken. The building settlement, however, will still increase with the increase in the excavation depth no matter which emergency measures are taken or auxiliary methods are employed, though the increased rate of settlement may be reduced with the implementation of emergency measures or auxiliary methods. Finally, the total settlement of the building at the final stage may exceed the settlement that the building can tolerate. Serious damage may occur accordingly. On the other hand, emergency or remedial measures that are taken during excavation are less effective and economical.

Though the above method is sometimes illogical, it is still applicable to most excavations. The main reason is that the alert value has the safety factor considered and there remains some room for the increase of deformation or stress when the measured value is over the alert value. Whether it is enough to last through the entire excavation process, however, is another question.

Considering the method above has its shortcoming, the author proposes conducting feedback analysis for the safety evaluation of excavations and adjacent buildings. Figure 12.29 diagrams the flow chart of feedback analysis in the evaluation of building safety. As shown in the figure, evaluate if the buildings are within the settlement influence zone before excavation; if not, the safety of the buildings can be guaranteed and excavation can continue. If they are, depending on the types of the buildings and their foundations, the allowable settlement of the buildings, δ_a, can be determined (see Section 11.2). Then predict the ground settlement for each excavation stage δ_i as well as that for the last stage δ_f. If δ_f derived from the analysis is larger than δ_a, it implies that the buildings will be subject to the settlement which engineers do not expect at the end of excavation and the buildings may be damaged. The excavation plan has to be re-evaluated. If necessary, building protection measures or control of the movement must be designed. Repeat the process till δ_f is smaller than δ_a, which means the safety of the buildings can be secured and the excavation can start.

If the measured value (δ_m) of ground settlement at the ith stage equals the predicted value, δ_i, it means that the prediction is accurate and the excavation can continue.

If δ_m is larger than δ_i, it means the prediction is inaccurate and the excavation may come out unsafe. A feedback analysis is now required. That is to say, the excavation sequence and soil parameters used in the analysis have to be adjusted till δ_i is about equal to δ_m. If δ_i is about equal to δ_m but δ_f is larger than δ_a, it means the buildings will be subject to the settlement which engineers do not expect at the end of excavation and the buildings may be damaged. Under such conditions, the excavation plan has to be re-evaluated and building protection measures or control of the movement must be designed. If δ_f is still smaller than δ_a, it means the excavation is within the safety range and excavation can be continued.

If δ_m is smaller than δ_i, it means that, though the prediction is not accurate, the prediction is conservative and the excavation-induced ground settlement will not cause damage to the buildings. The excavation can be continued or a re-examination of the excavation plan can be made to reduce the cost (for example, decrease the numbers of the strut levels).

Though Figure 12.29 takes ground settlement for example, it is the same principle for the lateral deformation of the retaining wall, strut load, and the heave of the excavation bottom.

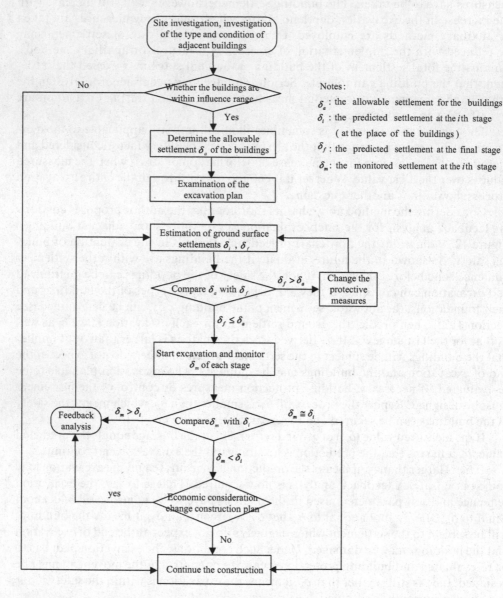

Figure 12.29 Feedback analysis using settlements as parameter.

PROBLEMS

12.1   What is the object of setting up the monitoring system in an excavation?

12.2   Explain the basic principle of the resistance type of strain gauges.

12.3   Elucidate the basic principle of the vibrating type of strain gauges.

12.4   Compare the strengths and shortcomings of the resistance type and vibrating type of strain gauges.

12.5   If the depth of the bottom of the inclinometer casing within the diaphragm wall is the same as that of the diaphragm wall, the displacement of the bottom of the diaphragm wall cannot be obtained and the true displacement curve is not to be obtained accordingly. How can we obtain the true displacement curve of the diaphragm wall? Please propose 2–3 methods.

12.6   The depth of a diaphragm wall is 21.0 m, an inclinometer casing is installed in the wall, and the depth of the inclinometer casing is the same as that of the diaphragm wall. The measurement method is to first place the inclinometer at the bottom of the casing, read the value from the indicator, and then lift the inclinometer 50 cm each time to read the value. Repeat the process until to the top of the casing. The value read from the indicator is the magnification value of the sine function, i.e., $\sin\theta$, of the tilt angle of the casing. It is known the setting of the indicator $\sin 30° = 12500$. Before excavation, use an inclinometer to measure the initial value of the inclinometer casing as shown in Table P12.6. As excavation advances to a depth of 5.0 m, the top displacement of the inclinometer is 1.5 cm, and the measured tilt values of the casing in the A+ and A− directions, respectively, are also shown in this table. Compute the displacements of the diaphragm wall at each depth, and plot the diagram of displacement versus depth.

12.7   Elucidate why the measured value on the tiltmeter on the wall of a building does not necessarily represent the angular distortion.

12.8   To monitor the ground settlement induced by excavation, what is the benefit of the settlement profile measured from the settlement marks which are perpendicular to the retaining wall and set in the central section of the excavation site, as far as excavation safety and property protection are concerned?

12.9   What are the objects of monitoring the heave of an excavation bottom and the uplift of central posts?

12.10   Elucidate the installation process of a rebar stress meter.

12.11   Describe the most suitable location and depth of a rebar stress meter on the retaining wall.

12.12   To measure the strut load, the commonly used methods are the hydraulic jack pressure, the load cell, and the strain gauge. Elucidate their strengths and shortcomings.

12.13   Describe the objective of monitoring the strut load.

12.14   Elucidate the installation process of an earth pressure cell.

12.15   Explain the objective of monitoring the lateral earth pressure on the retaining wall.

12.16   The commonly used devices to measure water pressures are the open standpipe piezometer and the electronic piezometer. Explicate their strengths and shortcomings separately.

12.17   What is the objective of installing piezometers within an excavation zone?

12.18   What then is the objective of installing piezometers outside the excavation zone?

12.19   Explicate the similarities and dissimilarities between an observation well and a piezometer in installation.

12.20   Judging on the basis of the installation process of an observation well, can it be used to monitor the water pressure?

Table PI2.6

Depth (m)	Initial value		Measured value after 5m excavation		depth (m)	Initial value		Measured value after 5m excavation	
	Direction of +A	Direction of -A	Direction of +A	Direction of -A		Direction of +A	Direction of -A	Direction of +A	Direction of -A
0.00					11.00	-461	443	-450	438
0.50	377	-389	266	-302	11.50	230	-253	261	-306
1.00	342	-356	299	-321	12.00	416	-428	443	-467
1.50	384	-400	343	-364	12.50	168	-186	216	-229
2.00	684	-700	650	-654	13.00	2	-20	41	-58
2.50	750	-766	710	-736	13.50	29	-46	67	-85
3.00	481	-498	439	-473	14.00	35	-50	74	-90
3.50	387	-404	346	-366	14.50	64	-79	105	-122
4.00	265	-276	222	-248	15.00	67	-79	108	-126
4.50	83	-101	51	-64	15.50	53	-68	94	-115
5.00	3	-19	-33	17	16.00	-256	236	-215	187
5.50	23	-39	-12	-5	16.50	-230	215	-181	174
6.00	150	-162	115	-123	17.00	-161	147	-119	101
6.50	109	-124	88	-106	17.50	-149	135	-107	90
7.00	118	-134	103	-120	18.00	-94	78	-64	42
7.50	25	-40	18	-41	18.50	28	-43	71	-89
8.00	-4	-8	-12	-5	19.00	-287	272	-249	213
8.50	179	-198	182	-203	19.50	-533	520	-491	474
9.00	64	-76	71	-91	20.00	-510	494	-463	445
9.50	118	-133	135	-145	20.50	-197	173	-168	141
10.00	-32	12	-31	-14	21.00				
10.50	-494	483	-454	436					

Appendix A

Conversion factors

Meter (symbol: m), gram (symbol: g), and newton (symbol: N) are the basic units of length, mass, and force for the SI system, respectively. Prefixes, as listed in Table A.1, are used to form multiples of SI units, for example, $km = 10^3$ m and $MN = 10^6$ N (Tables A.2–A.10).

Table A.1

Factor	Prefix	Symbol	Factor	Prefix	Symbol
10^{18}	exa	E	10^{-1}	deci	d
10^{15}	peta	P	10^{-2}	centi	c
10^{12}	tera	T	10^{-3}	milli	m
10^{9}	giga	G	10^{-6}	micro	μ
10^{6}	mega	M	10^{-9}	nano	n
10^{3}	kilo	k	10^{-12}	pico	p
10^{2}	hecto	h	10^{-15}	femto	f
10^{1}	deka	da	10^{-18}	atto	a

Length

Table A.2

SI Units to British Units	British Units to SI Units
1 km = 0.6214 mile (US)	1 mile (US) = 1.609 km
1 m = 1.094 yd	1 yd = 0.9144 m
1 m = 3.281 ft	1 ft = 0.3048 m
1 cm = 0.3937 in	1 in = 2.54 cm
1 mm = 0.03937 in	1 in = 25.4 mm

Note: 1 mile (US) = 1,760 yd = 5,280 ft, 1 ft = 12 in

Area

Table A.3

SI Units to British Units	British Units to SI Units
$1 \text{ km}^2 = 0.3861 \text{ mile}^2$	$1 \text{ mile}^2 = 2.59 \text{ km}^2$
$1 \text{ m}^2 = 1.196 \text{ yd}^2$	$1 \text{ yd}^2 = 0.8361 \text{ m}^2$
$1 \text{ m}^2 = 10.764 \text{ ft}^2$	$1 \text{ ft}^2 = 0.09290 \text{ m}^2$
$1 \text{ cm}^2 = 0.155 \text{ in}^2$	$1 \text{ in}^2 = 6.452 \text{ cm}^2$
$1 \text{ mm}^2 = 1.55 \times 10^{-3} \text{ in}^2$	$1 \text{ in}^2 = 645.2 \text{ mm}^2$

Volume or section modulus

Table A.4

SI Units to British Units	British Units to SI Units
$1 \text{ m}^3 = 1.308 \text{ yd}^3$	$1 \text{ yd}^3 = 0.765 \text{ m}^3$
$1 \text{ m}^3 = 35.32 \text{ ft}^3$	$1 \text{ ft}^3 = 0.0283 \text{ m}^3$
$1 \text{ m}^3 = 61,023 \text{ in}^3$	$1 \text{ in}^3 = 16.387 \times 10^{-6} \text{ m}^3$
$1 \text{ cm}^3 = 0.061024 \text{ in}^3$	$1 \text{ in}^3 = 16.387 \text{ cm}^3$
$1 \text{ mm}^3 = 0.061024 \times 10^{-3} \text{ in}^3$	$1 \text{ in}^3 = 16.387 \times 10^3 \text{ mm}^3$
$1 \text{ L} = 0.03532 \text{ ft}^3$	$1 \text{ ft}^3 = 28.32 \text{ L}$
$1 \text{ L} = 0.220 \text{ gal (BS)}$	$1 \text{ gal (BS)} = 4.546 \text{ L}$
$1 \text{ L} = 0.2642 \text{ gal (US)}$	$1 \text{ gal (US)} = 3.785 \text{ L}$

Note: $1 \text{ m}^3 = 1,000 \text{ L (L)}$

Moment of inertia

Table A.5

SI Units to British Units	British Units to SI Units
$1 \text{ mm}^4 = 2.403 \times 10^{-6} \text{ in}^4$	$1 \text{ in}^4 = 0.4162 \times 10^6 \text{ mm}^4$
$1 \text{ cm}^4 = 0.02403 \text{ in}^4$	$1 \text{ in}^4 = 41.62 \text{ cm}^4$

Mass

Table A.6

SI Units to British Units	British Units to SI Units
$1 \text{ Mg} = 0.9842 \text{ long ton}$	$1 \text{ long ton} = 1.016 \text{ Mg}$
$1 \text{ Mg} = 1.102 \text{ short ton}$	$1 \text{ short ton} = 0.9072 \text{ Mg}$
$1 \text{ kg} = 2.205 \text{ lb}$	$1 \text{ lb} = 0.4536 \text{ kg}$
$1 \text{ kg} = 0.06852 \text{ slug [lb-force/(ft.s}^2\text{)]}$	$1 \text{ slug} = 14.59 \text{ kg}$

Note: $1 \text{ Mg} = 1 \text{ ton}$, 1 long ton (US) $= 2,240 \text{ lb}$, 1 short ton (BS) $= 2,000 \text{ lb}$

Density

Table A.7

SI Units to British Units	British Units to SI Units
$1\ Mg/m^3 = 62.43\ lb/ft^3$	$1\ lb/ft^3 = 0.01602\ Mg/m^3$
$1\ kg/m^3 = 0.06243\ lb/ft^3$	$1\ lb/ft^3 = 16.02\ kg/m^3$
$1\ g/cm^3 = 62.43\ lb/ft^3$	$1\ lb/ft^3 = 0.01602\ g/cm^3$

Note: $1\ Mg/m^3 = 1\ ton/m^3 = 1\ g/cm^3 = 10^{-3}\ kg/cm^3$

Force or weight

Table A.8

SI Units to Other Units	Other Units to SI Units
$1\ N = 0.10197\ kg - force$	$1\ kg - force = 9.807\ N$
$1\ kN = 0.10197\ ton - force$	$1\ ton - force = 9.807\ kN$
$1\ N = 0.2248\ lb - force$	$1\ lb - force = 4.448\ N$
$1\ kN = 0.2248\ kip - force$	$1\ kip - force = 4.448\ kN$
$1\ kN = 0.1004\ long\ ton - force$	$1\ long\ ton - force = 9.964\ kN$
$1\ kN = 0.1124\ short\ ton - force$	$1\ short\ ton - force = 8.896\ kN$

Note: $N = kg \cdot m/sec^2$

Stress or pressure

Table A.9

SI Units to Other Units	Other Units to SI Units
$1\ MPa = 10.197\ kg - force/cm^2$	$1\ kg - force/cm^2 = 0.09807\ MPa$
$1\ MPa = 101.97\ ton - force/m^2$	$1\ ton - force/m^2 = 9.807 \times 10^{-3}\ MPa$
$1\ kPa = 0.010197\ kg - force/cm^2$	$1\ kg - force/cm^2 = 98.07\ kPa$
$1\ kPa = 0.10197\ ton - force/m^2$	$1\ ton - force/m^2 = 9.807\ kPa$
$1\ MPa = 0.06475\ long\ ton - force/m^2$	$1\ long\ ton - force/m^2 = 15.44\ MPa$
$1\ MPa = 0.145\ kip - force/in^2$	$1\ kip - force/in^2 = 6.895\ MPa$
$1 MPa = 145 lb - force/in^2$	$1\ lb - force/in^2 = 6.895 \times 10^{-3}\ MPa$

Note: $1\ Pa = 1\ N/m^2$, $1\ bar = 100\ kPa$, $1\ kg - force/cm^2 = 10\ ton - force/m^2$,

Atmospheric pressure: $1\ P_a = 1.033\ kg - force/cm^2 = 10.33\ ton - force/m^2 = 101.34\ kN/m^2$

Unit weight

Table A.10

SI Units to Other Units	Other Units to SI Units
1 MN/m^3 = 101.97 ton – force/m^3	1 ton – force/m^3 = 9.807 ×10^{-3} MN/m^3
1 kN/m^3 = 0.10197 ton – force/m^3	1 ton – force/m^3 = 9.807 kN/m^3
1 kN/m^3 = 6.366 lb – force/ft^3	1 lb – force/ft^3 = 0.1571 kN/m^3

Note: 1 kg – force/cm^3 = 1,000 ton – force/m^3

Unit weight of water γ_w = 1.0 ton – force/m^3 = 62.429 lb – force/ft^3

Soil properties at the TNEC excavation site

The construction of the Taipei National Enterprise Center (TNEC) project is a typical excavation using the top-down construction method. The author and his research group (Ou et al., 1998, 2000a, 2000b) carried out detailed studies on the excavation project. The construction of the TNEC excavation project is summarized in Section 3.6. The plane view of the site and the observation system are shown in Figure 3.34. The construction sequence of the foundation construction is shown in Figures 3.35 and 3.36 and Table 3.3.

The soil properties including small strain properties at the site were studied as shown in the related references (Ou et al., 1998; Kung, 2003, 2007; Liu et al., 1998; Teng, 2010; Teng et al., 2014), which are summarized below.

Figure B.1a shows the variation of SPT-N values of the construction project site. Figure B.1b shows the undrained shear strength obtained from the triaxial

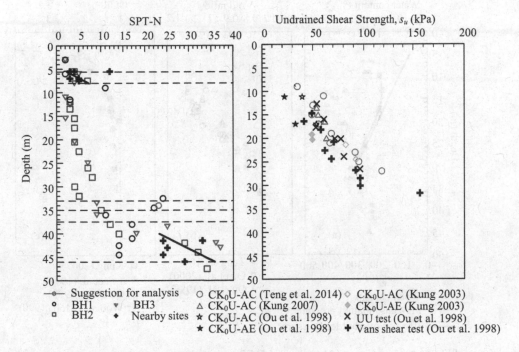

Figure B.1 SPT N and undrained shear strength.

unconsolidated undrained (UU) test, field vane shear test, triaxial axial compression undrained test under K_o consolidation (CK_0U-AC), and triaxial axial extension undrained test under K_o consolidation (CK_0U-AE). According to Figure B.1b, the ratios (s_u/σ_v') of the undrained shear strength (s_u) to the effective overburden pressure (σ_v') obtained from the field vane shear test and the UU triaxial test are 0.29 and 0.33, respectively. Those from the CK_0U-AC and CK_0U-AE tests are 0.32 and 0.19, respectively. Figure B.2a–c shows the water contents, void ratio, and overconsolidation ratio (OCR), respectively. Figure B.2a also shows the variation of the effective vertical overburden pressure (σ_v') and preconsolidation pressure with depth. As shown in Figure B.2a and c, the soil was overconsolidated within a depth of 15 m, which may be related to the existence of five-story buildings before the construction of the TNEC project. Figure B.3a and b shows the variation of the compression index (C_c) and swelling index (C_s) with depth.

According to the local geological history, site investigation, and Figure 3.36, the subsoil conditions at the site can be roughly divided into seven layers. Starting from the ground surface, they can be described as follows: the first layer, so-called Sungshan VI formation, is soft silty clay (CL), which ranges from GL-00 to GL-5.6 m and whose N-value is around 2~4. The second layer (Sungshan V formation), from GL-5.6 to GL-8.0 m, is loose silty fine sand with N-values between 4 and 11 and $\phi' = 28°$. The third layer (Sungshan IV formation), from GL-8.0 to GL-33.0 m, is again soft silty clay (CL) whose N-value is around 2–5 and *PI* is in the range of 9–23, with an average value of 17. This layer is the one that most affects the excavation behavior. The soil mainly contains 40%–55% of silt and 45%–60% of clay. The coefficient of permeability (k) is

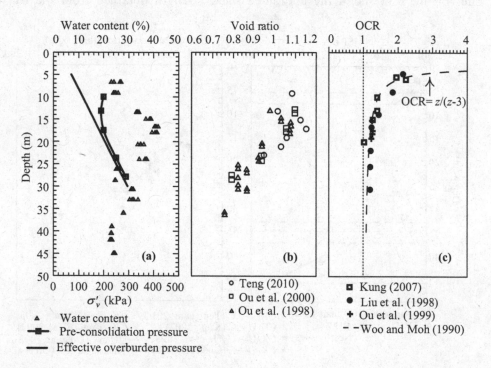

Figure B.2 Water content, void ratio and OCR.

around 4×10^{-6} cm / sec. The coefficient of consolidation ranges from 3×10^{-3} cm^2 / sec to 1.1×10^{-2} cm^2 / sec. The fourth layer (Sungshan III formation), ranging from GL-33.0 to GL-35.0 m, is medium dense silty fine sand with N-values between 22 and 24 and $\phi' = 32°$. The fifth (Sungshan II formation) is medium soft clay, which ranges from GL-35.0 to GL-37.5 m with $N = 9-11$. The sixth (Sungshan I formation) is medium dense-to-dense silt or silty fine sand, which ranges from GL-37.5 to GL-46.0 m with $N = 14-37$ and $\phi' = 32°$. Below the sixth layer, there is dense Chingmei formation (gravel soil) with N above 100.

The groundwater levels in Sungshan deposits were at around GL-2.0 m. Theoretically, as the Taipei Basin was formed, the groundwater level in the Chingmei formation should be in the hydrostatic condition, that is, the same as that in Sungshan formations (GL-2.0 m). Between 1950 and 1970, the residents of Taipei City pumped a large amount of groundwater from the Chingmei formation for people's livelihood, resulting in a sharp drop in the groundwater level in the Chingmei formation. After 1973, pumping began to be controlled so that the drop rate of the groundwater level in the Chingmei formation slowed down, but the groundwater level dropped to a historical low (about GL-42 m). After 1977, the groundwater level in the Chingmei gravel formation gradually recovered. By the time of TNEC excavation (1992), it is speculated that the groundwater level in the Chingmei formation had risen to about GL-12 m. Since the clay layer (CL) of the Sungshan II formation may be pinched out somewhere in the Taipei Basin, which connects Sungshan I and Sungshan III formations, the pumping of groundwater from the Chingmei gravel formation will also affect the groundwater level in the Sungshan III formation. The monitoring results showed that in 1992, the groundwater level of Sungshan III was about GL-30 m.

When the groundwater level in the Chingmei gravel formation dropped to GL-42 m, the effective stress of the Sungshan II clay formation reached the maximum due to the drop of the groundwater level in the Chingmei gravel formation (or Sungshan I formation) and Sungshan III formation. In 1992, the groundwater level in the Chingmei gravel formation rose back to GL-12 m, the pore water pressure of the Sungshan II clay formation increased, the effective stress decreased, and as a result, the clay was in the state of overconsolidation, which would affect the undrained shear strength of the clay and the coefficient of earth pressure at rest.

Similarly, as the groundwater level in Sungshan III sand formation dropped to GL-30 m, the effective stress in the Sungshan IV clay formation that connected the Sungshan III sand formation would be affected, which in turn influenced the undrained shear strength and other mechanical properties of the Sungshan IV clay formation.

Moreover, the soil below GL-5.6 and GL-37.5 m is silty sand (SM) (refer to Figure 3.34b). Therefore, the site was drilled with 3–4 deep wells to the SM, below GL-37.5 m. The groundwater in these two layers has been lowered before excavation. Due to the small number of deep wells, it is estimated that the dewatering effect is limited. When excavation was down to GL-5.6 m, the groundwater of this layer was still encountered. At this time, some sumps were installed for surface drainage.

The performance of the diaphragm wall and soil was reported in Ou et al. (1998, 2000a, 2000b), which is summarized here. The lateral deformation of the diaphragm wall and the ground surface settlement at each excavation stage are shown in Figure 6.10. Time-dependent movements of the diaphragm wall and surface settlement are shown in Figures 6.23–6.25. The soil displacements outside the excavation zone are

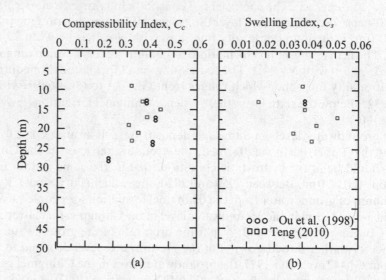

Figure B.3 Cc and Cs.

shown in Figure 6.19. The heaves and heaving rate of the excavation bottom are shown in Figure 6.22b. The variation of bending moment of the diaphragm wall is shown in Figures 7.10 and 7.11.

There were six buildings in the vicinity of the excavation site of TNEC. During the process of excavation, the variations of tilts of the buildings are described in Section 11.4.4 and shown in Figures 11.11 and 11.12.

Appendix C

Definition of plane strain

Figure C.1 illustrates an infinitely long diaphragm wall. Under the action of the earth pressure behind the wall, the wall moves forward. The lateral displacement of the wall and the surrounding soils (in x-direction) changes with the variation of the depth (in y-direction). There also exists some amount of vertical displacement of the wall and the surrounding soils. At the same depth, the lateral displacements of the wall and soils in different positions (along z-direction) are the same. That is, the displacement along z-direction is zero. Since displacement only occurs in two directions, the displacement behavior is called a two-dimensional behavior.

Because the displacement of the wall and soils only occurs in the x- and y-directions, the normal strains and shear strains in the z-direction are all none, so it follows that the strains are only found on the xy plane (i.e., the $z = 0$ plane). The behavior is called the plane strain or two-dimensional plane strain behavior.

Since the strain in the z-direction is zero, the σ_z-value at each section (Sections A-A and B-B, for example) is the same. That is to say, the stress condition at each section of the soil is the same. When performing an analysis, we can take any section to be a representative.

Analysis of structures subject to plane strain behavior is much simpler than that of those with three-dimensional behavior. Therefore, in most books on geological engineering or foundation engineering, three-dimensional problems are often simplified into plane strain problems, as best as possible. For example, the flow net, seepage, the bearing capacity of continuous or strip foundations, the safety factor for slope stability, the lateral deformation of a retaining wall, and the stability analysis of excavations are all analyzed on the basis of plane strain condition.

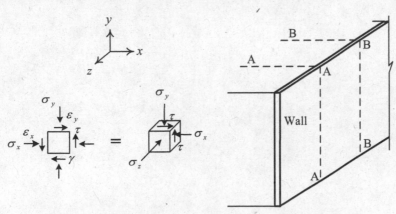

Figure C.1 Section of plan strain for a infinite long wall.

References

ACI (2019), *Building Code Requirements for Structural Concrete (ACI 319-19)*, American Concrete Institute, Farmington Hills, MI.

Alpan, I. (1967), The empirical evaluation of the coefficient k_0 and k_{0R}, *Soils and Foundations*, Vol. VII, No. 1, pp. 31–40.

American Association of Sate Highway and Transportation Officials (AASHTO) (1992), *Ground Anchors*.

Becker, P. (2008), Time and stress path dependent performance of excavations in soft soils, *Proceeding of the 19th European Young Geotechnical Engineers' Conference*, Gyor, Hungary, pp. 62–70.

Benz, T., Vermeer, P. A., and Schwab, R. (2009), A small-strain overlay model. *International Journal for Numerical and Analytical Methods in Geomechanics*, Vol. 33, pp. 25–44.

Bergado, D. T., Asakami, H., Alfaro, M. C., and Balasubramaniam, A. S. (1991), Smear effects of vertical drains on soft Bankok clay, *Journal of Geotechnical Engineering*, Vol. 117, No. 10, pp. 1509–1530.

Bjerrum, I. (1963), Allowable settlement of structures, *Proceedings of European Conference on Soil Mechanics and Foundation Engineering}, Weisbaden, Germany*, Vol. 2, pp. 35–137.

Bjerrum, L. (1971), *Recent Research on the Consolidation and Shear Behavior of Normally Consolidated Clays*, NGI Internal Report 50302, Imperial College, London, England.

Bjerrum, L. (1972), Embankments on soft ground, *Proceeding of Specialty Conference on Performance of Earth and Earth Supported Structures*, Vol. 2, pp. 1–54.

Bjerrum, L. (1973), Problems of soil mechanics and construction on soft clay, *State-of-the-Art Report, Session 4, Proceedings of Eighth International Conference on Soil Mechanics and Foundation Engineering*, Moscow, USSR, Vol. 3.

Bjerrum, L. and Eide, O. (1956), Stability of strutted excavation in clay, *Geotechnique*, Vol. 6, pp. 32–47.

Bolton, M. D. (1986), The strength and dilatancy of sands, *Geotechnique*, Vol. 36, No. 1, pp. 65–78.

Bonaparte, R. and Mitchell, J. K. (1979), The Properties of San Francisco Bay Mud at Hamilton Air Force Base, California, *Geotechnical Engineering Research Report*, Department of Civil Engineering, University of California, Berkeley.

Borja, R. I. (1990), Analysis of incremental excavation based on critical state theory, *Journal of Geotechnical Engineering, ASCE*, Vol. 116, No. 6, June, pp. 964–985.

Boscardin, M. D. and Cording, E. J. (1989), Building response to excavation induced settlement, *Journal of Geotechnical Engineering, ASCE*, Vol. 115, No. 1, pp. 1–15.

Bowles, J. E. (1988), *Foundation Analysis and Design*, 4th Ed., McGraw-Hill Book Company, New York.

British Standard Institute (BSI DD81)(1989), *British Standard Code of Pratice for Ground Anchorages*.

Burland, J. B. and Potts, D. M. (1981), The overall stability of free and propped embedded cantilever retaining walls, *Ground Engineering*, July, pp. 28–38.

Burland, J. B. and Wroth, C. P. (1974), Settlement of buildings and associated damage, *Proceeding of Conference on Settlement of Structures*, Pentech Press, London, England, pp. 611–654.

Calvello, M. and Finno, R. (2004), Selecting parameters to optimize in model calibration by inverse analysis, *Computer and Geotechnics*, Vol. 31, pp. 410–424.

Caquot, A. and Kerisel, J. (1948), *Tables for the Calculation of Passive Pressure, Active Pressure, and Bearing Capacity of Foundations*, Gauthier-Villars, Paris.

Caspe, M. S. (1966), Surface settlement adjacent to braced open cuts, *Journal of the Soil Mechanics and Foundations Division, ASCE*, Vol. 92, SM4, pp. 51–59.

Chang, R. F. (1991), *An introduction to geotechnical observational system of the mass rapid transit system*, MRT Technology, Bureau of TRTS.

Cho, W., and Finno, R. J. (2010). Stress-strain responses of block samples of compressible Chicago glacial clays, *Journal of Geotechnical and Geoenvironmental Engineering*, Vol. 136, No. 1, pp. 178–188.

Chou, H. L. and Ou, C. Y. (1999), Boiling failure and resumption of deep excavation, *Journal of Performance of Constructed Facilities, ASCE*, Vol. 13, No. 3, pp. 114–120.

Chu, H. and Chou, L. M. (1989), Building correction near the construction of Taipei Rapid Transit System, *Sino-Geotechnics*, Vol. 66, pp. 75–84.

Chu, J., Bo, M. W., Chang, M. F. and Choa, V. (2002), Consolidation and permeability properties of Singapore marine clay, *Journal of Geotechnical and Geoenvironmental Engineering*, Vol. 128, No. 9, pp. 724–732.

CICHE (1998), *Criteria and Descriptions for Design and Construction of Anchors*, Chinese Institute of Civil and Hydraulic Engineering, Taipei.

Clough, G. W. (1969), *Finite Element Analysis of the Soil-Structure Interaction in U-Frame Locks*, Ph.D Dissertation, University of California, Berkeley, CA.

Clough, G. W. and O'Rourke, T. D. (1990), Construction-induced movements of insitu walls, *Design and Performance of Earth Retaining Structures*, ASCE Special Publication, Vol. 25, pp. 439–470.

Coulomb, C. A. (1776), Essai sur une application des regles de maximis et minimis a quelques problemes de statique, *relatifs a l'architecture, Mem. Roy. des Sciences, Paris*, Vol. 3, pp. 38.

Culmann, C. (1875), *Die graphische statik*, Meyer and Zeller, Zurich.

Das, B. M. (1995), *Principles of Foundation Engineering*, PWS Publishing Company, MA.

Dawson, M. P., Douglas, A. R., Linney, L. F., Friedman, M. and Abraham, R., (1996) Jubilee Line Extension, Bermondsey Station box: design modifications, instrumentation and monitoring, *Geotechnical Aspect of Underground Construction in Soft Ground, London*, pp. 99–104.

Deutsche Industrie Norm (DIN) (1988), *Verpressanker fur Voruburgehende Zwecke im Lockergestein, Bemessung*, Ausfuhrung und Prufung.

Diaz-Rodriguez, J. A., Leroueil, S., Aleman, J. D. (1992), Yielding of Mexico city and other natural clays, *Journal of Geotechnical Engineering*, Vol. 118, No. 7, pp. 981–995.

Do, T. N. (2015), *A Study of Stability of Deep Excavations in Clay with Consideration of a Full Elastoplastic Support System*, PhD Dissertation, National Taiwan University of Science and Technology, Taipei, Taiwan.

Do, T. N., Ou, C. Y. and Chen, R. P. (2016), A study of failure mechanisms of deep excavations in soft clay using the finite element method, *Computers and Geotechnics*, Vol. 73, pp. 153–163.

Driscoll, F. G. (1989), *Groundwater and Wells*, Johnson Filtration Inc., St. Paul, MN.

Duncan, J. M. and Buchignani, A. L. (1976), *An Engineering Manual for Settlement Studies*, Department of Civil Engineering, University of California, Berkeley, 94 p.

Duncan, J. M. and Chang, C. Y. (1970), Nonlinear analysis of stress and strain in soils, *Journal of the Soil Mechanics and Foundations Division, ASCE*, Vol.96, No.5, pp.637–659.

Dunnicliff, J. (1988), *Geotechnical Instrumentation for Monitoring Field Performance*, John Wiley and Sons, New York.

Dupuit, J. (1863), *Etudes Theoretiques et Pratiques sur le Mouvement des eaux*.

Federation Internationale de la Precontrainte (FIP) (1982), *Recommendation for the Design and Construction of Prestressed Concrete Ground Anchors*.

Ferris, J. G., Knowles, D. B., Brown, R. H., and Stallman, R. W. (1962), *Theory of aquifer tests*, USGS, Water-Supply Paper 1536-E, 174 pp.

Finno, R. J., Bryson, S., Calvello, M. (2002), Performance of a stiff support system in soft clay, *Journal of Geotechnical and Geoenvironmental Engineering*, Vol. 128, No. 8, pp. 660–671.

Finno, R. J., Calvello, M., Bryson, S. (2002), *Analysis and Performance of the Excavation for the Chicago-State Subway Renovation Project and its Effects on Adjacent Structures – Final Report*, Department of Civil Engineering, Northwestern University.

Fredlund, D. G. (1997), An introduction to unsaturated soil mechanics, *Unsaturated Soil Engineering Practice, ASCE*, pp. 1–37.

Gerber, E. (1929), *Untersuchungen uber die Druckverteilung im Orlich belasteten Sand*, Technische Hochschule, Zurich.

Goodman, R. E., Taylor, R. L., and Brekke, T. L. (1968), A model for mechanics of jointed rock, *Journal of the Soil Mechanics and Foundation Division*, Vol. 94, No. 3, pp. 637–658.

Graf, E. D. (1992), Compaction grouting, *Grouting/soil improvement and geosynthetics, ASCE Special Publication*, Vol. 30, pp. 275–287.

Grant, R., Christian, J. T., and Vanmarcke, E. H. (1974), Differential settlement of buildings, *Journal of the Geotechnical Division, ASCE*, Vol. 100, No. 9, pp. 973–991.

Hanna, T. (1985), *Field Instrumentation in Geotechnical Engineering*, Trans. Tech Publication, Germany.

Hantush, M. S. (1962), Aquifer tests on partially penetrating wells, *Transactions, ASCE*, Vol. 127, Part 1, pp. 284–308.

Harza, L. F. (1935), Uplift and seepage under dams in sand, *Transaction*, ASCE.

Hausman, M. R. (1990), *Engineering Principles of Ground Modification*, McGraw-Hill Publishing Company.

Higgin, K. G., Mair, R. J. and Potts, D. M. (1996), Numerical modeling of the influence of the Westminster Station excavation and tunneling on the Big Ben clock tower, *Geotechnical Aspect of Underground Construction in Soft Ground, London*, pp. 525–530.

Hsieh, P. G. and Ou, C. Y. (1998), Shape of ground surface settlement profiles caused by excavation, *Canadian Geotechnical Journal*, Vol. 35, No. 6, pp. 1004–1017.

Hsieh, P. G. and Ou, C. Y. (2012), Analysis of deep excavations in clay under the undrained and plane strain condition with small strain characteristics, *Journal of the Chinese Institute of Engineer*, Vol. 35, No. 5, pp. 601–616.

Hsieh, P. G. and Ou, C. Y. (2016), Simplified approach to estimate the maximum wall deflection for deep excavations with cross walls in clay under the undrained condition, *Acta Geotechnica*, Vol. 11, pp. 177–189.

Hsieh, P. G. and Ou, C. Y., (2018), Mechanism of buttress walls in restraining the wall deflection caused by deep excavation, *Tunneling and Underground Space Technology*, Vol. 82, pp. 542–553.

Hsieh, P. G. and Ou, C. Y., and Hsieh, W. H. (2016), Efficiency of excavations with buttress walls in reducing the deflection of the diaphragm wall, *Acta Geotechnica*, Vol. 11, pp. 1087–1102. doi:10.1007/s11440-015-0416-6.

Hsieh, P. G., Ou, C Y. and Shih, C. (2012), A simplified plane strain analysis of the lateral wall deflection for excavations with cross walls, *Canadian Geotechnical Journal*, Vol. 49, pp. 1134–1146.

Huang (1992), Case histories of underpinning during the construction of the Singapore mass transit system, *Sino-Geotechnics*, Vol. 40, pp. 77–90.

ICE (1990), Geotechnical Instrumentation in Practice, Purpose, Performance and Interpretation, *Proceedings of the Conference on Geotechnical Instrumentation in Civil Engineering Projects*, Organized by the Institution of Civil Engineers, London.

Jacob, C. E. (1940), On the flow of water in an elastic artesian aquifer, *Transaction, American Geophysical Union*, pp. 574–586.

Jaky, J. (1944), The coefficient of earth pressure at rest, *Journal of the Society of Hungarian Architects and Engineers (in Hungarian)*, Vol. 8, No. 22, pp. 355–358.

James, R. G. and Bransby, P. L. (1970), Experimental and theoretical investigation of a passive earth pressure problem, *Geotechnique*, Vol. 20, No. 1, pp. 17–37.

Janbu, N. (1963), Soil compressibility as determined by oedometer and triaxial tests, *European Conference on Soil Mechanics and Foundation Engineering, Wissbaden, Germany*, Vol. 1, pp. 19–25.

JSA (1988), *Guidelines of Design and Construction of Deep Excavations*, Japanese Society of Architecture.

JSF (1990), *Design and Construction Criteria of Ground Anchors*, Japanese Soil and Foundation Society.

Khoiri, M. and Ou, C. Y. (2013), Evaluation of deformation parameter for deep excavation in sand through case histories, *Computers and Geotechnics*, Vol. 47, Jan. pp. 57–67.

Konder, R. L. (1963), Hyperbolic stress-strain response: cohesive soils, *Journal of the Soil Mechanics and Foundation Division, ASCE*, Vol. 89, No. 1, pp. 115.

Kozeny, J. (1953), *Hydraulik*, Springer, Verlag.

Kruseman, G. P. and de Ridder, N. A. (1990), A*nalysis and Evaluation of Pumping Test Data*, International Institute for Land Reclamation and Improvement, Wageningen, Netherlands.

Kung, G. T. C. (2003), *Surface Settlement Induced by Deep Excavation with Consideration of Small Strain Behavior of Taipei Silty Clay*, Ph.D. Dissertation, Department of Construction Engineering, National Taiwan University of Science and Technology.

Kung, G. T. C. (2007), Equipment and testing procedures for small strain triaxial tests, *Journal of the Chinese Institute of Engineers, Transactions of the Chinese Institute of Engineers, Series A*, Vol. 30, No. 4, pp. 579–591.

Ladd, C. C. and Foote, R. (1974), New design procedure for stability of soft clay, *Journal of Geotechnical Engineering Division, ASCE*, Vol. 100, GT7, pp. 763–786.

Ladd, C. C., Foote, R., Ishihara, K., Schlosser, F., and Poulous, H. G. (1977), Stress-Deformation and Strength Characteristics, State-of-the-ART Report, *Proceedings of the Ninth International Conference on Soil Mechanics and Foundation Engineering, Tokyo*, Vol. 2, pp. 421–494.

Ladd, C.C. and Lambe, T.W. (1963), The strength of undisturbed clay determined from undrained tests, *Laboratory Shear Testing of Soils, ASTM Special Technical Publication* No. 361, pp. 342–371.

Lade, P. V. and Duncan, J. M. (1973), Cubical triaxial tests on cohesionless soil, *Journal of the Soil Mechanics and Foundation Division, ASCE*, Vol. 99, No. 10, pp. 793–812.

Lee, F. H., Yong, K. Y., Quan, K. C. N., Chee, K. T. (1998), Effect of corners in strutted excavations: field monitoring and case histories, *Journal of Geotechnical and Geoenvironmental Engineering*, Vol. 124, No. 4, pp. 339–349.

Lim, A., Ou, C. Y. and Hsieh, P. G. (2010), Evaluation of clay constitutive models for analysis of deep excavation under undrained conditions, *Journal of GeoEngineering*, Vol. 5, No. 1, pp. 9–20.

Lim, A., Ou, C. Y. and Hsieh, P. G. (2016), Evaluation of buttress wall shapes to limit movements induced by deep excavation, *Computers and Geotechnics*, Vol. 78, pp. 155–170.

Lim, A., Ou, C. Y. and Hsieh, P. G. (2020), A novel strut-free retaining wall system for deep excavation in soft clay: numerical study, *Acta Geotechnica*, Vol. 15, pp. 1557–1576.

Lin, H. D. and Lin, C. B. (1999), A preliminary study for the frictional resistance and loading factor of the cast-in-place pile, *Journal of the Chinese Institute of Civil and Hydraulic Engineering*, Vol. 11, No. 1, pp. 13–31.

Littlejohn, G. S. (1970), Soil Anchors, *Proceedings of Conference on Ground Engineering, Institute of Civil Engineering*, London.

Liu, C. C., Chen, S. H. and Cheng, W. L. (1998), Undrained behavior of Taipei silty clay under simple shear condition, *Journal of the Chinese Institute of Civil and Hydraulic Engineering*, Vol. 10, No. 4, pp. 627–637. (in Chinese).

Lo, K. Y. (1962), Shear strength properties of a sample of volcanic material of the valley of Mexico, *Geotechnique*, Vol. 12, No. 4, pp. 303–318.

Mackey, R. D. and Kirk, D. P. (1967), At rest, active and passive earth pressures, *Proceedings of Southeast Asia Regional Conference on Soil Engineering*, Bangkok, pp. 187–199.

Mana, A. I. and Clough, G. W. (1981), Prediction of movements for braced cut in clay, *Journal of Geotechnical Engineering Division, ASCE*, Vol. 107, No.6, pp. 759–777.

Marino, M. A. and Luthin, J. A. (1982), *Seepage and Groundwater*, Elsevier Scientific Publishing Company, New York.

Mesri, G., Rokhsar, A., Bohor, B. F. (1975), Composition and compressibility of typical samples of Mexico city clay, *Geotechnique*, Vol. 25, No. 3, pp. 527–554.

Milligan, G. W. E. (1983), Soil deformation near anchored sheet-pile walla, *Geotechnique*, Vol. 33, No. 1, pp. 41–55.

Moh, Z. C., Chin, C. T., Liu, C. J and Woo, S. M (1989), Engineering correlations for soil deposits in Taipei, *Journal of the Chinese Institute of Engineers*, Vol. 12, No. 3, pp. 273–283.

Moh, Z. C., Nelson, J. D. and Brand, E. W. (1969), Strength and deformation behavior of Bangkok clay, *Proceedings of the 7th International Conference on Soil Mechanics and Foundation Engineering*, Mexico city, pp. 287–295.

Mohr, O. (1900), *Welche Umstande bedingen die Elastizitatsgrenze und den Bruch eines Materiales Zeitschrift des Vereines Deutscher Ingenieure*, Vol. 44, pp. 1524–1530; 1572–1577.

NAVFAC DM7.2 (1982), *Foundations and Earth Structures*, Design Manual 7.2, Department of the Navy, U.S.A.

Ng, C. W. W. (1992), *An Evaluation of Soil-Structure Interaction Associated with a Multi-Propped Excavation*, PhD Thesis, University of Bristol, UK.

Nicholson, D. P. (1987), *The design and performance of the retaining wall at Newton station, Proceeding of Singapore Mass Rapid Transit Conference*, Singapore, pp. 147–154.

Nonveiller, E. (1989), *Grouting Theory and Practice*, Elsevier, Amsterdam, pp. 225–227.

Ou, C. Y. and Chen, M. J. (2015), *Prediction of Diaphragm Wall Caused by Deep Excavation in Clays, Geotechnical Research Report No. GT1501*, Department of Civil and Construction Engineering, National Taiwan University of Science and Technology, Taipei, Taiwan.

Ou, C. Y. and Chen, S. H. (2010), Performance and analysis of pumping tests in a gravel formation, *Bulletin of Engineering Geology and the Environment*, Vol. 69, pp. 1–12.

Ou, C. Y. and Hsiao, J, L. (1999), *Stability Analysis of Deep Excavations in Sand, Geotechnical Research Report No. GT99005*, Department of Construction Engineering, National Taiwan University of Science and Technology, Taipei, Taiwan, R.O.C.

Ou, C. Y. and Hsieh, P. G. (2011), A simplified method for predicting ground settlement profiles induced by excavation in soft clay, *Computers and Geotechnics*, Vol. 38, pp. 987–997.

Ou, C. Y. and Hu, M. Y. (1998), *Stability Analysis of Excavations in Clay, Geotechnical Research Report No. GT99007*, Department of Construction Engineering, National Taiwan University of Science and Technology, Taipei, Taiwan, R.O.C.

Ou, C. Y. and Shiau, B. Y. (1998), Analysis of the corner effect on the excavation behavior, *Canadian Geotechnical Journal*, Vol. 35, No. 3, pp. 532–540.

Ou, C. Y. and Wu, C. H. (1990), Deformation behavior of excavations in sandy soils due to grouting, *Journal of the Civil and Hydraulic Engineering*, Vol. 2, No. 2, pp. 169–182.

Ou, C. Y. and Yang, L. L. (2011), Observed performance of diaphragm wall construction, *Geotechnical Engineering Journal, SEAGS*, Vol. 42, No. 3, pp. 41–49.

Ou, C. Y., Chiou, D. C. and Wu, T. S. (1996), Three-dimensional finite element analysis of deep excavations, *Journal of Geotechnical Engineering, ASCE*, Vol. 122, No. 5, pp. 337–345.

Ou, C. Y., Hsieh, P. G. and Chiou, D. C. (1993), Characteristics of ground surface settlement during excavation, *Canadian Geotechnical Journal*, Vol.30, pp.758–767.

Ou, C. Y., Liao, J. T. and Cheng, W. L. (2000a), Building response and ground movements induced by a deep excavation, *Geotechnique*, Vol. 50, No. 3, pp.209–220.

Ou, C. Y., Liao, J. T. and Lin, H. D. (1998), Performance of diaphragm wall constructed using Top-Down Method, *Journal of Geotechnical and Geoenvironmental Engineering*, ASCE, Vol. No.9, pp.798–808.

Ou, C. Y., Shiau, B. Y. and Wang, I. W. (2000b), Three-dimensional deformation behavior of the TNEC excavation case history, *Canadian Geotechnical Journal*, Vol. 37, No. 2, pp.438–448.

Ou, C. Y., Wu, T. S. and Hsieh, H. S. (1996), Analysis of deep excavation with column type of ground improvement in soft clay, *Journal of Geotechnical Engineering, ASCE*, Vol.122, No.9, pp.709–716.

Padfield, C. J. and Mair, R. J.(1984), *Design of Retaining Walls Embedded in Stiff Clay, CIRIA Report No. 104*, England, pp. 83–84.

Peck, R. B. (1969), Advantages and limitations of the observational method in applied soil mechanics, *Geotechnique*, Vol. 19, No. 2, pp. 171–187.

Peck, R. B. and Ireland, H. O. (1961), Full-scale lateral load test of a retaining wall foundation, *Proceedings of 5th International Conference on Soil Mechanics and Foundation Engineering*, Vol. 2, pp. 453–458.

Peck, R. B.(1943), Earth pressure measurements in open cuts Chicago (III) subway, *Transactions, ASCE*, Vol. 108, p.223.

Peck, R. B., Hanson, W. E., and Thornburn, T. H. (1977), *Foundation Engineering*, John Wiley and Sons, New York.

Peck, R.B. (1969), Deep Excavation and Tunneling in Soft Ground, *Proceedings of the 7th International Conference on soil Mechanics and Foundation Engineering*, Mexico City, State-of-the-Art Volume, pp. 225–290.

Phuoc, D. H. (2014), *Study Three-dimensional Excavation Behavior and Adjacent Structure Responses Using Advanced Soil Model and Inverse Analysis Technique*, PhD Dissertation, Department of Civil and Construction Engineering, National Taiwan University of Science and Technology, Taipei, Taiwan.

Post-Tensioning Institute (PTI) (1980), *Recommendations for Prestressed Rock and Soil Anchors.*

Potyondy, J. G. (1961), Skin friction between various soils and construction materials, *Geotechnique*, Vol. 11, pp. 339–353.

Powers, J. P. (1992), *Construction Dewatering*, John Wiley and Sons, New York.

Pratama, I.T., Ou, C. Y. and Ching, J. (2020), Calibration of reliability-based safety factors for sand boiling in excavations, *Canadian Geotechnical Journal*, Vol. 57, No. 5, pp. 742–753.

Quinion, D. W. and Quinion, G. R. (1987), *Control of Groundwater*, Thomas Telford, London.

Rankine, W. M. J. (1857), On stability on loose earth, *Philosophic Transactions of Royal Society*, London, Part I, pp. 9–27.

Reddy, A. S. and Srinivasan, R. J. (1967), Bearing capacity of footing on layered clay, *Journal of the Soil Mechanics and Foundations Division, ASCE*, Vol. 93, No.2, pp.83–99.

Rehnman, S. E. and Broms, B. B. (1972), Lateral pressures on basement wall: results from full-scale tests, *Proceedings of 5th European Conference on Soil Mechanics and Foundation Engineering*, Vol. 1, pp. 189–197.

Robertson, P. K. and Campanella, R. G. (1989), *Guidelines for Geotechnical Design Using the Cone Penetrometer Test and CPT with Pore Pressure Measurement*, Hogentogler Co., Inc.

Roscoe, K. H. and Burland, J. B. (1968), *On the generalized stress-strain behavior of 'wet' clay, Engineering Plasticity*, Cambridge University, pp. 535–609.

Rowe, P. W. (1962), The stress-dilatancy relation for static equilibrium of an assembly of particles in contact, *Proceedings of the Royal Society, Series A*, No. 269, pp. 500–527.

Rowe, P. W. and Peaker, K. (1965), Passive earth pressure measurements, *Geotechnique*, Vol. 15, No. 1, pp. 57–78.

Schanz T., Vermeer P. A., and Bonnier P. G. (1999), The hardening soil model: formulation and verification. In R.B.J. Brinkgreve, *Beyond 2000 in Computational Geothechnics*, Balkema, Roterdam.

Schmidt, B. (1967), *Lateral stresses in unaxial strains, Bulletin, No. 23*, Danish Geotechnical Institute, pp. 5–12.

Schweiger, H. F. (2010), Design of deep excavations with FEM - Influence of constitutive model and comparison of EC7 design approaches, *Proc. of the 2010 Earth Retention Conference* (Finno, R.J., Hashash, Y.M.A., Arduino, P., eds.), ASCE, Bellevue, Washington, USA, pp. 804–817.

Seed, H. B., Noorany, I., and Smith, I. M. (1964), *Effects of Sampling and Disturbance on the Strength of Soft clay, Report TE-1*, University of California, Berkeley.

Sharma, K. G. and Desai, C. S. (1992), Analysis and implementation of thin-layer element for interfaces and joints, *Journal of Engineering Mechanics*, Vol. 118, No. 12, pp. 2444–2461.

Shen, M. S. (1999), *Hazard Mitigation Technology of Construction*. Wen-Sen Publisher, Taipei.

Sichart, W. (1928), *Das Fassungsvermogen von Rohrbrunnen*, Julius Springer, Berlin.

Sichart, W. and Kyrieleis, W. (1930), *Grundwasser Absekungen bei Fundierungsarbeiten*, Berlin.

Simpson, B., (1992), Retaining structures: displacement and design, *Geotechnique*, Vol. 42, No. 4, pp. 541–576.

Skempton, A. W. (1951), The bearing capacity of clays, *Proceeding of Building Research Congress*, Vol. 1, pp.180–189.

Skempton, A. W. and McDonald, D. H. (1957), Allowable settlement of buildings, *Proceedings, Institute of Civil Engineers*, Part III, Vol. 5, pp.727–768.

Smith, P.R., Jardine. R.J. and Hight, D.W. (1992), The yielding of Bothkennar clay, *Geotechnique*, Vol. 42, No. 2, pp. 257–274.

Son, M. and Cording, E. J. (2005), Estimation of building damage due to excavation-induced ground movements, *Journal of Geotechnical and Geoenvironmental Engineering*, Vol. 131, No. 2, pp. 162–177.

Spangler, M. G. (1938), *Horizontal pressures on retaining walls due to concentrated surface loads*, Iowa State University Experimental Station, Bulletin, No. 140.

Tan, T. S., Lee, F. H., Chong, P. T. and Tanaka, H. (2002), Effect of sampling disturbance on properties of Singapore clay, *Journal of Geotechnical and Geoenvironmental Engineering*, Vol. 128, No. 11, pp. 898–906.

Tatsuoka, F. and Ishihara, K. (1974), Yielding of sand in triaxial compression, *Soils and Foundations*, Vol. 14, No. 2, pp. 63–76.

Teng, F. C. (2010), *Prediction of ground movement induced by excavation using the numerical method with the consideration of inherent stiffness anisotropy*, PhD Dissertation, Department of Civil and Construction Engineering, National Taiwan University of Science and Technology.

Teng, F. C., Ou, C. Y. and Hsieh, P. G. (2014), Measurements and numerical simulations of inherent stiffness anisotropy in soft Taipei clay, *Journal of Geotechnical and Geoenvironmental Engineering, ASCE*, Vol. 140, No. 1, pp. 237–250.

Terzaghi, K. (1922), Der Grundbrunch on Stauwerken und Seine Verhutung, *Die Wasserkraft*, Vol. 17, pp. 445–449. Reprinted in *From Theory to Practice in Soil Mechanics*, John Wiley and Sons, New York, pp. 146–148, 1961.

Terzaghi, K. (1943), *Theoretical Soil Mechanics*, John Wiley & Sons, Inc., New York.

Terzaghi, K. and Peck, R. B. (1967), *Soil Mechanics in Engineering Practice*, John Wiley and Sons, New York.

Terzaghi, K., Peck, R. B. and Mesri, G. (1996), *Soil Mechanics in Engineering Practice*, 3rd., John Wiley and Sons, New York.

TGS (2001), *Design Specifications for the Foundation of the Building*, Taiwanese Geotechnical Society.

Theis, C. V. (1935), The relation between the lowering of the piezometric surface and the rate and discharge of a well using ground water storage, *Transactions of the American Geophysical Union. 16th Annual Meeting*.

Thiem, G. (1906), *Hydrologische Methoden*, JM Gephardt, Leipzig.

Vermeer, P.A. and Wehnert, M. (2005), Beispiele von FE-Anwendungen – Man lernt nie aus. In: *FEM in der Geotechnik – Qualität, Prüfung, Fallbeispiele*, Hamburg, pp. 101–119.

Vesic, A. B. (1961), Bending of beams on isotropic elastic solid, *Journal of the Engineering Mechanics Division, ASCE*, Vol. 87, No. 2, pp.35–53.

Wahls, H. E. (1981), Tolerable settlement of building, *Journal of the Geotechnical Division*, ASCE, Vol.107, No. 11, pp.1489–1504.

Waterman, D., 2009. Personal communication.

Winkler, E. (1867), Die Lehre Von Elasticitaet Und Festigkeit, Pray (H. Dominicus), pp. 182–184.

Wong, K. S. and Broms, B. B. (1989), Lateral deflection of braced excavation in clays, *Journal of Geotechnical Engineering, ASCE*, Vol. 115, No. 6, June, pp. 853–870.

Wong, L. W., Shau, M. C. and Chen, H. T. (1996), Compaction grouting for correcting building settlement, *Grouting and Deep Mixing,* Edited by Yonekura, Terashi and Shibazaki, A. A. Balkema, Rotterdam, The Netherlands.

Woo, S. M. (1992), Method, design and construction for building protection during deep excavation, *Sino-Geotechnics*, No. 40, pp.51–61.

Woo, S. M. and Moh, Z. C. (1990), Geotechnical characteristics of soils in the Taipei Basin, *Proceedings of the Tenth Southeast Asian Geotechnical Conference*, Vol. 3, Taipei, pp.51–65.

Wu, W. T. (1987), Engineering characteristics of subzones of Taipei Basin, *Sino-Geotechnics*, No. 22, pp. 5–27.

Yang, K. H. (2014), Personal communication.

Yen, D. L. and Chang, G. S. (1991), A study of allowable settlement of buildings, *Sino-Geotechnics*, No. 22, pp. 5–27.

Index

Printed in the United States
by Baker & Taylor Publisher Services

Printed in the United States
by Baker & Taylor Publisher Services